PHYSICS: Fundamentals & Frontiers

JAE R. BALLIF
and
WILLIAM E. DIBBLE
Physics Department
Brigham Young University

JOHN WILEY & SONS, INC.
New York London Sydney Toronto

Library of Congress Catalog Card Number: 70-37429
ISBN 0-471-04635-3

Printed in the United States of America.

10 9 8 7 6 5 4 3 2 1

Preface

In this book we present, in a nonmathematical way, the fundamental principles of physics. These are principles which have caused revolutionary changes in the way men think of the natural world which surrounds them. From a practical point of view, a knowledge of these principles is needed to understand the changing technology of modern civilization.

The book is designed for either a one-semester or a one-quarter course. All of the material can be covered in a one-semester course. In a shorter course it might be desirable to omit some topics.

We have profited from the responses of many of the thousands of students on whom this material has been tested. We are indebted to several of our colleagues, particularly B. Kent Harrison, for many helpful discussions. Suggestions from David Perry, Robert Smith, and Donald Wieber were very useful. We acknowledge the assistance and encouragement of the staff of John Wiley & Sons, particularly Donald H. Deneck, Gerhard R. Brahms, Bernard Scheier, Linda Riffle, Phyllis Niklas, Elodie Sabankaya, Stella Kupferberg, and John Balbalis.

Provo, Utah
January 1972

JAE R. BALLIF
WILLIAM E. DIBBLE

Contents

1 Introduction

Science is the creation of concepts and their exploration in the facts. It has no other test of the concept than its empirical truth to fact. Truth is the drive at the center of science; it must have the habit of truth, not as a dogma but as a process.

J. Bronowski[1]

Throughout history, men and women have tried to understand their environment—an environment into which birth so suddenly and so strangely places them. Sometimes necessity forces this attempt at understanding. Food, shelter, protection from enemies, all must be wrested from an often stubborn and unforgiving earth. Sometimes, however, it appears to be a natural and driving curiosity that prompts a person to seek for the fundamental principles, forces, and elements that form, structure, and order the changing appearances and perplexing variety of nature.

Science and Technology

When we try to apply scientific information to achieve a specific purpose, to obtain food, shelter, protection, or some

[1]Reprinted from SCIENCE AND HUMAN VALUES by J. Bronowski by permission of Julian Messner, A Division of Simon & Schuster, Inc. Copyright © 1956, 1965 by J. Bronowski.

FIGURE 1-1 Courtesy of Syd Greenberg, DPI.

other desired product or result, we are engaging in technology. All of us have some interest in technology, if for no other reason than that we need food to eat, water to drink, and air to breathe. Logically speaking, however, before we can have any technology, our present technology or any other that we decide upon, we must have information to apply. This is one place where science comes into the picture. The activities of science are the search for reliable information and its organization into fundamental laws and principles. On the basis of these laws and principles one can then predict how nature will behave in a wide variety of circumstances. Thus one can create technologies. But science is not technology, rather it is more aptly described, as J. Bronowski has put it, as "the search to discover unity in the wild vaariety of nature—or more exactly, in the variety of our experience."[2] In this book we shall consider many of the fundamental laws which help to bring such unity to our understanding of nature.

Physics is generally considered to be the most fundamental of the sciences, providing much, if not all, of the scientific basis for chemistry, geology, and the various fields of engineering. Physics may be said to be the study of space, time, and matter—a rather all inclusive subject.

Indeed, physics *is* concerned with the elementary particles and fields of matter and energy, their behavior under various forces, and the structure, properties, and behavior of the larger aggregates of matter composed of many particles. Today even systems as complicated as living cells and the nervous systems of animals can be profitably studied by the methods of physics.

In the course of our discussion of the principles and development of physics, we shall have occasion to discuss and comment on the methods of science, and the way in which scientific principles are verified. At the outset, however, we should realize that although correct and clear reasoning is very important in physics, physics is an *experimental* science. This means that its principles and laws should be tested by experiment, and that if they fail to hold they should be rejected or modified. In general, a theory may be proposed to explain a number of facts. Predictions are then made on the basis of the theory, and experiments are performed to see if the predictions of the theory actually hold true. If they do, then the theory will be gener-

[2]Reprinted from SCIENCE AND HUMAN VALUES by J. Bronowski by permission of Julian Messner, A Division of Simon & Schuster, Inc. Copyright © 1956, 1965 by J. Bronowski.

ally accepted. As we shall see by specific examples later, however, it may happen that a theory passes many tests, becomes widely and solidly accepted, and then finally fails to meet the test of some new experiment. Thus the theory is never absolutely proved, because new evidence may disprove it. Several times in the development of physics, new evidence has required revolutionary changes, changes which seemed and still seem to require new attitudes toward nature and even toward truth itself. One of the particular purposes of this book is to discuss two such revolutions.

Two Scientific Revolutions

The first revolution is associated with the names (among others) of the Italian Galileo Galilei (1564–1642), his German contemporary Johannes Kepler (1571–1630), and the man who tied it all together, the Englishman Isaac Newton (1642–1727). The revolutionary idea that the earth moved caused major revisions in physics, astronomy, philosophy, and theology. Those holding to traditional views became offended and even enraged over the new philosophy. In the ensuing battle, the scientific revolution gained its heroes, its villains, and its martyrs. The new ideas have shaped our present view of the cosmos, and the new methods of thought have shaped much of what is thought of as "scientific" even today. The cosmos was, as it were, cut from its moorings, and a whole world view was changed.

The impact of the second revolution, it appears to us, is less widely appreciated than that of the first. Neither Albert Einstein (1879–1955), Max Planck (1858–1947), Niels Bohr (1885–1962), Louis de Broglie (1892–), Erwin Schrodinger (1887–1961), Werner Heisenberg (1901–), nor Paul Dirac (1902–) were burned at the stake or imprisoned as far as we know (though some people might not have objected). Yet, the principles of relativity and quantum physics appear to be as challenging to traditional ideas of nature and truth as were the ideas of the first scientific revolution.

At the risk of oversimplification, we should like also to contrast two technologies. (See Figures 1-2 and 1-3). The technology of the last two hundred years or so created mechanical devices, and eventually electrical and electromechanical devices. It exploited "conventional" energy sources in coal, petroleum, and water power. With these devices and energy sources it established rapid transportation over land, water, and through the air; indeed through space itself. With these devices and energy sources it increased production of

FIGURE 1-2 Discoveries of the first scientific revolution have long had their application in the design of heavy equipment. Courtesy of David Crofoot, DPI.

food, manufactures, and more machines to add to the "muscle power" of the human race.

The new physics, of the second revolution, has already its technological applications as well. Many of these applications arise from facts and principles discovered from the study of the invisible—particles and systems much too small to be seen even in a light microscope. Nuclear processes offer a vast new source of energy, when properly harnessed. The electron microscope and the laser operate on "quantum" principles. The transistor and related devices have made electronic circuits much more compact. The computer, which only a few years ago took up an entire room, can now be made as small as a suitcase. A true technological revolution, the information revolution, appears to be well started. Computers and other information processing equipment are taking over more and more of the mental drudgery of the human race. Information processing, statistical prediction and decision making, and automatic control may be as characteristic of our time as steel production and railways were of another.

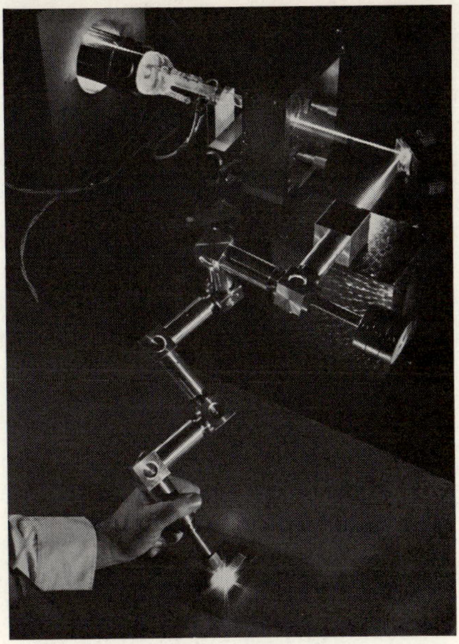

FIGURE 1-3 **An example of modern technology: Laser "knife" developed at Bell Laboratories has an articulated arm that allows the beam from a stationary laser to be moved freely for use in surgery, microcircuit fabrication or many other applications. Courtesy of William Vandivert.**

Again, however, we must emphasize that technology is not science, although science opens new possibilities for technology. A scientist may tell us that a nuclear energy project is possible, an engineer may tell us what it will cost, and together they can perhaps tell us what side effects it will have and how safe it will be. Neither of them can tell us whether we want the new project or not, nor would we want them making that decision for us. That is the responsibility of the citizen, not of any special class or expertise. Nor is it necessary to be able to build or operate a nuclear power plant yourself in order to make a reasonably intelligent decision. It would be useful, however, to know what energy is, and what a nucleus is.

Increasingly, political decisions are going to require scientific evaluations. We hope that the subjects we will discuss will be of some help in building a general understanding of physical principles that may be of use in such decisions. Of course, we are also interested in the subject for its own sake, and hope to appeal to your natural curiosity as well.

Quantities

Many of the laws of physics are expressed in terms of physical quantities, such as distance and time duration. We shall begin the discussion of the laws of physics in the next chapter. Here, however, we would like to discuss some of the quantities with which we shall be concerned.

Quantities and Measurement

A *quantity* may be defined, roughly, as an observable property or process in nature with which a number may be associated. This number is obtained by the operation of *measurement*. The number may be obtained directly by a single measurement, or indirectly, as, for example, by multiplying together two numbers obtained in separate operations of measurement. In this section we will first be concerned with the quantities *distance* and *time duration*. We obtain distances by making measurements of space. We obtain time durations by making measurements of time. Almost everyone has had practical experience making measurements of distance using rulers and making measurements of time duration using clocks. Of course measurements of these quantities are made only to a certain degree of accuracy. Greater accuracy generally requires greater care or more elaborate equipment.

Fundamental quantities are those which we choose *not* to define in terms of other quantities. For example, we choose distance and time duration to be fundamental; space and time seem very basic to us. Other quantities, which we can then define in terms of the fundamental quantities, are called *derived* quantities.

Scalar quantities can be specified using a single number. For example, the length of a rope is a scalar quantity. Some quantities cannot be specified completely without a direction as well as a number. For example, we cannot completely describe the motion of an object without giving the direction of its motion. A quantity that has a direction as well as a number is called a *vector* quantity. We must distinguish between scalar and vector quantities because they obey different rules of arithmetic, as we shall see in Chapter 4 where vector quantities are more precisely defined.

The Unit and Standard of Distance

We measure a distance in terms of some other distance. We say, for example, that the distance from Chicago to New York is 713 miles. The number 713 alone is not enough; we need the whole term, 713 *miles*, in order to indicate the distance correctly. The *mile*, which is itself a distance, is the unit in terms of which we usually measure the distance from Chicago to New York. We may thus define a *unit* of distance as that distance in terms of which we measure other distances. We could, of course, choose any distance we liked for our unit: a mile, a foot, an inch. In fact, we shall choose a unit commonly used in scientific work—the *meter*.

In order to define the meter (or any unit of distance), we need to have some sample distance we can rely on. We need some material object that will give us a distance to which we can refer. This material object, properly used, is the *standard* of distance. For a long time the standard of distance was a rigid bar, made of a corrosion-resistant metal alloy, which was kept in a controlled environment at a constant temperature and pressure (see Fig. 1-4). This bar had a thin scratch near each end. The distance between those scratches was defined to be one meter. Rulers one meter long were constructed by comparing them with the bar, and other distances were measured in meters using these rulers. A meter is exactly equal to 39.37 inches.

The rigid metal bar has been supplanted by another standard of distance.[3] However, it illustrates the use of standards and units in physics: quantities are measured in terms of units. These units can be defined by using suitable material objects or processes as standards.

The Unit and Standard of Time Duration

For our unit of time duration we shall choose the familiar *second*. There is, of course, no direct way to compare a second of time that

[3]In this book we shall use standards which are accessible and convenient for the purpose at hand. Sometimes, therefore, we shall not use the most precise standard currently adopted by scientists. We note here, for completeness, without describing the matter in detail, that currently the meter is defined in terms of the wavelength associated with a certain spectral line. The second is defined in terms of a particular atomic clock.

FIGURE 1-4 **Meter No. 27 at the National Bureau of Standards. This bar is similar to the original standard bar in Paris. Courtesy of the National Bureau of Standards.**

elapsed twenty years ago with a second of time today. Thus there is a problem in deciding just what should be called a second of time. The solution is to consider some process which happens over and over again in essentially the same way, such as the swinging of a pendulum or the rotation of the earth. We then agree that each seemingly identical process takes the same time. For our purposes here, we may consider that the earth takes the same time for each rotation (which is very nearly true) and obtain the *second* as our unit of time duration as follows: the second is defined as one 86,400th part of the earth's daily rotation period. In other words, there are 86,400 seconds in a day. (There are 60 seconds in a minute, 60 minutes in an hour, and 24 hours in a day. $60 \times 60 \times 24 = 86,400$.)

Derived Quantities

In our considerations of matter in motion, we will need to define many other quantities. However, we do not consider most of these quantities fundamental since they can be derived from other quantities. Area, speed, and velocity are examples of such *derived quantities*.

Area. If we have a rectangular region six meters long and three meters wide, we obtain the area of the region by multiplying these two distances together:

$$\begin{aligned} 6\text{ meter} \times 3\text{ meter} &= 18 \times (\text{meter}) \times (\text{meter}) = 8\text{ meter}^2 \\ &= 18\text{ square meters} \end{aligned}$$

Note that we can multiply units together just as if they were numbers. We multiply meter times meter and get meter squared, or, as we call it, "square meters." The unit of area is the square meter, or meter^2.

Just as we obtained an area by multiplying two distances together, so we can obtain a volume by multiplying three distances together. To find the volume of a rectangular object—for example, a brick—we multiply together its length, width, and height. The unit of volume is the cubic meter, or meter^3.

Speed and Velocity. A useful quantity associated with the motion of an object is its *speed*. The speed is determined by the amount of distance traveled by the object in a given time duration. In order to derive the quantity speed, we divide the fundamental quantity distance by the fundamental quantity time duration. For example, suppose that an object travels 10 meters in two seconds. We divide 10 meters by two seconds and obtain a speed of five meters per second:

$$\frac{10\text{ meter}}{2\text{ second}} = 5\,\frac{\text{meter}}{\text{second}} = 5\text{ meters per second}$$

(In this example, we are assuming that the speed does not change in these two seconds.) Thus, in general we obtain the speed of a moving object by dividing the distance traveled by the time duration. The unit of speed is

$$\text{meter per second} = \frac{\text{meter}}{\text{second}}$$

This unit is obtained by dividing the unit of distance by the unit of time duration.

Speed alone does not completely describe the motion of an object. We also need to know the direction in which the object is moving. Therefore, it is convenient to define a new quantity, which includes both the speed and the direction of a moving object. This quantity is called *velocity*. In order to specify the velocity of an object, we must give both the speed of the object and the direction in which it is traveling. Because it includes the direction, velocity is a vector quantity. Speed, on the other hand, is a scalar quantity and it omits the direction.

FUNDAMENTAL IDEAS

I. GENERAL INTRODUCTION TO SCIENCE
   - A. SCIENCE AND TECHNOLOGY AND THE DISTINCTION BETWEEN THEM
   - B. TWO SCIENTIFIC REVOLUTIONS

II. QUANTITIES. These are obtained by measurement or by combining other quantities that are obtained by measurement.
   - A. FUNDAMENTAL QUANTITIES
     1. Distance, a scalar.
     2. Time duration, a scalar.
   - B. DERIVED QUANTITIES
     1. Area, a scalar.
     2. Volume, a scalar.
     3. Speed, a scalar.
     4. Velocity, a vector.

QUESTIONS

General Introduction to Science

1. Explain what is generally meant by science, physics, and technology.
2. Which of the following activities are more nearly science and which are more nearly technology: repairing a television set, designing a television set, searching for the cause of cancer, searching for a cure for can-

Galileo ridiculed those followers of Aristotle who claimed that an arrow, after it leaves the bowstring, continues to move forward only because the air stirred up by the bowstring keeps pushing the arrow along. Galileo pointed out that if this were true, the arrow would point feathers first, as does an arrow blown by the wind.[3] In fact, the air exerts a retarding force, called friction.

Friction

There are nearly always forces that cause moving objects to slow down and stop. This is why people did not discover the first law of motion sooner. Objects that are sliding, swimming, or flying are rubbing aginst something—a solid surface, the water, or the air. The rubbing causes a force which tends to slow the objects down and eventually stops them. Such retarding forces are called friction.

Illustrations of the First Law of Motion

Our experience with automobiles furnishes us with many illustrations of the first law of motion. Once a car is rolling along, a force is required to stop it. If we wait long enough, friction will stop it; if we are unwilling to wait, we increase the frictional force by stepping on the brakes. If we jam on the brakes, the frictional forces stop the car quickly. But the brakes do not act directly on the passengers inside the car. Therefore, they continue their forward motion in accordance with the first law—right on through the windshield if nothing prevents it. Hence the use of seat belts. If the car turns suddenly, the occupants continue in the original direction and, therefore, slide across the seat unless friction prevents it.

The first law of motion also helps us to understand what produces "whiplash" injury in an automobile collision. When a car is struck from the rear, the seat, which is firmly attached to the car itself, pushes the occupant's body forward. But his head, which is not rigidly fastened to his body or supported by the car seat, tends to remain where it was. The motorist therefore suffers an injury at the top of the spine, where the head and body meet. The situation is illustrated in Figure 2-3, for one common posture. The unsuspecting driver nearly loses his head, so to speak.

[3]Galileo Galilei, *Dialogue Concerning the Two Chief World Systems, Ptolemaic & Copernican*, translated by Stillman Drake, Berkeley: University of California Press, 1962, p. 152–153. (See credit in footnote 2.)

FIGURE 2-3 A whiplash situation—Why do they tend to lose their heads?

A sky diver falling from an airplane provides another illustration of Newton's first law. The sky diver does not fall straight down; he has the same forward speed as the plane when he is dropped, and he preserves that forward motion—somewhat retarded by air friction—as he falls. Therefore, he keeps moving forward at the same time as he moves downward (see Fig. 2-4).

Galileo's Reasoning[4]

For most physical principles, the process of discovery is either unknown or mathematically inacessible to the general reader. Galileo, however, often presented the development of his ideas in ways that were "elegant" and simple.

Consider the inclined planes shown in Figure 2-5*a*. We allow a ball to roll up and down the planes. You will notice that we have rounded off the corners where the planes meet the level; this is done so that the ball will roll smoothly. If a ball struck a corner, it might bounce.

[4]Galileo Galilei, *Discourses and Mathematical Demonstrations Concerning Two New Sciences;* see *Dialogues Concerning Two New Sciences*, translated by Henry Crew and Alfonso de Salvio, New York: McGraw Hill, 1963, pp. 164–166. (We have adopted the alternate title suggested by I. B. Cohen, *The Birth of a New Physics*, New York: Doubleday Anchor Books, 1960, p. 193.) We have not been able to find this argument in Galileo, but it is ascribed to him by Ernst Mach, in *The Science of Mechanics*, LaSalle, Ill.: Open Court, 1960, pp. 168–169; see also Galileo Galilei, *Dialogue Concerning the Two Chief World Systems, op. cit.*, pp. 146–147.

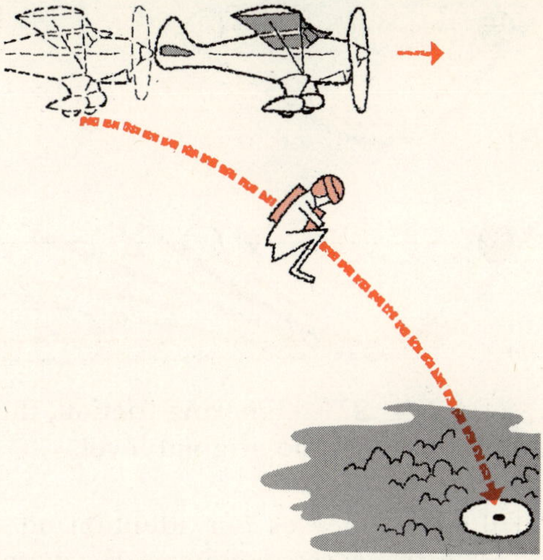

FIGURE 2-4 **The sky diver continues to move forward even after he jumps from the plane.**

We release a ball from the left-hand plane, letting it roll down that plane, across the level, and then up the other inclined plane. How far up the other plane will it go? Intuitively we answer: as high above the level as it was when it started on the left-hand plane. Then we think more carefully; if there is some friction, the ball will probably not go quite so far up as it was when it started. So we idealize the experiment we are thinking about. In practice we cannot remove friction entirely. However, the smoother and harder the surface and the ball, the less friction there will be and the more nearly the ball will come back to the same height.

Now we lower the plane on the right, still repeating the same experiment (see Fig. 2-5*b*). The ball should still come up to the same height, but the same height will put the ball farther and farther to the right as the plane is made less steep. In the limiting case, when the plane on the right is horizontal, the ball should keep going forever. Thus we have arrived at the first law of motion: the ball, being neither retarded nor pushed ahead by any force, keeps moving uniformly forever.

In the preceding paragraphs we performed what is sometimes called a "thought experiment." We "talked" an experiment without actually doing it. (There are those who accuse Galileo of doing too much of this.) We assumed certain ideal conditions that would not exist in actuality—for example, the absence of all friction. This type

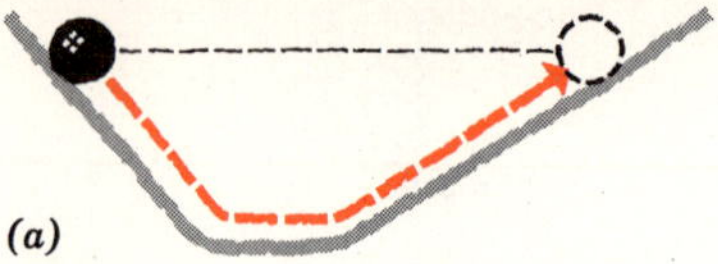

(a)

(b)

FIGURE 2-5 Ignoring friction, the ball keeps going until it reaches the original level.

of thinking focuses our attention on certain ideas and helps us to work out their consequences. We were in this way led to the first law of motion, which gives the behavior expected of objects with no forces acting on them. Actually, of course, there do not appear to be any such force-free objects—there always appear to be some forces acting on an object. Thus the first law of motion describes an idealized condition that never occurs. Yet, after making this idealization, we are able to understand real processes better than if we merely made detailed observations of the real processes themselves. Experimental science abstracts and idealizes from experience to arrive at principles, which form a sort of model of reality. It is amazing how precisely this abstract model of reality can predict the behavior of real objects.

Of course principles such as the first law of motion are not established by thought experiments alone, just by thinking, without actual observations. Since we only talked about the experiment, we need to check its conclusions by actual experiment, as Galileo clearly pointed out.[5] Also, some of our experience went into the discussion of the thought experiment. What really convinced us that the ball could rise no higher than it started was our general experience that balls do not run uphill by themselves. Similar notions were used by other early scientists in developing theories.

Later we shall fasten our attention again on the interesting fact that the ball returns to the same height it started from. It suggests that something remains the same, or is conserved, as the ball moves on the inclined planes.

[5]Galileo Galilei, *Dialogues Concerning Two New Sciences, op. cit.*, pp. 163–166.

The New Philosophy

And new Philosophy calls all in doubt,
The Element of fire is quite put out;
The Sun is lost, and th' earth, and no mans wit
Can well direct him where to looke for it.

. . .

'Tis all in peeces, all cohaerence gone;
All just supply, and all Relation.

John Donne[6]

The New Astronomy

At the time of Galileo, tradition favored the physics of Aristotle and the astronomy of Ptolemy (2nd century A.D.). This old astronomy and physics had not gone unchallenged, but it had been reconciled with Christian theology in the so-called Scholastic philosophy of Thomas Aquinas (1225?–1274?). Thus it had great authority for both philosophical and theological reasons.

The old physics and astronomy divided space into two regions. Both the nature of objects and their motion were different in the two realms. All objects higher than the moon were celestial. They were thought to be light, unchanging, perfectly smooth, and spotless spheres. Many thought that the moon also was perfectly smooth and spherical. Furthermore, these celestial spheres were assumed to move in perfectly circular paths. In the terrestrial realm below the moon, heavy objects were imperfect, changeable, and had a natural motion toward the center of the earth. Objects such as arrows and cannonballs could also have an unnatural motion in other directions if they were continually pushed.

According to Aristotle and Ptolemy, the earth itself was motionless in the center of the universe. The heavens revolved about it once a day, causing sunrise, sunset, and the motion of the stars seen in the night sky. The revolution of objects in the solar system was troublesome in this scheme since these objects appear to change their position in the sphere of stars.

The planets in particular presented a formidable problem. When viewed against the starry background at the same hour on many consecutive nights, their motion is seen to be eastward, with occasional temporary reversals that for a time have a westward direction (see

[6]John Donne, as quoted in I. B. Cohen, *The Birth of a New Physics*, *op. cit.*, p. 89.

FIGURE 2-6 **As seen from the earth, a planet regularly appears to reverse the direction of its path among the stars for a period of time.**

Fig. 2-6). It was considered essential for all the heavenly bodies to move in circular motion, for this was supposed to be the natural motion of perfect celestial objects. Since the planets did not appear to revolve in this simple way, it was necessary to invent a scheme or trick to save the theory. Ptolemy's scheme (and that of others) was to use epicycles. To explain what an epicycle is, let us imagine a certain point moving around the earth in a perfect circle (see Fig. 2-7). Instead of letting that point represent a planet, however, we let it be the moving center of another circle. It is around this moving center that the planet is supposed to revolve. In a sense, then, an epicycle is a circle on another circle. If necessary, other epicycles could be added, until the theoretical model fit the observed movement—that is, until the calculated motion agreed with the recorded observations and thus "saved the appearances." Actually, dozens of epicycles were necessary, as well as other tricks, to make the whole system agree with the observed appearances.

Copernicus (1473–1543) suggested that the sun is the center of the universe, and that the planets, including the earth, revolve around the sun (see Fig. 2-8). He further suggested that the earth also rotates on its axis once a day, making the heavens appear to go around the earth. He could then account for the temporary reversal in the motion of the planets. For example, when the fast-moving earth passes closest to the slower-moving Mars, Mars appears to go backward in its orbit (see Fig. 2-8*b*). Since the planets do not go around the sun in perfect circles, it was necessary for Copernicus also to use epicycles to account exactly for the observed planetary motion and thus save the appearances.

Copernicus knew that he was reviving the doctrine of Aristarchus and certain other ancient Greek philosophers and astronomers when he suggested that the earth moved. These doctrines, however, had

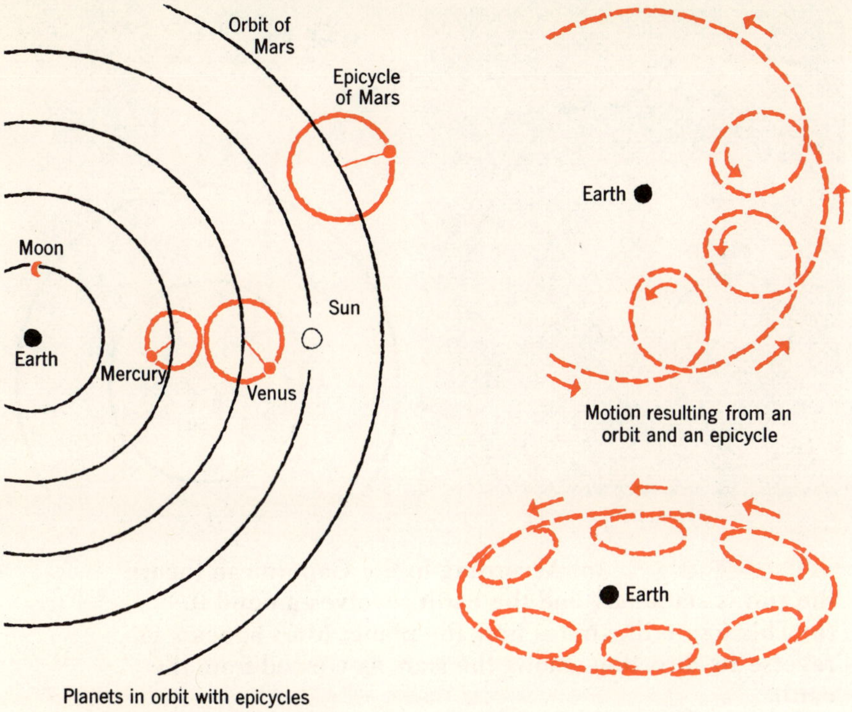

FIGURE 2-7 **According to the Ptolemaic theory the earth is stationary. Epicycles were used to account for the observed motion of other planets.**

long been out of favor. Osiander, in an unsigned preface to Copernicus's main treatise, made the doctrine more palatable to some by emphasizing that the suggestion was merely a mathematical technique to save appearances. Johannes Kepler, convinced of the reality of the sun-centered system, said that this preface was "written by a jackass for the use of other jackasses."[7]

It was Kepler who finally discovered that the planets did not travel in perfect circles but in *ellipses*. The ellipse is an oval curve that can be more or less like a circle. An ellipse can be drawn as shown in Figure 2-9 by looping a string around two nails or pins and then pulling a pencil around the loop. The ellipse can be made very similar to a circle by putting the two nails close together. The position of the

[7]Quoted in Giorgio de Santillana *The Crime of Galileo*, Chicago: University of Chicago Press, 1955, p. 101. Copyright © 1955 by the University of Chicago. All rights reserved. Copyright © 1955 under the International Copyright Union.

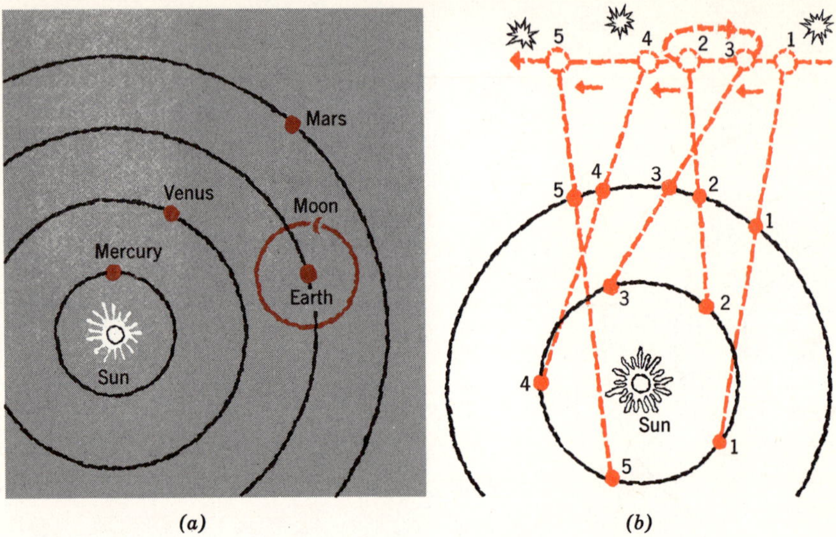

FIGURE 2-8 (*a*) According to the Copernican theory the sun is stationary and the earth revolves around it. (*b*) This figure illustrates how the planet Mars appears to reverse its direction among the stars as viewed from the earth.

FIGURE 2-9 An ellipse may be drawn using a loop of string.

two nails gives the position of the foci of the ellipse. Of course, if the two foci are at the same point, the ellipse is a perfect circle. The planets go around the sun in ellipses that are very close to circles.

Kepler worked laboriously for a long time, using the accurate data on the orbit of Mars which the Danish astronomer Tycho Brahe (1586–1601) had obtained. No matter what scheme he used in putting together an orbit from perfect circles, he could never quite match the observed orbit of Mars. He was always left with a very small error in angles of observation. However, he found that a single ellipse matched the observations to their limit of accuracy:

> But for us, who by divine kindness were given an accurate observer such as Tycho Brahe, for us it is fitting that we should acknowledge this divine gift and put it to use. . . . For if I had believed that we could ignore these eight minutes [a very small angle] I would have patched up my hypothesis accordingly. But since it was not permissible to ignore them, those eight minutes point the road to a complete reformation of astronomy. . . .[8]

And to much else besides, we might add.

Kepler arrived at three famous laws of planetary motion from his analysis of Tycho Brahe's data.

1. The planet moves in an ellipse about the sun, with the sun at one focus (Fig. 2-10).

2. The line between the sun and the planet sweeps across equal

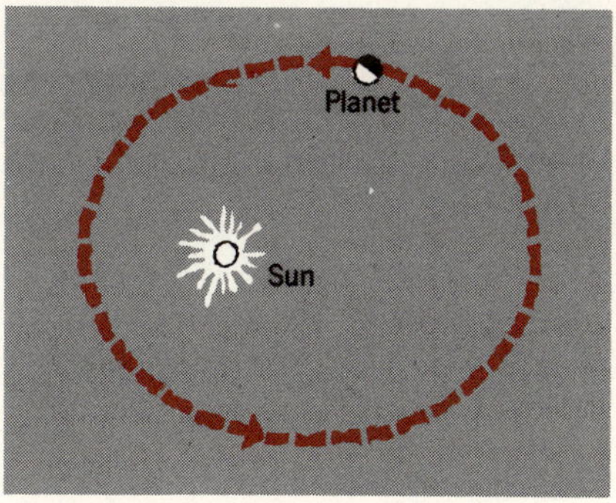

FIGURE 2-10 Kepler's first law—the planet moves in an ellipse about the sun, with the sun at one focus.

[8]Johannes Kepler, *Astronomia Nova, II*, Chap. 19, as quoted in Arthur Koestler, *The Sleepwalkers*, New York: Macmillan, 1959, p. 322. © Arthur Koestler 1959.

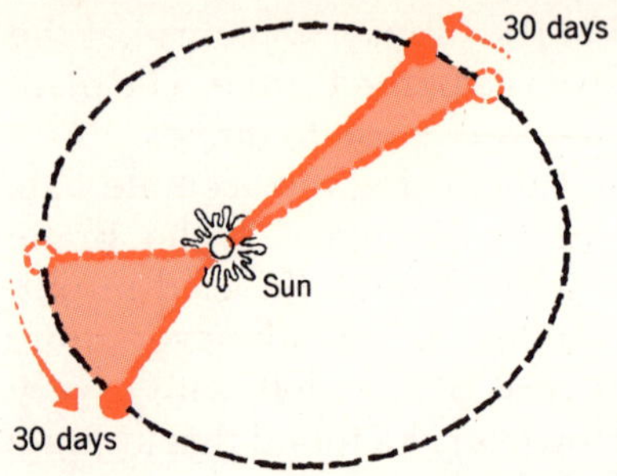

FIGURE 2-11 **Kepler's second law—the planet sweeps out equal areas in equal time intervals. The two orange regions are equal in area. The planet moves slower when it is farther away.**

amounts of the area of the ellipse in equal amounts of time (Fig. 2-11).

3. The square of the period of revolution of the planet around the sun is proportional to the cube of its average distance from the sun (Fig. 2-12).

We do not wish to enter into a detailed explanation of these laws; it should be noted, however, that they gave a precise mathematical description of planetary motion. If Galileo had known and appreciated Kepler's laws he would have had an easier time defending the motion of the earth. Kepler's book, *Astronomia Nova* (The New As-

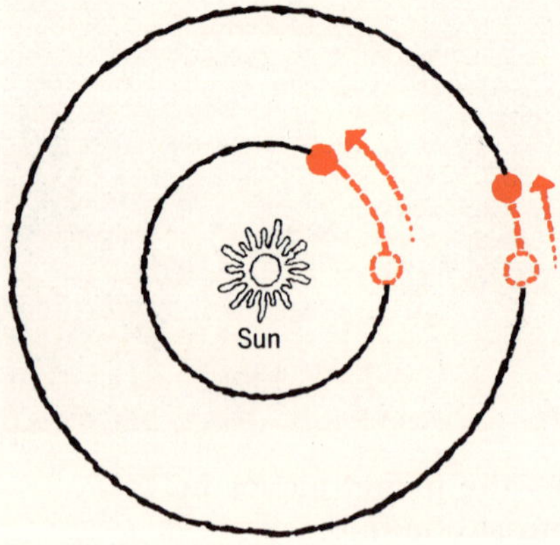

FIGURE 2-12 **Kepler's third law implies that the farther the planet is from the sun the longer it takes to go around and the slower it moves in its orbit.**

tronomy), containing his first two laws, may even have been sitting on Galileo's shelf unheeded.

Galileo improved the telescope, which had just been invented, and made many astronomical discoveries that upset the old ideas. He saw the planet Jupiter with its moons going back and forth in the vicinity of the planet, and concluded that the moons were revolving around the planet. Here was a solar system in miniature, for all to see. He observed that the planet Venus showed changes like the moon, appearing at various times as full, half full, and crescent, in a way that would not occur in the Ptolemaic system. Then he saw mountains and valleys on the moon—a supposedly smooth celestial body. He saw spots on the sun, a supposedly spotless celestial body. These were not the first observational assaults on the old astronomy. Even before this time, in 1572 and 1604, two new stars, the very rare supernovae, had been seen in the supposedly unchangeable heavens. Tycho Brahe showed that they were much further away than the moon because they showed no measurable parallax.

The Conflict with Authority

Traditional philosophers and conservative theologians were disturbed by the new astronomy. At least one Aristotelian philosopher refused to look through the telescope. Another claimed that "geometry is of the Devil" and that "mathematicians should be banished as the authors of all heresies."[9] Still another charged that "to see the satellites of Jupiter, men had to make an instrument that would create them."[10]

Even before Galileo's time, responsible religious leaders were concerned because they thought the astronomy of Copernicus conflicted with the sacred scriptures. John Calvin (1509–1564) opposed it: "Who will venture to place the authority of Copernicus above that of the Holy Spirit?"[11] Martin Luther (1483–1546) wrote:

> People gave ear to an upstart astrologer who strove to show that the earth revolves, not the heavens or the firmament, the sun and the moon. . . . Whoever wishes to appear clever must devise some new system, which of all systems is of course the very best. This fool wishes to reverse the entire science of astronomy; but sacred scrip-

[9]A. D. White, *A History of the Warfare of Science with Theology in Christendom*, Vol. I, Dover Publications Inc., New York, 1960, p. 133. Reprinted through permission of the publisher.

[10]Quoted in *ibid.*, p. 132.

[11] *Ibid.*, p. 127.

tures tell us that Joshua commanded the sun to stand still and not the earth.[12]

St. Robert Bellarmine, in Galileo's day, wrote:

> . . . To say that on the supposition of the Earth's movement and the Sun's quiescence all the celestial appearances are explained better than by the theory of eccentrics and epicycles is to speak with excellent good sense and to run no risk whatever. Such a manner of speaking is enough for a mathematician. But to want to affirm that the Sun, in very truth, is at the center of the universe and only rotates on its axis without going from east to west, is a very dangerous attitude and one calculated not only to arouse all Scholastic philosophers and theologians but also to injure our holy faith by contradicting the Scriptures. . . .
>
> If there were a real proof that the Sun is in the center of the universe, that the Earth is in the third heaven, and that the Sun does not go around the Earth but the Earth around the Sun, then we should have to proceed with great circumspection in explaining passages of Scripture which appear to teach the contrary, and rather admit that we did not understand them than declare an opinion to be false which is proved to be true. . . .[13]

Galileo could not prove that the earth moved but only that the arguments against its motion were invalid. In 1623, after conferring with his apparent admirer,[14] the newly elected Pope Urban VIII, Galileo published his *Dialogue Concerning the Two Chief World Systems*, which presented an apparently strong argument for the Copernican system, although one of his arguments was erroneous.

Apparently Galileo's argument was stronger than the Pope had expected, and the book, though very popular, was banned. Galileo was tried, convicted of heresy, and forced to renounce his theory. Although three cardinals refused to sign the condemnation, and the doctrine of Copernicus was never formally declared heretical, Galileo was sentenced to imprisonment by the Inquisition.[15] The sentence includes the following charge:

> The proposition that the Sun is the center of the world and does not move from its place is absurd and false philosophically and for-

[12] *Ibid.*, p. 126.

[13] As quoted in Santillana, *op. cit.*, pp. 99–100.

[14] *Ibid.*, pp. 156, 157, and Chap. VIII.

[15] For a detailed discussion of the political and theological motives for this action against Galileo and his book, see Santillana, *op. cit.*, Chaps. VIII, XV.

mally heretical, because it is expressly contrary to the Holy Scripture.

The proposition that the Earth is not the center of the world and immovable but that it moves, and also with a diurnal motion, is equally absurd and false philosophically and theologically considered at least erroneous in faith.[16]

What were some of the Scriptures which the new philosophy supposedly contradicted?

Then spake Joshua to the Lord in the day when the Lord delivered up the Amorites before the children of Israel, and he said in the sight of Israel, Sun, stand thou still upon Gibeon; and thou, Moon, in the valley of Ajalon. (Joshua 10:12)

. . . the world also is stablished, that it cannot be moved. (Psalm 93:1)

Galileo himself felt that there was no conflict when the Scriptures were properly understood, but he feared that ignorant and antiscientific theologians would bring the scriptures into disrepute.[17] In a famous letter to the Grand Duchess Christina (1615),[18] he wrote that if the Bible had spoken openly of the earth's motion no one would have understood. He also stated:

. . . I think that in discussions of physical problems we ought to begin not from the authority of scriptural passages, but from sense-experiences and necessary demonstrations; for the holy Bible and the phenomena of nature proceed alike from the divine Word. . . . It is necessary for the Bible, in order to be accomodated to the understanding of every man, to speak many things which appear to differ from the absolute truth so far as the bare meaning of the words is concerned. But Nature, on the other hand, is inexorable and immutable; she never trangresses the laws imposed upon her, or cares a whit whether her abstruse reasons and methods of operation are understandable to men. For that reason it appears that nothing physical which sense-experience sets before our eyes, or which necessary demonstrations prove to us, ought to be called in question (much less condemned) upon the testimony of biblical passages which may have some different meaning beneath their words.[19]

[16]As quoted in *ibid.*, p. 307.

[17]From *Discoveries and Opinions of Galileo*, by Stillman Drake. Copyright ©, 1957 by Stillman Drake. Quotations to follow used by permission of Doubleday & Company, Inc., p. 168.

[18] *Ibid.*, pp. 175–216.

[19]*Ibid.*, pp. 182–183.

At the very heart of the new philosophy was the importance of "sense experience" and necessary reasoning, particularly mathematical reasoning. This means that things are not to be taken on authority but must be checked by observation and calculation. Galileo emphasized that we really know very little and cannot, at least at present, answer all questions. He ridiculed those who appealed to authority alone, citing the experience of a certain expert in anatomy:

Turning to a gentleman whom he knew to be [an Aristotelian] philosopher, and on whose account he had been exhibiting and demonstrating everything with unusual care, he asked this man whether he was at last satisfied and convinced that the nerves originated in the brain and not in the heart. The philosopher, after considering for a while, answered: "You have made me see this matter so plainly and palpably that if Aristotle's text were not contrary to it, stating clearly that the nerves originate in the heart, I should be forced to admit it to be true."[20]

And again:

. . . what is more revolting in a public dispute, when someone is dealing with demonstrable conclusions, than to hear him interrupted by a text (often written to some quite different purpose) thrown into his teeth by an opponent? If, indeed, you wish to continue in this method of studying, then put aside the name of philosophers and call yourselves historians, or memory experts; for it is not proper that those who never philosophize should usurp the honorable title of philosopher.

. . . So put forward the arguments and demonstrations, Simplicio—either yours or Aristotle's—but not just texts and bare authorities, because our discourses must relate to the sensible world and not to one on paper.[21]

Although Galileo did not reject the civil and religious authority of civil and religious leaders, he did reject claims to authority where scientific truth is concerned. Indeed, he made it clear that truth is not subject to authority. Instead, he argued for a point of view usually regarded as inherent in the so-called scientific method, as in the following excerpts from his reply to "Sarsi" (the pseudonym of the noted Jesuit astronomer, Horatio Grassi):

[20]Galileo Galilei, *Dialogue. . .*, *op. cit.*, p. 108. Originally published by the University of California Press. Reprinted by permission of the Regents of The University of California.

[21]*Ibid.*, p. 113.

In Sarsi I seem to discern the firm belief that in philosophizing one must support oneself upon the opinion of some celebrated author, as if our minds ought to remain completely sterile and barren unless wedded to the reasoning of some other person. Possibly he thinks that philosophy is a book of fiction by some writer, ... productions in which the least important thing is whether what is written there is true. Well, Sarsi, that is not how matters stand. Philosophy is written in this grand book, the universe, which stands continually open to our gaze. . . .

I cannot but be astonished that Sarsi should persist in trying to prove by means of witnesses something that I may see for myself at any time by means of experiment. . . . So when Sarsi brings in this multitude of authors it appears to me that instead of strengthening his conclusion he merely ennobles our case by showing that we have outreasoned many men of great reputation. . . .

Sarsi says he does not wish to be numbered among those who affront the sages by disbelieving and contradicting them. I say I do not wish to be counted as an ignoramus and an ingrate toward Nature and toward God; for if they have given me my senses and my reason, why should I defer such great gifts to the errors of some man? Why should I believe blindly and stupidly what I wish to believe, and subject the freedom of my intellect to someone else who is just as liable to error as I am?[22]

Galileo is still a controversial figure. His biographer, Santillana, contends that Galileo was trying to keep the authorities from making fools of themselves and his Church, since he could see "the three forces of mathematics, physics, and astronomy converging rapidly toward a junction which would make them irresistible."[23] Arthur Koestler severely criticizes Galileo for his fight with the Jesuits and seems to blame Galileo's arrogance for facilitating the split between the scientific and spiritual sides of our culture.[24] Scientists generally consider Galileo as a champion of freedom of scientific investigation.

We are not historians, but we find in the writings of Galileo the same scientific attitude that we are used to in this modern age. True, our attitudes toward the immutable mathematical truth of nature have altered somewhat since Galileo's day. Still, the combination of the practical, the workable, with the intellectual and the mathematical seems very much the scientific point of view of our own day. In

[22]Drake, *op. cit.*, pp. 237–238, 271–272.
[23]Santillana, *op. cit.*, p. 106.
[24]Koestler, *op cit.*, pp. 425–426.

 the new philosophy we are compelled to be as accurate in practical matters as in logic. It is indeed the possibility of practicable operation that determines truth, according to one point of view.

FUNDAMENTAL IDEAS

I. FUNDAMENTAL LAWS AND PRINCIPLES

A. THE FIRST LAW OF MOTION. An object continues in its state of rest, or of uniform motion in a straight line, unless it is compelled to change that state of motion by forces impressed upon it.

II. APPLICATIONS

A. FRICTION

III. METHODS OF SCIENCE. We discuss Galileo, Kepler, and others and the place in scientific thought of:

A. REASON: including thought experiments.

B. EXPERIMENT: necessary to verify laws and theories.

C. AUTHORITY: not acceptable as proof of scientific ideas.

QUESTIONS

The First Law of Motion

1. Can something really keep moving without being "pushed"?
2. If an elephant and an ant were both moving at different speeds on a level, frictionless surface, which would stop first? (See Fig. 2-13)

FIGURE 2-13 See question 2.

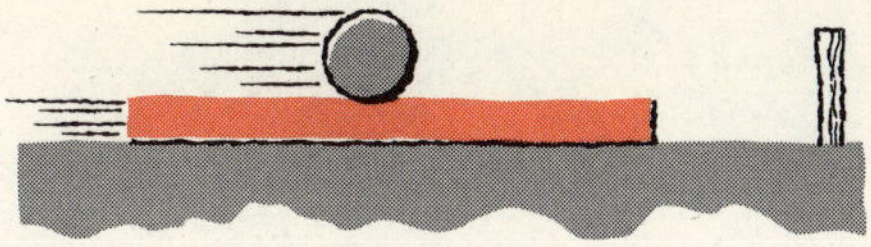

FIGURE 2-14 See question 4.

3. List some activities that would be impossible or very difficult if there were no such thing as friction. List some that would be possible.
4. A ball is placed on a board that is resting on a frictionless surface. If we suddenly move the board, what happens to the ball? If the ball and board are moving across the surface as illustrated in Figure 2-14, and the board strikes a post, what happens to the ball?
5. Galileo considered the thought experiment shown in Figure 2-15. If we put the peg in as shown, where will the pendulum bob swing?
6. How did people think that the universe was arranged before the new philosophy was developed?
7. What is the apparent motion of the planets relative to the stars as observed from the earth?
8. How was the apparent motion of the planets explained by Ptolemy? by Copernicus?
9. What did Galileo and others observe that could not be explained by the old physics and astronomy?
10. What did eight minutes of arc have to do with the history of astronomy?
11. What, according to Galileo, was the way to find out the truth about nature?

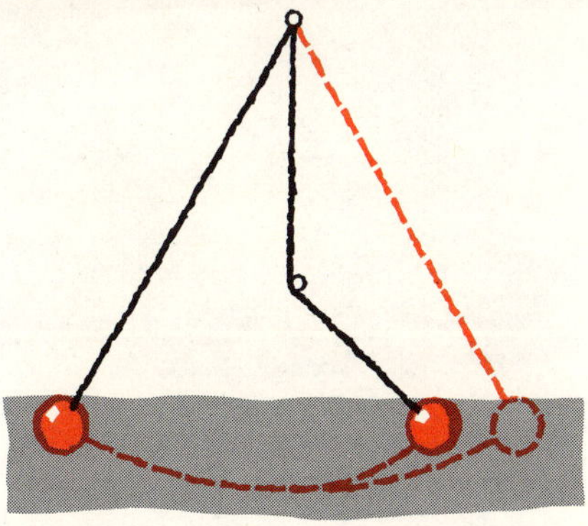

FIGURE 2-15 See question 5.

3 Systems in Motion

Odd as it may seem, most people's views about motion are part of a system of physics that was proposed more than 2000 years ago and was experimentally shown to be inadequate at least 1400 years ago. It is a fact that presumably well-educated men and women tend even today to think about the physical world as if the earth were at rest, rather than in motion. By this I do not mean that such people really believe the earth is at rest; if questioned, they will reply that of course they "know" that the earth rotates once a day about its axis and at the same time moves in a great yearly orbit around the sun. Yet when it comes to explaining certain common physical events, these same people are not able to tell you how it is that these everyday phenomena can happen, as we see they do, on a moving earth.

I. Bernard Cohen[1]

[1]I. Bernard Cohen, *The Birth of a New Physics*, Anchor Books, Garden City, New York: Doubleday & Company, Inc., 1960, p. 13.

FIGURE 3-1 The earth viewed from the moon. What is really moving? Courtesy of NASA.

Most people in Galileo's day thought that the earth was motionless in the center of the universe and that the sun and other heavenly bodies revolved around it. Galileo, on the other hand, was convinced that the earth itself moved around the sun, even though the motion was not apparent to people on the earth. Lacking any direct evidence that the earth was actually rushing through space at a speed of some nineteen miles per second, he had to persuade his opponents by explaining how the earth could appear to be stationary when in fact it was not. Such rapid motion of the earth, his antagonists claimed, would leave the air behind. Even a bird perched on a branch would be left behind in space the instant it let go of the branch, as the earth rushed on — or so it was argued (see Fig. 3-2). How can we account for the fact that this does not occur?

Galileo pointed out that the bird was moving with the earth while it was on the branch and retained that motion when it let go. Thus the relative motion of the bird with respect to the tree would be the same as if the earth were at rest. According to Galileo, it is the relative motion that is important. Before discussing relative motion in greater detail, we need to understand what is meant by a reference frame.

FIGURE 3-2 **The bird is *not* left behind by the moving earth, despite this figure.**

Reference Frames

Anyone who has followed a map to find his way from one place to another has used what we call a *reference frame.* You cannot begin such a trip until you first "get oriented." This orientation amounts to establishing a reference frame. You must know your present position with respect to some object (such as a tree, a rock, or a city), and you must also be able to establish the direction of your goal. To establish direction on a map, we ordinarily imagine two intersecting lines, one pointing north and the other east. The first can be established with a compass needle, and the other is a right angles to it. These imaginary lines give us an example of a reference frame. (Other kinds of reference frames are also used.) Now travel instructions have some meaning. For example, we can go 8000 meters north from the rock, 10,000 meters east, and another 16,000 meters north and find the treasure (see Fig. 3-3*a*).

The two imaginary lines that intersect at right angles to make up the reference frame are called *axes*. The point at which these axes intersect is called the *origin*. In our example, the axes intersect at the rock. The rock, then, is our origin.

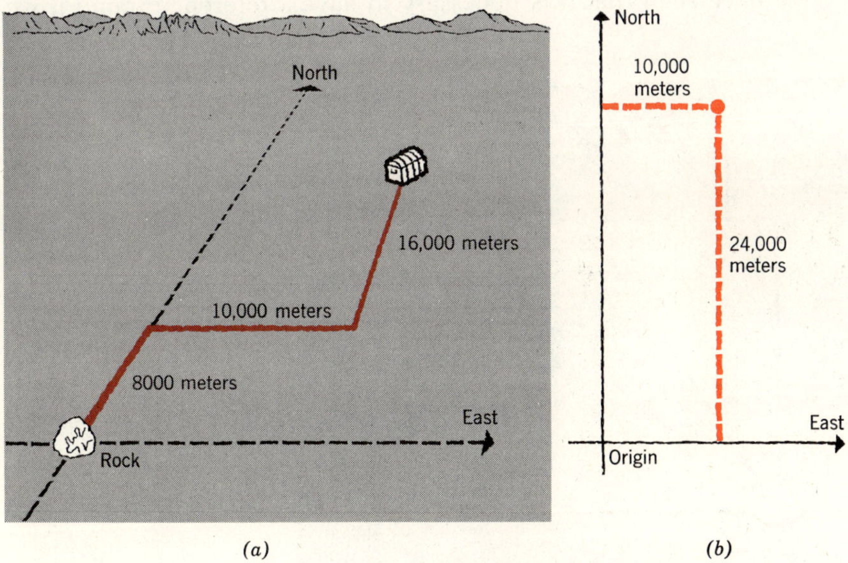

FIGURE 3-3 **A reference frame being used to locate a treasure.**

Many different sets of instructions could be made up that would eventually take us from the rock to the treasure. Once we have set up the reference frame, the simplest instructions would require just two statements. The only instructions actually needed to go to the treasure from the rock (origin) are the following ones:

(a) Go 10,000 meters east, parallel to the east-west axis.

(b) Go 24,000 meters north, parallel to the north-south axis.

The order in which we follow these instructions does not matter; either way, we will arrive at the same spot (see Fig. 3-3*b*).

In using just two lines in our reference frame, we have assumed that we were on a surface. If the treasure were buried underground or hidden in an old oak tree, we would need another instruction in order to locate it. This instruction would tell us how far above or below the surface we must go to get to the position of the treasure. As this example suggests, to locate objects in ordinary space we need three reference lines or axes in our reference frame, all intersecting at the same point (the origin) in such a way that each is perpendicular to the other two (Fig. 3-4). To locate any point in ordinary space, three instructions, or items of information, are necessary. For this reason we refer to ordinary space as three-dimensional.

We have seen that it is necessary to have a reference frame if we

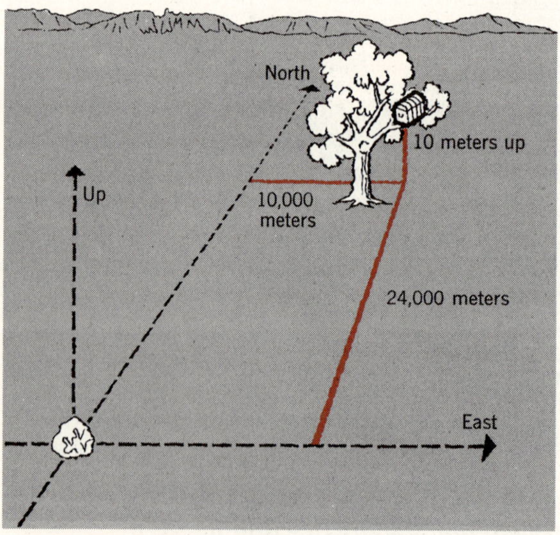

FIGURE 3-4 **A reference frame in three dimensional space being used to locate a treasure in an old oak tree.**

want to locate a position in space from a set of instructions. It is also necessary to use a reference frame in describing the motion of an object. The velocity (speed and direction) of an object should be given in reference to something else. Most often, the reference frame is similar to that described above for the location of a point or position. For example, the train in Figure 3-5*a* is moving to the northeast at 30 meters/second.

We are so accustomed to using compass directions fixed to the earth for our reference frame that we may not realize that reference frames attached to other objects and oriented in different directions are just as valid and sometimes more useful. Figure 3-5*b* shows the same train moving on the surface of the earth from a different point of view. The earth with the train on it is itself rotating and moving through space around the sun. What, we might ask, is the "real" motion of the train—its motion across the earth's surface, or its motion that includes the motion of the earth around the sun? Actually, either motion is real—the difference is in the reference frame with respect to which the motion is considered. In the former case we use the conventional reference frame fixed to the earth and describe the motion of the train as observed from the earth. In the latter case we view the motion as though we were located on the sun and refer it to a reference frame centered in the sun, with axes pointed outward toward selected stars. Although the motion looks very different from

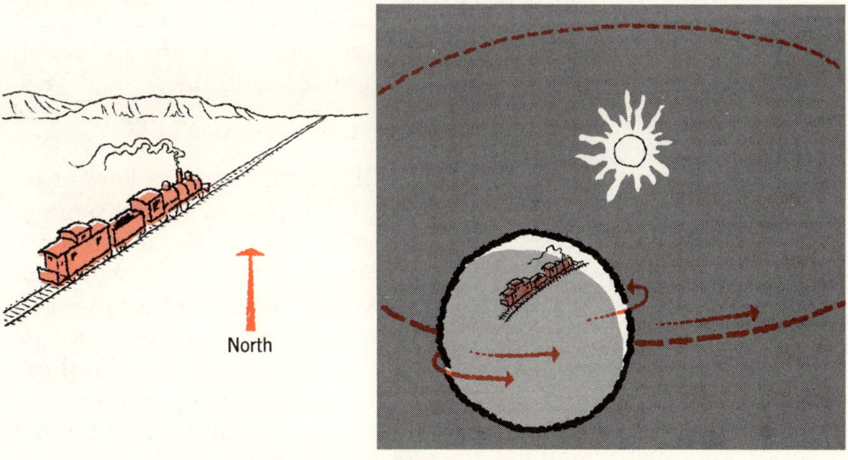

(*a*) (*b*)

FIGURE 3-5 **The motion of the train appears different as observed from different reference frames.**

 different reference frames, it is the same physical situation that is being observed and described.

Another example of the fact that the same motion can be described from different reference frames with equal validity is afforded by the motion of a stewardess in an airplane. Although the airplane may be traveling 600 miles per hour, to the passengers the stewardess appears to move around normally. That is, relative to a reference frame attached to the airplane, her motion appears no faster than it would be if she were walking on the ground. Relative to a reference frame fixed to the earth, however, she is rushing by at about 600 miles per hour. Since the same motion appears different from different reference frames, we cannot completely describe speed or velocity unless we tell what reference frame we are using.

The Special Principle of Relativity

The fundamental principle which we shall use in discussing the relative motion of different systems is called *the special principle of relativity*. This principle may be stated as follows:

All the laws of physics will be observed to be the same in all reference frames moving at a constant speed in a straight line.

A reference frame, or object, moving at a constant speed in a straight line is said to be in *uniform motion*, or *moving uniformly*. In the following paragraphs we shall discuss the meaning of the special principle of relativity, discuss its relation to the first law of motion, and illustrate it with several examples.

The special principle of relativity tells us that under a certain restricted condition, the laws of physics will appear to be the same in different reference frames. The restriction is that these reference frames must be either at rest or moving in a straight line with constant speed, and not rotating. In other words, they must be moving in the same way that a non-rotating object moves when there are no forces applied to it, according to the first law of motion. Under these conditions each reference frame, when viewed from another, will appear to be moving in a straight line with constant speed, and not rotating. An example will help us to understand this principle.

Suppose that we are in a railway car, but that the train has not yet started to move. It is sitting in the station (see Fig. 3-6*a*). If we drop a ball, it will fall straight down. Now suppose that we have left the sta-

(a) (b)

FIGURE 3-6 **As viewed from inside the train, the motion of the ball appears the same whether the train is at rest or in uniform motion.**

tion, and that the train is going eighty miles an hour along a straight smooth track (Fig. 3-6*b*). Once again, we drop the ball. What happens? To us in the train it looks as though the ball drops straight down. In other words, the result looks to us in the train exactly as it did when the train was at rest with respect to the ground. To an observer outside the train, however, the ball moves forward with the train as it drops.

We can account for this phenomenon by using the first law of motion: since the ball was moving with the train *before* it was dropped, it continues to move forward with the train *as* it drops. But an observer in the train is also moving forward with the train. Thus it looks to him as if the ball falls straight toward the earth. In fact, if we were to set up a scientific laboratory on the train and test all the known laws of physics, they would be just exactly the same as if we tested them in a laboratory that was at rest. The fact that the laws are tested in a moving reference frame fixed to the train makes no difference whatsoever. We say that the laws of physics are *invariant*, regardless of the straight uniform motion of the reference frame. This is another way of stating the special principle of relativity. Invariant means unchanged. The laws are not changed by using another reference frame that is moving uniformly (at a constant speed in a constant direction).

If all the laws are the same in the moving system as in a stationary system, then there is no way of telling that the system is moving without looking outside the system at something else. From the train we must look outside and see the trees and houses going by; then we know that we are in motion relative to them. There is no way to tell which is "really" moving, the train or the trees. Uniform motion is relative. That is, uniform motion is meaningful only as a relation between one frame of reference and another, or one object and another. Because all the laws of physics are unaffected by uniform motion, there is no way of finding any "real" or "absolute" uniform motion that a particular reference frame may have. We can measure relative motion only with respect to other objects or frames.

In all of this, of course, we are talking only about reference frames moving uniformly. If a frame is in nonuniform motion, we can detect it. If we are spinning, speeding up, or otherwise changing motion, we can feel or detect it without looking outside our system. If the train is vibrating, or if it turns a corner, we can tell it without looking out the window. In fact, of course, the train is moving on the earth, which is rotating and revolving. The effects of this nonuniform motion are too small for us to feel, but they can be detected by means of instruments. We can sometimes feel the vibrations of a moving train or automobile, but we do not feel its uniform motion.

Another demonstration of the special principle of relativity is provided by the standard laboratory toy illustrated in Figure 3-7. The "cannon" fires a ball straight up while the cannon itself is in motion to the right. Where will the ball fall? Assuming that we can disregard friction and that the cannon does not change its speed or direction, the ball will fall back into the mouth of the cannon. This, of course, is just where the ball would fall if the cannon were not moving. As far as the view from the cannon is concerned, the uniform motion of the cannon makes no difference.

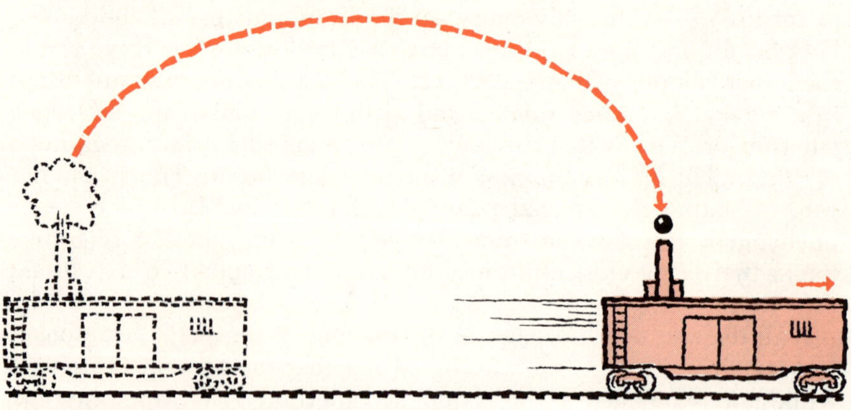

FIGURE 3-7 **Except for air friction, the cannonball fired vertically would fall back into the cannon's mouth if the train remains in uniform motion.**

An amusing example of this fact is given in Galileo's *Dialogue*. The *Dialogue* describes an imaginary conversation among three people, one of whom, Simplicio by name, opposes the idea that the earth moves. In the course of the argument, Simplicio claims that sailors measure the speed of a boat by dropping a rock from a mast. If the boat is in motion, he erroneously states, it will move forward (or backward) past the rock as the rock falls, and therefore the rock will strike the deck at some distance from the bottom of the mast. The faster the boat is moving, he claims, the greater this distance. Actually, this method will not work (see Fig. 3-8). Galileo criticizes Simplicio, and the "authorities" who made such claims, for merely quoting each other and for putting:

> . . . faith in their predecessors, right on back without ever arriving at anyone who had performed it. For anyone who does will find that the experiment shows exactly the opposite of what is written; that is, it will show that the stone always falls in the same place on the ship, whether the ship is standing still or moving with any speed you please.[2]

Thus, if the ship is in uniform motion, the experiment will come out the same, regardless of the speed of the ship.

Care must be taken in using the principle of relativity. In Figure 3-9 we illustrate two ways in which a badminton game can be played on a moving train. If the car is enclosed, as in Figure 3-9*a*, the game

FIGURE 3-8 **Except for the wind, the ball strikes the deck in the same place, whether the ship is at rest or in uniform motion.**

[2]*Galileo Galilei, Dialogue Concerning the Two Chief World Systems, Ptolemaic & Copernican*, translated by Stillman Drake, Berkeley: University of California Press, 1962, p. 144. Originally published by the University of California Press; reprinted by permission of The Regents of the University of California.

FIGURE 3-9 **The wind makes a difference in these two situations.**

will proceed just as if the train were at rest. If the game is played on an open flatcar, however, the "wind" generated by the movement of the train might spoil the game. The principle of relativity does not guarantee that the physical situation will be the same in a moving frame of reference as in a stationary frame. It *does* guarantee that the fundamental laws and principles of physics will be the same in two uniformly moving reference frames.

The Motion of the Earth

The Foucault Pendulum. We have said that the uniform straight-line motion of a system cannot be determined by measurements made within the system but must be determined by looking outside the system. The earth, however, is not moving in a straight line; it is spinning and revolving in near circles. These circles are large, so that the circular paths are very close to straight lines for short time periods. Still, there should be and are small effects. There are indeed ways to detect the earth's motion without looking out to see the stars go by.

One such method involves the use of the Foucault pendulum, which detects the rotation of the earth. In Figure 3-10 we show such a pendulum set up at the North Pole. The pendulum bob is not in rigid contact with the earth, since it has connection with the earth only through the cord and the framework that supports it. When the pendulum is set to swinging back and forth, it tends to go in as straight a line as possible, in accordance with the first law of motion.

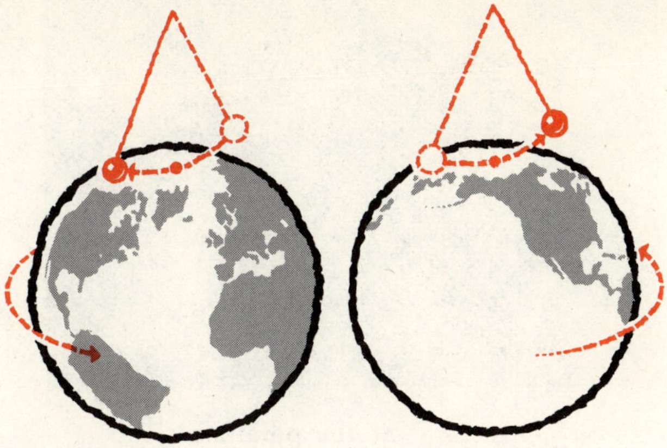

FIGURE 3-10 **The Foucault pendulum can be used to detect the rotation of the earth.**

As the earth turns under it, the pendulum bob tends to keep swinging back and forth in the same direction in space—that is, to keep swinging in the general direction of the same stars. To a person on earth, the pendulum appears to be mysteriously changing its direction of swing. Actually it is the earth that is moving, rotating underneath the pendulum. If the pendulum is not at the North Pole, its behavior is less simple to explain. What happens is that the direction of swing changes more slowly if a Foucault pendulum is moved to a location closer to the equator. (This pendulum effect was observed by J. L. Foucault in 1851.)

Even if the earth were covered with clouds so that we could not see the sky, we could still measure the rotation of the earth by means of such a pendulum. There are also other mechanical effects of the earth's rotation, small effects in the motion of bodies on the earth, but we shall not consider them here.

Stellar Parallax. In 1838 F. W. Bessel succeeded in obtaining a long-expected piece of evidence for the motion of the earth: the parallax of a star. Figure 3-11 illustrates what is meant by parallax. Since the earth is in different places in space at opposite points of its orbit around the sun, the relative position of stars should be slightly different as observed from the earth at these two points. This effect is extremely small because of the great distance of the stars, but Bessel finally observed it.

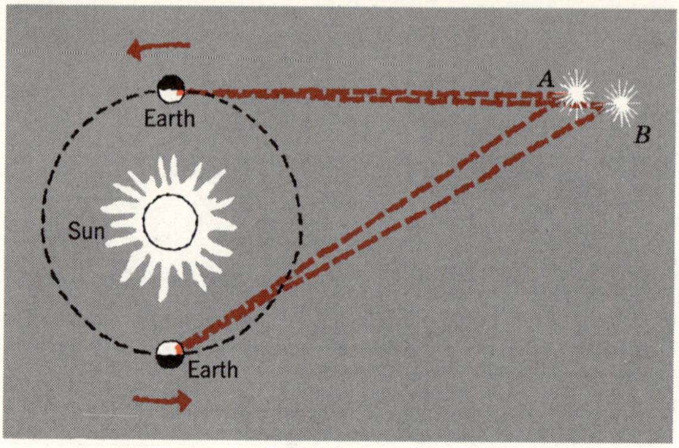

FIGURE 3-11 Stellar parallax.

Symmetry

The special principle of relativity tells us that the same laws of physics are obtained in all reference frames in uniform straight-line motion. That is to say, the laws of physics are not changed when a change is made from one of these reference frames to another. There are many other principles that tell us that we can change the reference frame without changing the way in which we express the laws of physics. These principles are called *symmetry principles* and they are also found to be of fundamental importance in physics. Before describing them, let us first explain in greater detail why they are called symmetry principles.

Figure 3-12*a* shows a square, with the corners labeled for convenience. We pretend that the letters are not there in reality but that we have imagined them there in order to keep track of the corners. Now suppose that the square has been rotated counter-clockwise through one quarter of a full revolution so that it appears as in Figure 3-12*b*. Without looking at the letters, you cannot tell that it has been moved. It looks exactly the same in its new position as it did in the old one. The square is a symmetric figure—there are certain things we can do to it without changing the way it looks. Something is said to be symmetric if it can be changed in such a way that its new condition is not detectibly different from its original condition. Any operation which makes such a change is called a symmetry operation. The operation that rotates the square through one quarter of a revolution is one of

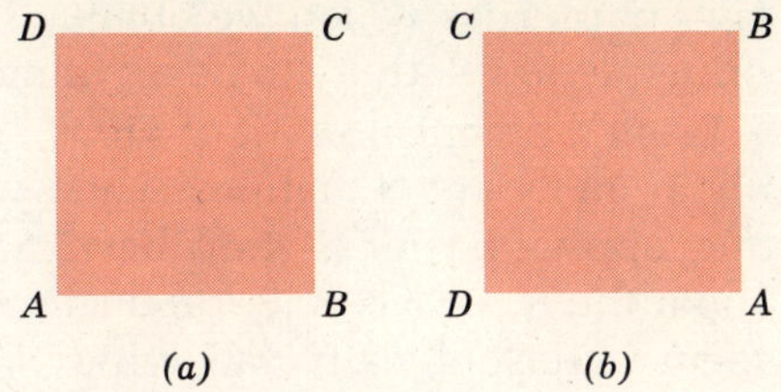

FIGURE 3-12 **After being turned a quarter of a revolution, the square still looks the same.**

the symmetry operations of the square. There are, of course, others.

We can study the symmetry of the square by changing the position of the observer instead of rotating the square. In Figure 3-13, for example, we have observers viewing the square from four different positions, yet the square looks the same to all of them. Thus, there are certain different ways of looking at the square for which the square still looks the same.

We are interested in the laws of physics. We look at them by using some reference frame in which we study the universe. There are certain ways of changing to new reference frames in which the laws of physics still look the same. That is, there are certain symmetry operations that will change our reference frame without changing the laws of physics. For example, according to the special principle of relativity, we can change to a frame in uniform motion without changing the

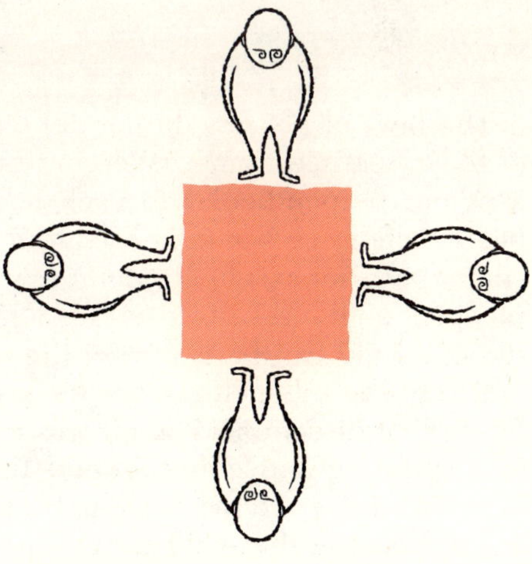

FIGURE 3-13 **The square looks the same from all four sides.**

laws of physics which we obtain. Therefore, we consider the special principle of relativity to be a symmetry principle.

From a practical point of view, symmetry principles are often useful in analyzing situations when we do not know the precise laws that apply. Besides, they have a certain elegance and simplicity about them, which is aesthetically appealing, and some of them also seem philosophically satisfying. They help us to express our faith that the universe is the same for all observers even though each may see it from a different vantage point. A law that can be expressed in the same way in different frames of reference seems likely to be an expression of some actual outside reality, something that is not completely dependent upon the situation of the individual observing mind.

Position Symmetry

The laws of physics do not depend upon the position of the origin of the reference frame. This means that the laws do not depend on the position in space at which we perform our experiments. The laws should turn out to be the same whether we make our experiments in New York, London, Moscow, or on the moon (see Fig. 3-14). In Galileo's day, the laws were supposed to be different on earth from those on the moon or above it. The heavens were thought to be different from the earth, and to obey different laws. Galileo's work helped to change that idea.

Directional Symmetry

The laws of physics do not depend upon the direction of the axes of the reference frame. (We continue to assume, however, that the axes are perpendicular to each other. It does not matter how the whole reference frame is oriented in space.) As far as the laws are concerned, our axis might point toward either Chicago or Miami with no effect on the results (see Fig. 3-15). In Galileo's day, however, the direction toward the center of the earth was regarded as very different from the other directions because heavy bodies would naturally fall in that direction. When it was realized that the earth was in motion, spinning and going around the sun, this idea had to change. What is "up" or "down" at one time may not be "up" or "down" later on, because the earth turns. The center of the earth also moves; therefore the direction to that center is not always the same.

We can appreciate the philosophical shock Galileo's contemporar-

FIGURE 3-14 **The laws of physics do not depend upon position in space.**

ies experienced when the "new physics" which implicitly contained these principles became widely known. The earth, which they had believed to be steady, was now flying through space. "Up to the heavens" and "down to earth" were now merely relative terms. The heavens and the earth obeyed the same laws. The earth was an arbitrary planet, moving through arbitrary space, in an arbitrary direction. We cannot blame them for feeling "lost in space."

Time Symmetry

We can set up a reference frame for time somewhat as we do for space. A point in time corresponding to some event is taken as the

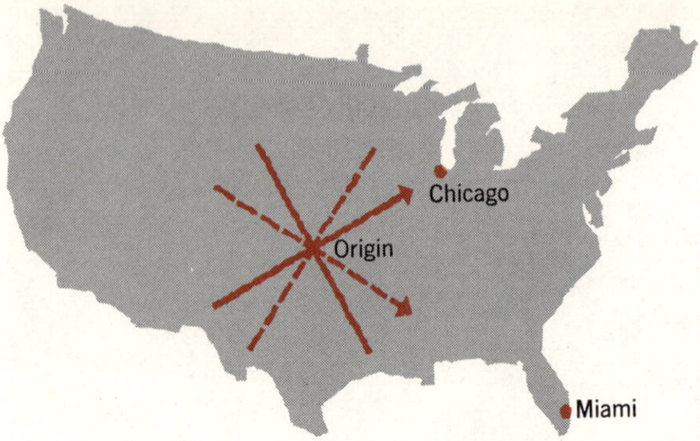

FIGURE 3-15 The laws of physics do not depend upon direction in space.

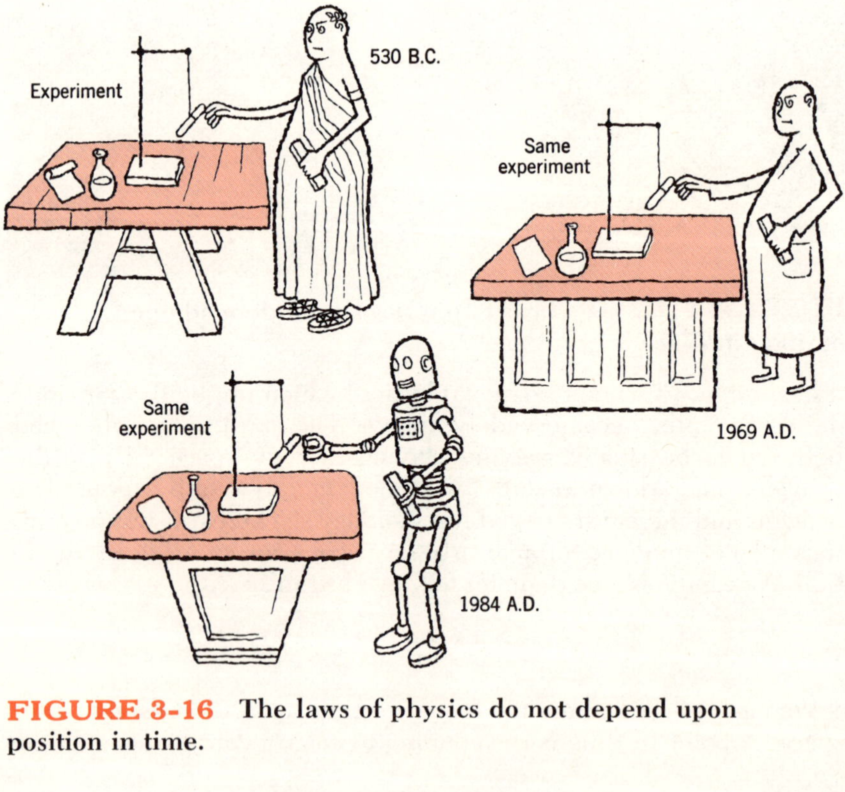

FIGURE 3-16 The laws of physics do not depend upon position in time.

reference point or origin in time, and events are dated forward and backward with respect to this point of reference. Our ordinary habit of referring to events as A.D. or B.C. uses the birth of Christ as such a reference point in time. As far as we know, it does not matter which instant is chosen. This gives us another symmetry principle: the laws of physics do not depend upon our chosen point of reference in time. Therefore, the laws alone can give us no reason for using one such reference point rather than another. It does not matter whether our experiments are done today, yesterday, or tomorrow; if we have taken everything into account, the laws should be the same (see Fig. 3-16). We may, of course, find out that we have been wrong, that our present laws are incorrect, but we believe that valid laws should be independent of time.

Time-Reversal Symmetry

The laws of physics should be the same whether time is running forward or is reversed to run backward. This may sound strange, since we cannot actually make time go backward. The easiest way to understand this principle is to suppose that we take motion pictures of all our scientific experiments. Then we determine the laws of physics from what we see when the film is run backward. If the laws turn out to be the same, then we say that the laws of physics are time-reversible, or that they have *time-reversal symmetry.* Although many of the laws of physics are reversible in this way, it is not known at present whether or not they all are. People, for example, are *not* time-reversible. If we take a movie of a man diving into a swimming pool and run it backward, we see the man come out of the water back onto the diving board (see Fig. 3-17). This reversed proc-

FIGURE 3-17 **Time reversal.**

ess does not happen in real life. The growth of a child is another example of a process that never occurs in reverse. Some processes in this world, therefore, do not appear to be time-reversible. We shall discuss such apparent violations of this symmetry principle in Chapter 10.

Mirror Symmetry

We might expect the laws of physics to be the same even if they were obtained by looking at our experiments in a mirror. This means that the laws we would discover by regarding the universe through a looking glass would still be the same (see Fig. 3-18). Thus we would have no way of telling the real universe from its mirror image as far as the laws of physics are concerned. This also means that if any process can occur, a mirror image of the process can also occur. This symmetry principle fails for certain situations, as far as we know at present. We shall describe one such situation in Chapter 19.

FUNDAMENTAL IDEAS

I. FUNDAMENTAL LAWS AND PRINCIPLES

A. THE SPECIAL PRINCIPLE OF RELATIVITY. All laws of nature are the same in all reference frames moving uniformly.

B. OTHER SYMMETRY PRINCIPLES. The laws of nature are not changed if:

FIGURE 3-18 Mirror reflection.

1. We observe nature from a different position in space.
2. We change our position in time.
3. We change our orientation in space.
4. We observe the world in a mirror.[3]
5. We reverse time.[3]

II. APPLICATIONS

A. DETECTING THE MOTION OF THE EARTH

1. Foucault pendulum.
2. Stellar parallax.

QUESTIONS

Reference Frames

1. Why do we use reference frames?
2. How many instructions, or items of information, are needed to locate a point on a plane surface? in space?
3. Sketch the motion of a ball released a few meters above the ground as it appears from a reference frame that is (a) fixed to the earth, (b) fixed to a flatcar in uniform motion passing nearby, (c) fixed to the sun.

The Special Principle of Relativity

4. Why would an object dropped from an airplane in horizontal flight remain directly under the plane as it fell? What could prevent the object from remaining directly under the plane?
5. If you were playing table tennis on a moving train, would you have to allow for the motion of the train in making your shots? Explain.
6. Can you detect uniform motion by performing an experiment? Explain.
7. It is often said that "motion sickness" is responsible for making people carsick, airsick, etc. Is the term "motion sickness" precisely correct?
8. If you set up your laboratory on a merry-go-round, would all the laws of physics be the same? What would happen to the first law of motion?

The Motion of the Earth

9. Make the best case you can for and against the motion of the earth.
10. You are suddenly transported to a new planet that is shrouded in clouds. How would you determine its speed of rotation?

[3]There are possible exceptions to principles 4 and 5.

11. Find all the symmetries of an equilateral triangle. (An equilateral triangle has all sides equal in length.) That is, find all the ways in which one can change the rigid triangle by turning or flipping it so that it will still look the same (see Fig. 3-19).
12. List some processes that are not time-reversible. List some that you think might be time-reversible.
13. Would the first law of motion still be true in a time-reversed world?
14. What does a right-hand glove look like when it is reflected in a mirror?
15. Would you expect the laws of physics to turn out to be the same if you did your experiments on Mars? in a diving airplane? in a rocket being propelled upward at an increasing speed? in an airplane flying smoothly on the level? in a laboratory fixed upside down on the ceiling of your room? if you looked at the results of the experiments in a mirror? if the results were reflected in two mirrors? in three mirors?

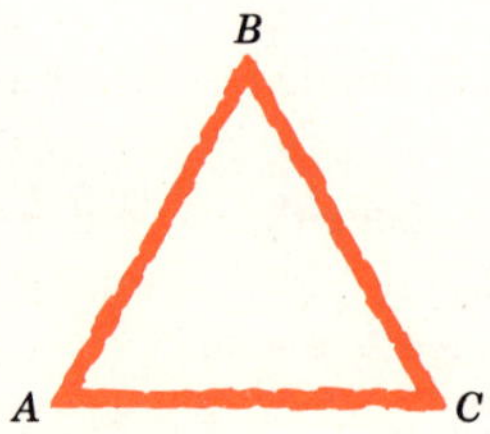

FIGURE 3-19 **See question 11.**

TOM HOOVER

4 The Second Law of Motion

The first law of motion, which we have already discussed, tells us that in the absence of force there is no change of motion. However, in the example illustrated in Figure 4-1, as in most cases, we are interested in *changes* in motion. Exactly how are such changes brought about? The effect of a force in changing motion was precisely stated by Isaac Newton in his second law of motion. Before discussing this law, however, we need to understand the quantities in terms of which the law is stated.

Acceleration

In Chapter 1 we defined *velocity* as the vector quantity that specifies both the speed and the direction of the motion of an object. The unit of velocity that we use is the *meter per second* (meter/second). The velocity of an object remains constant when no external influences are exerted on the object. External influences may cause the velocity to change in speed or direc-

FIGURE 4-1 What determines who will win a drag race? Courtesy of Adolph F. Rohrer, DPI.

tion or both. The quantity which describes such changes is called *acceleration.* The acceleration is the rate at which a change in velocity takes place.

First let us consider motion along a straight line. Let us assume that a certain object—for example, a car—is moving along a straight road (see Fig. 4-2). It can speed up, stop, slow down, or back up. For the moment, however, we will not allow it to turn any corners. In this case the acceleration is familiar to us because of our experience with cars. We say that a car with great acceleration is one that reaches a high speed very rapidly. That is, it changes its speed at a rapid rate. Most people do not regard braking to a stop as an acceleration, but in fact it is. There is an acceleration whenever the speed of the car is changing. The acceleration expresses the rate at which the speed is changing.

Now suppose that the speed of the car is constant, but its direction of motion is changing. Is is accelerating? According to our definition, it is. We have defined acceleration as the rate of change of velocity, and velocity is a directional quantity. Therefore, when the direction of motion is changing, there is an acceleration. Often the motion of an object changes in both speed and direction simultaneously.

The unit of acceleration is *meter per second per second.* The occurrence of "per second" twice in this expression requires some explanation. Let us again consider the special case of motion in a straight line, as illustrated in Figure 4-2. Suppose that a car initially at rest acquires a speed of three meters per second after one second. If it continues to accelerate—that is, to change its speed—at the same rate, it will be going six meters per second after two seconds, nine meters per second after three seconds, and so on. In this example, the speed changes by three meters per second *each second.* We express this by saying that the acceleration is "three meters per second per second." As long as the acceleration has this value, we know that no matter what speed the object has at a given time, one second later

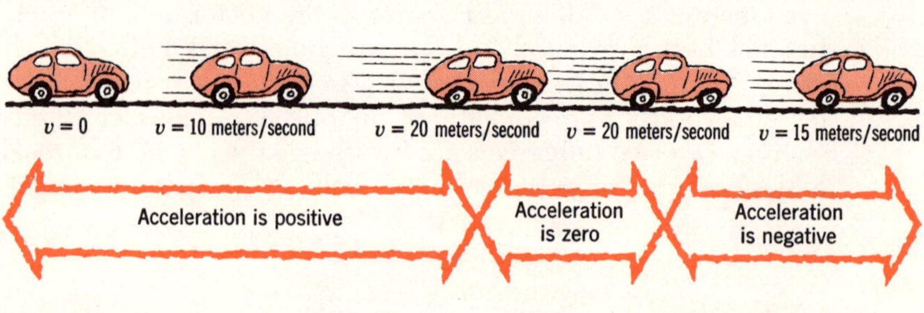

FIGURE 4-2 **A car experiencing different accelerations.**

its speed will have increased by three meters per second. The quantity acceleration describes how much the quantity velocity (which already involves time) changes in a certain amount of time. Therefore, the time must occur twice in the unit of acceleration. In our discussion of the unit of velocity or speed, we saw that the phrase "meter per second" is expressed mathematically as meter/second. The unit of acceleration, then, can be expressed mathematically as follows:

$$\text{Meter per second per second} = \frac{\text{meter/second}}{\text{second}}$$

According to the rule for performing mathematical operations with fractions, this is equal to

$$\frac{\text{meter}}{\text{second}} \cdot \frac{1}{\text{second}} = \frac{\text{meter}}{(\text{second})^2}$$

Mass

The fundamental quantity *mass* may be defined roughly as the amount of matter contained in an object. More precisely, this quantity is a measure of the response of an object to an interaction with another object or system of objects. Figure 4-3 shows what happens to a young man who wanders down the street kicking cans. Unknown to him, one of the cans is filled with lead. The interaction of his foot with this lead-filled can is much different from the interactions with the cans that are "empty" (i.e., filled with air). There is, in a certain

(*a*) (*b*)

FIGURE 4-3 **The can with more mass offers more resistance to acceleration.**

 sense, more matter in the lead-filled can, as is shown by the fact that it has lesser response to the same kick. Thus we say that the lead-filled can has more of the quantity mass than the other cans.

A Standard and Unit of Mass

In order to accurately define the quantity mass, we must first choose some object as a standard, establish a unit of measurement, and then devise some method of comparing the mass of other objects with that of the standard. We choose for a standard of mass the mass of a certain volume of pure water.[1] The volume we choose is one thousandth of a cubic meter. This is the volume, for example, of a cube that has sides one tenth of a meter long (see Fig. 4-4). We call the mass of this standard one *kilogram*, and adopt the kilogram for our unit of mass. Mass is a scalar quantity. (Actually we must specify the temperature and pressure of the water, as well as its volume, to specify the standard exactly. Temperature and pressure will be discussed in later chapters.)

The Measurement of Mass

There are several procedures that might be used to compare the mass of an object with the mass of the standard. The method that we shall now describe is not so practical as others, but it helps to make the concept of mass clear.

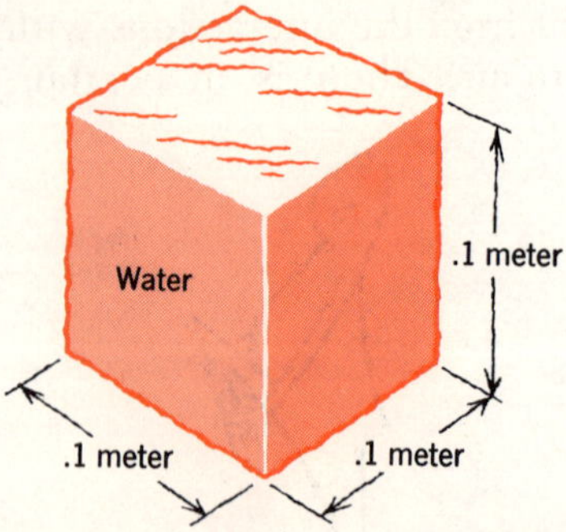

FIGURE 4-4 **A cube of water which has a mass of one kilogram.**

[1]The kilogram is currently officially defined as the mass of a certain cylinder made of platinum-iridium alloy kept at Sévres, France, by the International Bureau of Weights and Measures.

As illustrated in Figure 4-5, we imagine a kilogram of water in a very thin, rectangular container. We assume that the mass of the container is so small compared to the mass of the water that it can be disregarded. The container is resting on a perfectly smooth, frictionless surface. A string is attached to the container so that we can pull on it. We wish to pull on the container in such a way that we can continue to pull exactly as hard even while the container is moving. The exact device by which we achieve this constant pull is not important. However, to show that this is feasible we have illustrated a method by which this may be done. We have inserted a spring in the string as shown. We pull on the string in such a way that the spring remains stretched to the same length while the block is moving. Our basic operation is to pull on the block with an *unchanging* pull. The result of this pull is that the block accelerates.

As long as the pull on the block is the same, the acceleration is also the same. If, however, we change the mass of the object by pouring water into the container, the acceleration will no longer be the same. In fact, it will be less. The more mass (matter) there is in an object, the smaller the resulting acceleration if the pull remains the same. In other words, the mass is inversely related to the acceleration. A two-kilogram block of water (or the equivalent amount of matter in any shape or form) will accelerate at half the rate of our standard, the one-kilogram mass. An object with 1/3 kilogram of mass will have three times the acceleration as the standard, etc. Therefore, the mass of an object can be measured by comparing its acceleration to the acceleration of our standard mass. Any two objects of equal mass will receive the same amount of acceleration from the same amount of pull.

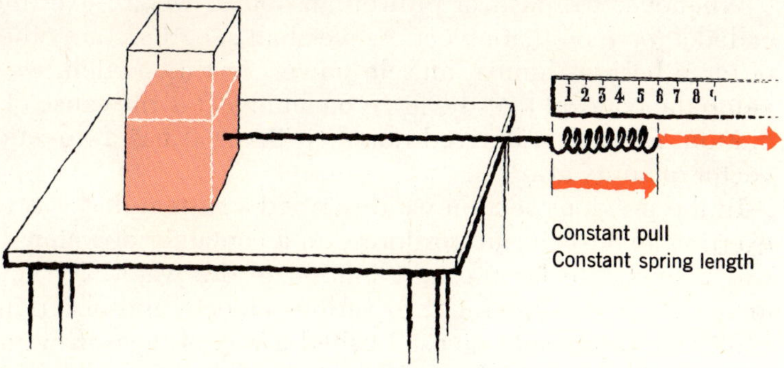

FIGURE 4-5 **A kilogram of water being accelerated by a constant pull.**

We wish to make it perfectly clear that the mass of an object is not dependent upon its volume alone but also depends on the type of material of which the object is composed. When we are dealing with only one kind of matter, such as the water in our previous example, we can increase the mass by increasing the volume of the material used. However, equal volumes of different kinds of material can have greatly different masses. For example, a cubic meter of lead has more than eleven times the mass of a cubic meter of water.

The Conservation of Mass

Mass is considered to be one of the basic realities of the universe. Mass can be neither created nor destroyed. Although matter can be changed from one form to another, the total amount of its mass does not change. According to all known experiments relating to this subject, there is just as much mass after any change or transformation as before. We express this fact by saying that *mass is conserved.*

If we boiled away a kettle-full of water, for example, it might appear that the mass of the water had disappeared. In fact, of course, if we were to collect all of the steam that had come from the kettle we would find that it had the same mass as the missing water. Although the form of the matter can be changed, the mass remains the same. Mass is one of several quantities in physics that are conserved.

Force

Whenever we push or pull on an object, we are exerting what is called a *force* on that object. As we shall see, there are other sources of force besides human muscle power. In this section, we shall develop the concept that the force on an object is the cause of its acceleration. Force is a derived quantity. Since it has a direction, it is a vector quantity also.

In the previous section we described a system that was capable of exerting a constant pull, or force, on a container of water. This force was responsible for the acceleration of the water. By keeping the force the same, applying it to various objects, and observing the acceleration produced, we established a way of measuring mass. Now we wish to compare the accelerations of the same object caused by different amounts of force.

Once again, our object is a container of water resting on a level frictionless surface (see Fig. 4-6). The mass of the water is the same in all three parts of the figure. Each of the strings shown pulls with the same force F. This means that the force in each string, acting alone, would produce the same acceleration of the container of water as the force in any other string, acting alone. We observe that doubling the force, as in Figure 4-6*b* doubles the acceleration also. Tripling the force, as in Figure 4-6*c*, triples the acceleration etc., thus, the acceleration is proportional to the force acting on the object. Therefore, one way of measuring a force is to measure the acceleration caused by that force on an object of standard mass.

The unit of force is called a *newton*. It is defined as the force needed to cause one kilogram of mass to accelerate at the rate of one meter per second per second. Two newtons of force give a one-kilogram mass an acceleration of two meters per second per second, and so on.

The Second Law of Motion

The two relations among mass, force, and acceleration discussed above are both contained in one comprehensive law: Newton's second law of motion. For the quantities and units we have discussed, this law may be stated as follows.

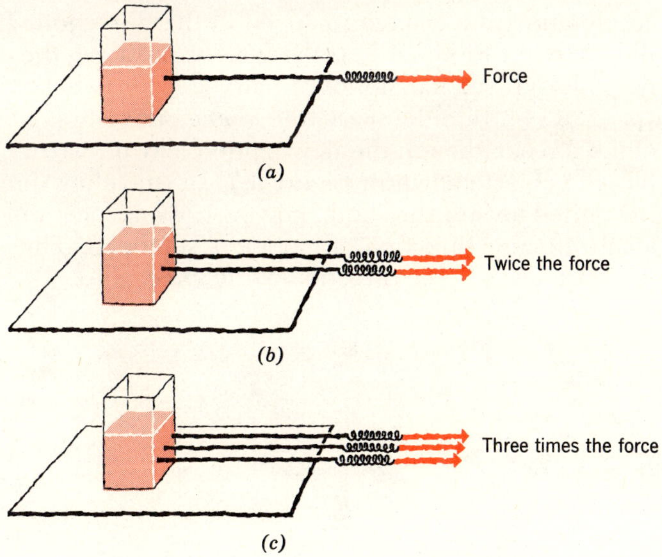

FIGURE 4-6 **A kilogram of water being accelerated by different forces. The greater the force, the greater the acceleration.**

The force on an object is equal to its mass multiplied by its acceleration.

(We are assuming that the mass of the object does not change while it moves.)

It should be noted that both force and acceleration are vector quantities, and that the acceleration of the object is in the direction of the force applied to it. We can write the second law of motion in equation form as follows:

$$F = ma \tag{1}$$

in which

F is the force on the object
m is the mass of the object
a is the acceleration of the object

The second law of motion tells us that a force causes an object to accelerate. It is a force, a push or a pull, from an external source that is responsible whenever an object undergoes a change in its motion. If there are no forces acting on the object, there is no acceleration.

The law also implies that the direction of the acceleration is the same as that of the applied force. If the force is in the direction of motion, the acceleration is also in that direction, and we observe the acceleration as a change in speed in that direction. This situation is illustrated in Figure 4-5. If the force is not along the direction of motion, the object will follow a curved path, whether it changes its speed or not. In either case, the *change* in motion is in the direction of the force, although the new motion may or may not be. For example, the object shown in Figure 4-7 was traveling due east. After being pulled toward the south, however, it has some motion toward the south. Its new direction of motion is southeast. Thus, the *change* in its motion was in the direction of the force.

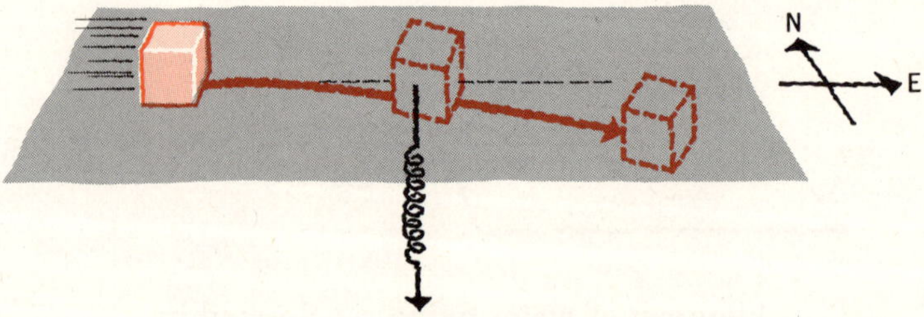

FIGURE 4-7 **A force is required to change the direction of an object's motion, as well as to change its speed.**

We can use the second law of motion to express the unit of force in terms of the units of mass, length, and time. We recall from the last section that one newton of force will give one kilogram of mass an acceleration of one meter per second per second. We simply insert these values for force, mass, and acceleration in the second law of motion, as follows:

$$\text{Force} = \text{mass} \times \text{acceleration}$$
$$\text{Newton} = (\text{kilogram}) \times (\text{meter/second}^2)$$

Therefore

$$\text{Newton} = \frac{\text{kilogram meter}}{\text{second}^2}$$

Since we are able to use the second law of motion to obtain the unit of force in this way, we may regard force as a derived quantity. We have derived it from the fundamental quantities mass, length, and time by using the second law of motion.

Vectors

Vectors and Scalars

We have already distinguished between the quantity speed (a scalar) and the quantity velocity (a vector). You will recall that the vector quantity, velocity, could not be specified without giving a direction, while the scalar quantity merely required a number. The number, with its units, is called the *magnitude* of a quantity. Thus velocity necessarily has a magnitude and a direction, while speed has a magnitude alone. In fact, the scalar quantity speed is just the magnitude of the vector quantity velocity. Many, but not all, scalar quantities are magnitudes of vector quantities. In general, then, we here define a vector quantity as a quantity that has both a *magnitude* and a *direction* and obeys the special rules of arithmetic discussed below. A *scalar* quantity is here defined as a quantity that has a magnitude only and obeys ordinary rules of arithmetic.

Distance and Displacement

We now define and discuss the vector quantity that we call *displacement.* This quantity is a measure of the difference in position between two points or places. The magnitude of this vector quantity is the straight-line distance between these two points. The direction of the displacement is the direction from one of these points (usually the first one mentioned) to the other. In Figure 4-8 for example, two points P and Q, are shown that are some distance apart. We draw a straight line from P to Q. We then draw an arrowhead so that the line points to Q as shown. The length of the straight line from P to Q gives the magnitude of the displacement, which is, of course, just the ordinary (straight-line) distance between the two points P and Q. The direction of the arrow is the direction of the displacement from P to Q. If we had chosen the opposite direction—that is, the direction from Q to P—we would not have the same vector. The magnitude would be the same, but not the direction. The displacement from P to Q is not equal to the displacement from Q to P even though the distance is the same, because the direction is different. We can generalize this principle as follows: two vectors are equal to each other if and only if they have the same magnitude and the same direction.

The displacements shown in Figure 4-9 are all equal to each other; they have the same length, and they are pointing in the same direction. In order for the direction of two vectors to be the same, it is not necessary that they lie on the same straight line, but merely that they lie on parallel straight lines, and that the arrows point in the same direction.

Figure 4-10, which contrasts the flight of a butterfly with that of a conventional bee, illustrates the displacement vector and permits us to compare distance and displacement. Both insects leave flower P and travel to flower Q. The butterfly, depending on the wind and his whim, might travel a very long path en route. Furthermore, he could make many flights of exactly the same path length and arrive at a different destination each time, possibly even right back where he started from. The "distance" traveled, the total length of the path, is a scalar; taken by itself it does not determine the end point of a particular journey.

The bee, on the other hand, lives up to his reputation for making "beelines" by traveling the shortest path between the two flowers.

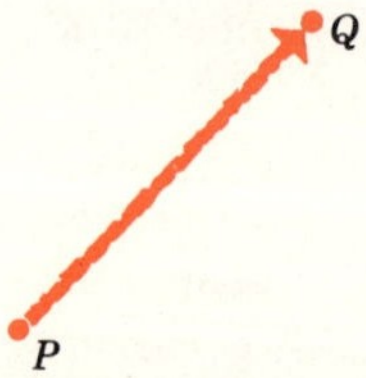

FIGURE 4-8 A displacement vector.

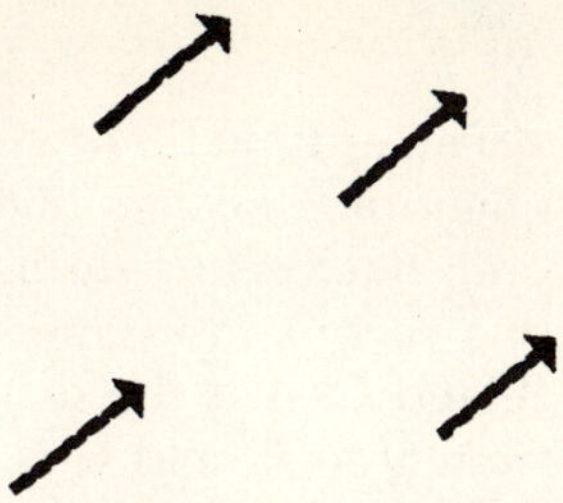

FIGURE 4-9 Equal displacements.

This directed quantity, a distance in a specific direction, is the displacement from P to Q. As before, we represent the displacement from P to Q by an arrow drawn in the direction of the line joining the two points with a length equal to the beeline distance. By traveling along this beeline, the bee travels a much shorter path than the butterfly. Nevertheless, the displacements shown in the diagram for the two insects are exactly the same, since they both start at the same point and end at the same point. Displacement is concerned only with the direction and the straight-line distance between two points—that is, with an overall change in position. The actual path length traveled in between is irrelevant as far as displacement is concerned. This is the reason we can represent the displacement vector as a straight arrow between the two points, pointing in the proper direction.

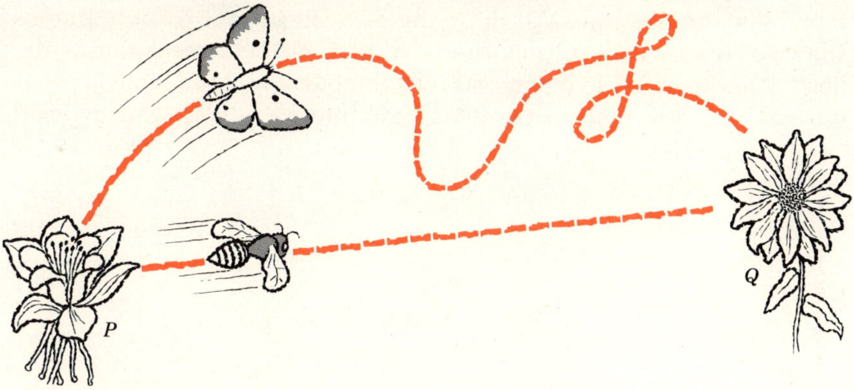

FIGURE 4-10 The bee and the butterfly experience equal displacements, but travel different distances.

Addition of Vectors

The arithmetic of vectors is different from the arithmetic of scalar quantities. Scalars obey the familiar rules of addition and subtraction, but vectors do not. We illustrate the rules for adding together two vectors in Figure 4-11, which shows the three displacements **A**, **B**, and **C**. **A** is the displacement from *P* to *Q*, **B** is the displacement from *Q* to *R*, and **C** is the displacement from *P* to *R*. Of course, the displacement **C** may be obtained by going directly from *P* to *R*. However, we can also get from *P* to *R* by going first from *P* to *Q* and then from *Q* to *R*. The result obtained by joining displacement **A** to displacement **B** is exactly the same as if we had taken the displacement **C** alone. In each case we have started at *P* and ended at *R*. Since the displacement depends only on the starting point and the end point, we have accomplished the same displacement in two different ways. We say, then, that the displacement **C** is the sum of the two displacements **A** and **B**, or

$$\mathbf{C} = \mathbf{A} + \mathbf{B}$$

We now present the rule for adding two vectors together. (We illustrate this addition for displacements, but the procedure can easily be adapted for vectors in general.) Consider the two vectors **A** and **B** in Figure 4-12. We wish to obtain the sum **A** + **B**. The first step is to construct a vector equal to **B** beginning at the arrowhead of vector **A**, as shown in Figure 4-12*b*. This step is sometimes described as moving vector **B** to the end of vector **A**. Care must be taken to see that the direction of **B** is the same after we move it as before. Next we connect the beginning point of **A** with the ending point of **B** in it's new position by a straight line. We draw the arrowhead on this new line as shown. This procedure puts the head of **C** at the same point as the head of **B** the second vector, and puts the tail of **C** at the same point as the tail of **A**, the first vector. This new line with its new arrowhead

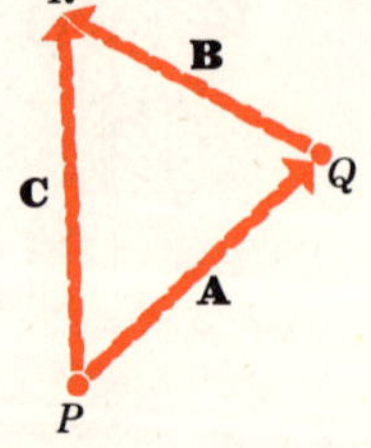

FIGURE 4-11 **Addition of displacements.**

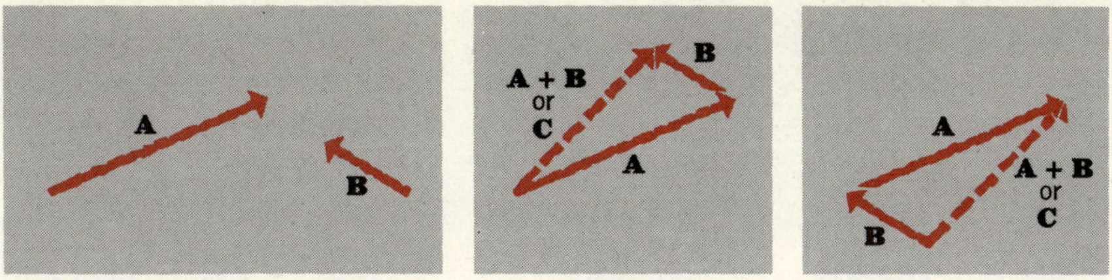

(a) (b) (c)

FIGURE 4-12 **Addition of vectors.**

gives us a new vector. This new vector **C** is called the *resultant* or *sum* of **A** and **B**.

In the above procedure, we brought the tail of **B** to the head of **A**. If we had instead brought the tail of **A** to the head of **B** as illustrated in Figure 4-12*c*, and then found our new vector, we would have obtained an equal resultant. It does not matter in what order the two vectors are added. In other words, we have the familiar-looking fact that

$$\mathbf{A} + \mathbf{B} = \mathbf{B} + \mathbf{A}$$

When we add vectors together, the magnitude of the resultant is not necessarily equal to the sum of the magnitudes of the individual vectors from which we obtained the resultant. Figure 4-13 gives

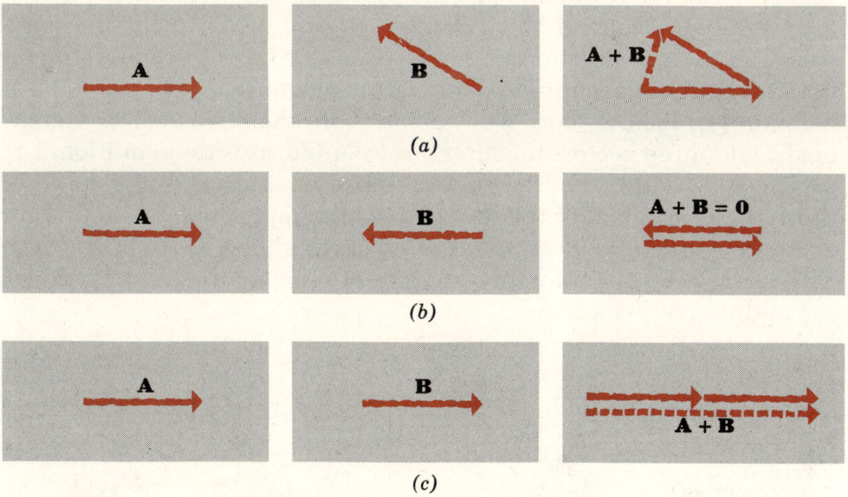

FIGURE 4-13 **The magnitude of the resultant depends upon the directions of the two vectors and not just on their magnitudes.**

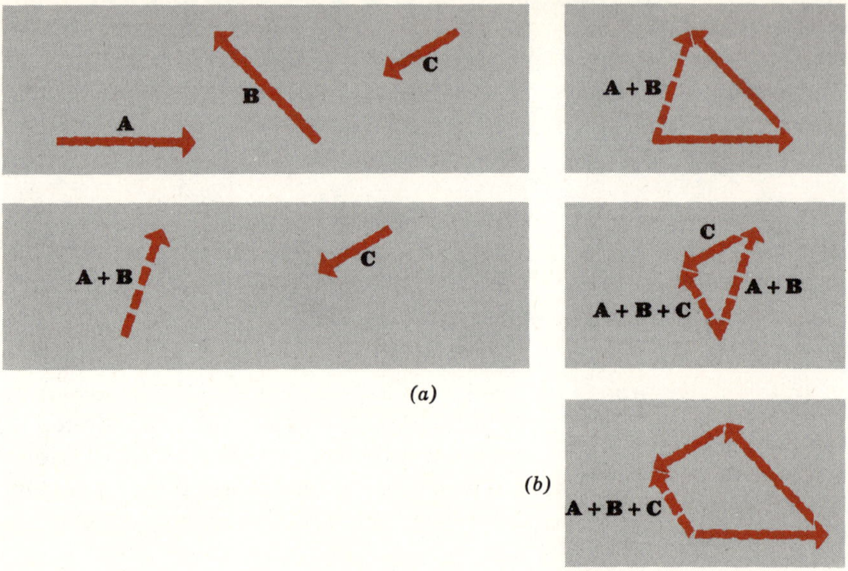

FIGURE 4-14 The addition of three vectors.

some examples of this fact. In Figure 4-13*a* the resultant has a magnitude that is even less than either the magnitude of **A** or **B** alone. In Figure 4-13*b* the resultant is zero, although neither **A** or **B** has a zero magnitude. In Figure 4-13*c* **A** and **B** are pointing in exactly the same direction; in this situation only, the resultant has a magnitude equal to the sum of the magnitude of **A** and the magnitude of **B**.

Figure 4-14 shows how we can add more than two vectors together. To add three vectors together, for example, as is done in Figure 4-14*a*, we first add two of the three vectors together, and then add the third to the resultant of the first two. This simply amounts to putting all three of them in a head-to-tail chain as shown in Figure 4-14*b*. This same rule works for any number of vectors; Figure 4-15 shows the addition of five vectors.

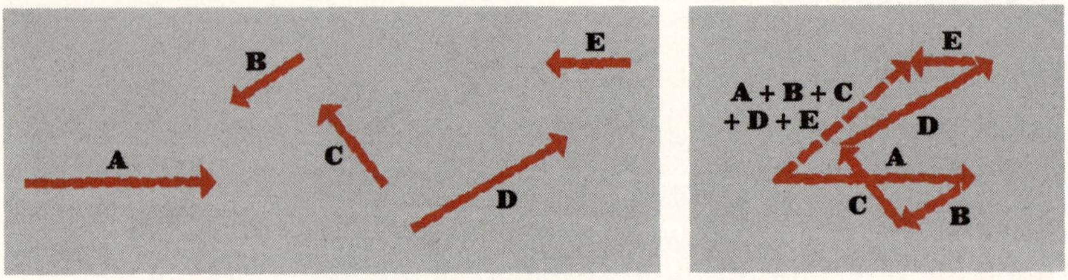

FIGURE 4-15 The addition of several vectors.

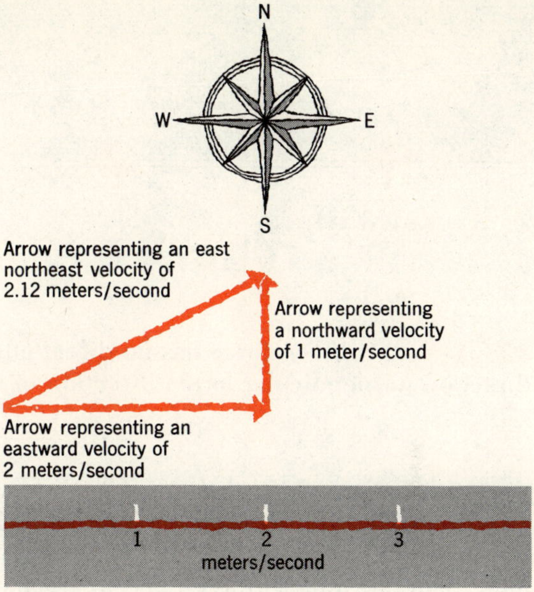

FIGURE 4-16 **The addition of velocity vectors.**

All vector quantities can be added by this same technique. For example velocity may also be represented on a diagram (see Fig. 4-16). We simply let one unit of velocity, one meter per second, be represented by some convenient length, shown by a scale included on the diagram. We then construct arrows representing each of the velocity vectors to be added. The length of each arrow is determined from the scale on the diagram and is therefore proportional to the magnitude of the vector velocity represented by the arrow. Then we add the vectors in the usual way and obtain the resultant. The magnitude of the resultant velocity is obtained by using the scale to convert the length of the resultant arrow on the diagram to meters per second. The direction of the resultant velocity is the same as the direction of the resultant on the diagram. By using such scale drawings, we may also find the resultants for other vector quantities besides velocity and displacement.

Resultant Force and the Second Law of Motion

In our earlier discussion of the second law of motion, we pointed out that it is force that causes an object to accelerate, and we noted

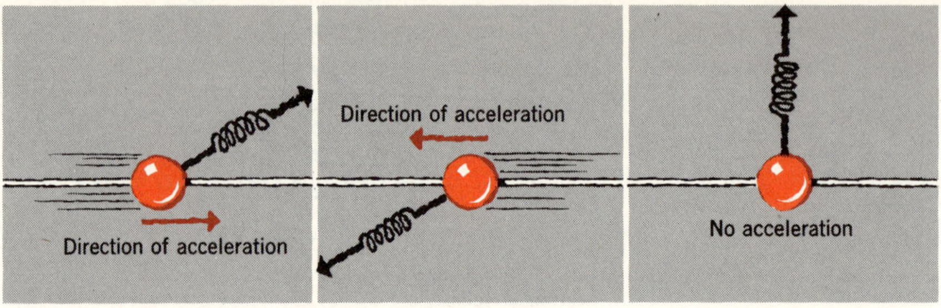

FIGURE 4-17 **A force has no direct effect on motion in a direction perpendicular to the direction of the force.**

that both force and acceleration are vector quantities. We described the second law of motion for the case in which a single force is acting on an object. However, it is possible for an object to be acted on by more than one force at the same time. In such a case, the acceleration of the object is produced by the *resultant* of all the forces acting on the object, and this resultant force is found by the method of vector addition outlined above.

If the resultant of all the external forces acting on an object is zero, there is no acceleration of the object as a whole and it is said to be in motion equilibrium.[2] This condition will be discussed further in Chapter 7.

Let us consider another concept. Figure 4-17 shows a bead which is free to slide on a stiff frictionless rod. If we pull it by the string it will move along the rod in one direction or the other as shown. The rod exerts forces which prevent the bead from moving in any direction except along the rod. Suppose that we pull with the string in a direction exactly perpendicular to the rod. Now which way will the bead move? Intuitively we guess that the bead will not move at all, since there would be just as much reason for it to move to the right as to the left. The symmetry of the situation indicates that no motion will result. This turns out to be correct, and it illustrates an important principle: a force has no direct effect on motion in a direction perpendicular to the direction of the force.

[2]We are here concerned with the acceleration of the object as a whole, and not with changes in its rate of rotation.

FUNDAMENTAL IDEAS

I. FUNDAMENTAL LAWS AND PRINCIPLES
   A. THE SECOND LAW OF MOTION.
   B. THE LAW OF CONSERVATION OF MASS

II. QUANTITIES
   A. FUNDAMENTAL QUANTITIES
      1. Mass, a scalar. The unit of mass is the kilogram.
      2. Displacement, a vector, whose magnitude is distance.
   B. Derived Quantities
      1. Acceleration, a vector. The unit of acceleration is the meter/second^2. Acceleration is the rate of change of velocity of an object.
      2. Force, a vector. The unit of force is the newton, which is equal to a kilogram meter/second^2
   C. VECTOR ADDITION

QUESTIONS

Acceleration

1. A car is in an intersection facing east. Describe its motion under each of the following sets of conditions. In this question, vectors pointing east are considered positive and vectors pointing west are considered negative.
   (a) Acceleration is zero; speed is positive.
   (b) Acceleration is zero; speed is negative.
   (c) Acceleration is zero; speed is zero.
   (d) Acceleration is positive; speed is zero.
   (e) Acceleration is negative; speed is positive.
   (f) Acceleration is positive; speed is negative.

Mass

2. Is mass related to size? Explain.
3. Describe a way to find the mass of an object. Show that your method is valid.
4. Suppose, for the moment, that all matter was made up of identical particles. How would you explain the fact that an object made of lead is more massive than an object of the same size made of wood?
5. Mass is conserved. List some processes in which mass appears to disappear as matter disappears or changes its form. Can you explain such situations?

Force and the Second Law of Motion

6. List as many ways as you can think of to exert a force on an object.
7. An object is accelerated by the only force acting on it until it reaches a speed of 20 meters per second. Then the force is removed. What would you expect the object to do after that?
8. A certain automobile can reach a speed of 100 miles per hour from a stop in one minute if the automobile is pulling a trailer of equal mass. Estimate the time it will take the automobile to reach this speed if it is not pulling the trailer, assuming that the automobile exerts a constant force regardless of its speed or load. Justify your answer.
9. A rocket out in space becomes easier and easier to accelerate as it travels along. Why? (*Hint*. The same thing is true of an automobile traveling on a level road between stops for gasoline, although the effect is less marked.)

Vectors

10. How does the displacement from one position to another differ from the distance between the two positions?
11. Find the sums of the following sets of vectors graphically.
    (a) A vector southwest of magnitude 6 inches.
        A vector northwest of magnitude 7 inches.
    (b) A displacement north of magnitude 10 meters.
        A displacement east of magnitude 5 meters.
    (c) A force of 10 newtons north.
        A force of 5 newtons north.
    (d) A velocity of 9 meters per second east.
        A velocity of 5 meters per second north.
        A velocity of 7 meters per second north, northwest.
12. An object is subject to three forces: a force f to the east, a force $2f$ to the

north, and an unspecified force. The object does not change its motion as a result of the forces. Draw a diagram more or less to scale illustrating how you would find the third force, and estimate its magnitude and direction.

13. Why is acceleration a vector quantity?

14. How is it possible for two vectors to add up to zero if neither of them is equal to zero?

5 The Third Law of Motion

Interactions

Whenever one object disturbs the motion of another object, it is itself disturbed. The *action* of one object on another is always accompanied by a *reaction* of the second object on the first; thus, we have an *interaction* between the two objects and both are disturbed. The action and reaction are simultaneous; we speak of the "first" and "second" object only for the sake of convenience. In this chapter we shall study the third law of motion, which governs such interactions. We will also introduce a new, derived, vector quantity, *momentum*, and show that it is conserved.

We all know better than to double up our fist and punch a brick wall. The action of our fist against the wall may or may not cause some damage to the wall, but the reaction of the wall on our fist will certainly give us some discomfort (see Fig. 5-2). When we push on a wall or any other object, the wall or object pushes back on us. For another example, consider our can-kicking friend illustrated in Figure 4-3 (p. 57). He cannot kick the can without receiving some reaction on his foot. If the can is massive and he kicks it hard, he may injure his foot or at least stub his toe. He cannot disturb the can with his foot without having the can disturb his foot in return.

☞ **FIGURE 5-1** Where does the force come from that knocks the man down?

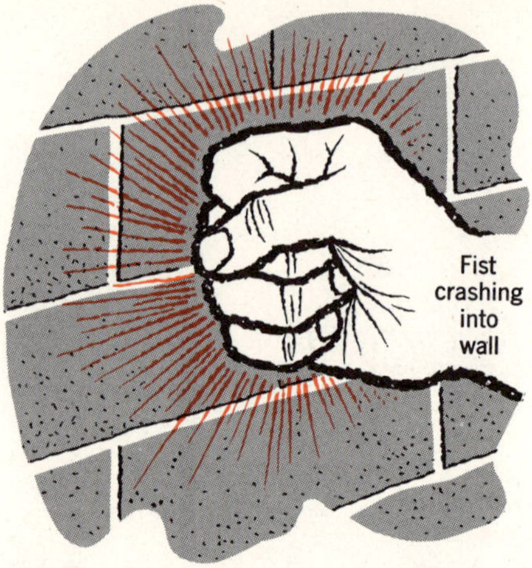

FIGURE 5-2 **When the fist strikes the wall, a force is exerted on both the fist and the wall.**

As another example, suppose that there are two balls moving on a smooth (frictionless) surface, as shown in Figure 5-3. The balls would continue to move along straight lines if they did not collide. However, when the balls collide, the motion of both balls changes, not just the motion of one alone. The motion of each ball changes because of a force exerted on it by the other ball. In each interaction, then, there must be at least two forces.

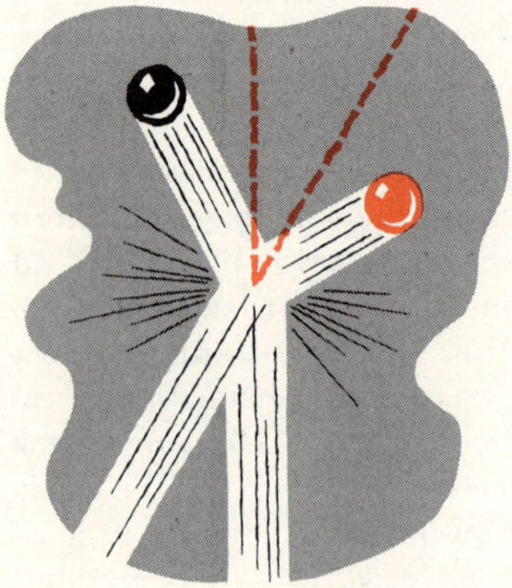

FIGURE 5-3 **The motion of both balls changes in the collision.**

The Third Law of Motion

Newton's third law of motion tells us exactly how the forces in an interaction are related. He stated this law as follows.

> To every action [force] there is always opposed an equal reaction [force]: or, the mutual actions of two bodies upon each other are always equal, and directed to contrary parts.[1]

This law tells us that if two objects interact, the forces they exert on each other are precisely the same in magnitude and exactly opposite in direction. The forces in an interaction never occur alone but always in pairs. We call such pairs *action-reaction force pairs.* The two forces in an action-reaction pair are equal in magnitude and opposite in direction. It is important to realize that when we are considering the forces of interaction between two objects, we never find that both forces of the action-reaction pair act on just one of the two objects. One of the forces acts on one object and the other acts on the other object. When we are concerned with the entire interaction it does not matter which of the two forces in the pair we call the action force and which the reaction force.

In summary, for two forces to make up an action-reaction pair they must fulfill the following requirements.

(a) The two forces must arise in the same interaction.
(b) The two forces must be of the same magnitude.
(c) The two forces must act in opposite directions.
(d) The two forces must act on different objects in the interaction.

Figure 5-4 illustrates these concepts in an interaction between two canoes, each bearing a passenger. In this case, each of the two objects is a system that includes both a canoe and its occupant. The two canoes in our illustration are of equal mass. A man is in one, and a girl of equal mass is in the other. Therefore, both objects or systems have the same mass. Suppose that the man reaches out his arm and pushes the other canoe away from him. This sets up an interaction between the two canoes. As a result of this interaction, not only the girl's canoe but the man's canoe also begins to move. The canoes move in opposite directions but at the same speed. In applying a force to the girl's canoe to set it in motion, the man automatically sets up an additional force which sets his canoe in motion at the same

[1]Isaac Newton, *The Mathematical Principles of Natural Philosophy and His System of the World*, translated by Andrew Motte in 1729, rev. by Florian Cajori, Berkeley: University of California Press, 1934, p. 13. Originally published by the University of California Press; reprinted by permission of The Regents of the University of California.

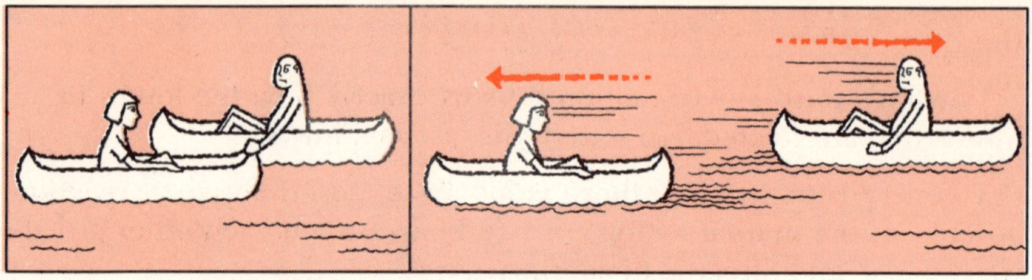

(a) (b)

FIGURE 5-4 He cannot push her canoe away without his canoe being pushed in the opposite direction at the same time.

time. We can tell that the two forces are in opposite directions because the canoes accelerate in opposite directions. Therefore, these two forces fulfill the conditions of an action-reaction pair. They arise in the same interaction, they have the same magnitude, they are in opposite directions, and they act on different objects.

Suppose that a young man is throwing packages from a boat, as illustrated in Figure 5-5. This, of course, sets up an interaction between the man and the package. Assuming that the young man does not lose his balance and fall overboard, we can consider the man and boat as one object (system) and the package as the other. In order to throw the package to the dock, he must exert a force on it and thus set up an interaction. In this interaction his force on the package must be one of the two forces in an action-reaction pair. Therefore, there must be another force by which the package pushes back on him as he throws it. He feels this force in his hand as he throws the

(a) (b)

FIGURE 5-5 Why does he move backwards when he throws the package?

package, and it causes him (and the boat) to start moving away from the dock. As is required in any action-reaction force pair, the two forces act in opposite directions and on different objects. It is also true, as must be the case, that the magnitudes of the two forces are equal. The reason that the motion of the package is changed more than that of the man and boat is that the package has much less mass.

This same principle can be used to explain the operation of a rock-

FIGURE 5-6 The rocket carrying the Apollo 15 astronauts David R. Scott, Alfred M. Worden, and James B. Irwin to the moon, July 26, 1971. Courtesy of NASA.

 et. The rocket, of course, does not throw packages out behind it. It does, however, throw out large quantities of gas through the nozzle at the rear (see Figure 5-6). To expel the gas, the rocket must exert a force on it. Since this force must be part of a force pair, the gas exerts a similar force on the rocket, which pushes the rocket in the opposite direction from the gas. Again we have a typical situation. One force is exerted on the gas by the rocket to expel it. Another force of equal magnitude is exerted by the gas on the rocket in the opposite direction.

Momentum

Now we define a new, derived, vector quantity, called *momentum.* The momentum of any object is its mass multiplied by its velocity. The magnitude of the momentum of an object is its mass multiplied by its speed. The direction of the momentum of any object is the same as the direction of its velocity.

The letter **p** is commonly used to represent the momentum of an object. Using m to represent the mass of an object and **v** to represent its velocity, we may write

$$\text{momentum} = \mathbf{p} = m\mathbf{v}$$

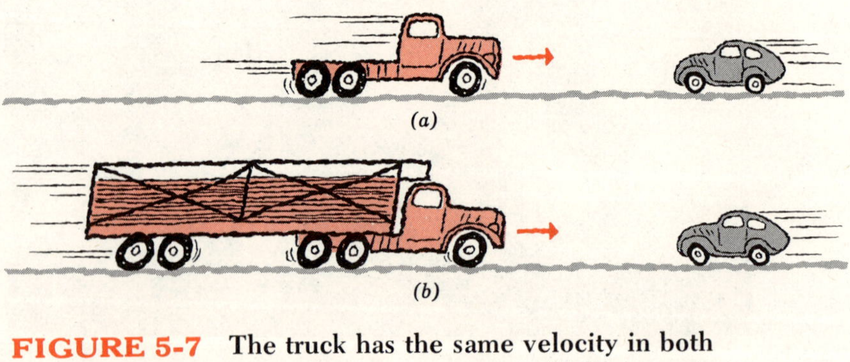

FIGURE 5-7 The truck has the same velocity in both (a) and (b), but it has greater momentum in (b) because it has greater mass.

Momentum is closely related to velocity, but it is more useful in some situations. Figure 5-7 illustrates a circumstance in which the momentum of an object is more important than its velocity alone. A small car is facing immediate interaction with a runaway truck. (A collision is a form of interaction.) In both of the cases illustrated, the velocity of the truck is the same. In the second case, however, a trailer carrying 10 tons of steel is still attached to the cab. We recognize this case as far more frightening than the first. This illustrates that momentum, mass times velocity, is an important quantity in a collision.

Conservation of Momentum

If we add up the momentum of all the objects involved in an interaction, we find that this sum is the same before, during, and after the interaction. The *total momentum* of the system of objects is *conserved* in an interaction. To illustrate this principle, let us refer again to Figure 5-4. Before the man exerts a force to push the girl's canoe away, both canoes are motionless in the water. In this condition each canoe has a momentum equal to zero because the velocity of each is zero. The total momentum for the overall system, which includes both canoes (and both people), is also zero, since it is obtained by adding together the momenta of the two objects. After the interaction, each canoe is moving, and therefore each has a nonzero momentum. Since the two systems have the same mass and speed, their momenta have the same magnitude. Since the canoes move in opposite directions, the resultant total momentum is zero. Thus momentum is conserved in this simple example.

We could, of course, do the same experiment with unequal masses, and momentum would still be conserved. Suppose, for example, that the girl in the second canoe has a mass several times the mass of the man. If the man then pushes the canoes apart, using the same force for the same length of time as he did in the first example, his canoe will move with the same speed as before, but the other canoe, since it is carrying more mass, will be accelerated to a lower speed. Nevertheless, its momentum will still be equal in magnitude to the momentum of the man's canoe. The mass is greater, but the final speed is proportionally less, so that the momentum, the mass times the velocity, still has the same magnitude. Thus the total momentum of the whole system still adds up to zero. In this special case, the two objects in interaction have zero speed and therefore zero momentum to begin with. Had the canoes been moving before they were pushed apart they would probably have had some total momentum to begin with. In this case the total momentum after the interaction would not have added up to zero but, rather, to the same total momentum the two canoes had to begin with. The total momentum remains the same in an interaction.

Conservation of Momentum and the Laws of Motion

Newton originally stated the second law of motion in terms of momentum rather than acceleration. The law read as follows.

The change of momentum is proportional to the motive force impressed; and is made in the direction of the straight line in which that force is impressed.[2]

This means that the force on an object is proportional (in our system of units, actually *equal*) to the rate of change of momentum of the object. Using $\mathbf{f}$ for force, m for mass, and $\mathbf{v}$ for velocity, we may express the law as follows:

$$\mathbf{f} = \text{rate of change of } (m\mathbf{v})$$

For most of the circumstances we have been discussing, the mass itself changes very little, if at all. Therefore, we have been concerned with changes of velocity. Under these circumstances, since the acceleration $\mathbf{a}$ is just the rate of change of the velocity $\mathbf{v}$, we have:

$$\begin{aligned} \text{rate of change of } (m\mathbf{v}) &= m \times \text{rate of change of } \mathbf{v} \\ &= m\mathbf{a} \end{aligned}$$

Therefore, once again we obtain the law in the form more usually seen: $\mathbf{f} = m\mathbf{a}$.

It is easy to prove that momentum must be conserved in interactions if Newton's laws hold. The third law of motion tells us that forces of interaction always occur in pairs. The two forces in such a

[2] *Ibid.* Motte's translation uses the word *motion* rather than *momentum*. However, the quantity referred to is defined as we define momentum. We also use the term *straight line* rather than Motte's *right line*. See footnote 1, p. 77.

pair have the same magnitude but opposite directions. According to the second law, each of these forces is equal to the rate at which it is changing the momentum of some object. Since the forces in a pair are equal in magnitude, they must produce a change of momentum of equal magnitude in the two objects they act on. However, the forces in a pair are in opposite directions. Therefore, the change of momentum caused by one of the forces is in the opposite direction to the change of momentum caused by the other force. The total momentum is not changed at all. Thus, Newton's laws of motion imply that momentum is conserved in interactions.

FUNDAMENTAL IDEAS

I. FUNDAMENTAL LAWS AND PRINCIPLES

- A. THE THIRD LAW OF MOTION. Forces in an interaction occur in pairs. The forces in a pair are equal in magnitude, opposite in direction, and are exerted on different objects.
- B. CONSERVATION OF MOMENTUM

II. QUANTITIES

- A. MOMENTUM; a derived vector quantity, equal to mass times velocity.

QUESTIONS

The Third Law of Motion

1. Why can I hurt my foot when I kick a rock?
2. Why does a gun recoil?
3. If the man in Figure 5-4 were able to push very quickly on the other canoe, could he exert a greater force on it than it exerts on him? Explain.
4. In the example illustrated in Figure 5-5, in which a man is throwing a package from a boat, explain what forces are operating in that act and how they are related.

Third Law of Motion or Conservation of Momentum

5. Explain the operation of a rocket.
6. A man, fully dressed, is standing in the middle of a frictionless ice

sheet. How can he get off the ice?

7. Would you expect a rocket to work better in an atmosphere or outside an atmosphere? Why?
8. Why would it not be possible for a rocket to have some device to catch the escaping gases after they had slowed down and use them over again, so that it would not have to take so much fuel along?
9. An astronaut drifting in space has a lead weight fastened to a rope. Can he move himself through space by throwing out the weight and then hauling it back in again by the rope? Explain.
10. A man claims to have perfected a device which, when operating on internal batteries, will rise off the ground, hover, and move around. The box is sealed, nothing enters or leaves it, and it does not exert any forces on its surroundings. Would it be wise to invest money in this invention? Explain.

FIGURE 5-8 Gospel According to Peanuts ©, 1960. Robert L. Schultz, United Features Syndicate. See question 12.

11. A bullet strikes a block of wood resting on a frictionless surface. Under which of the following conditions will the block receive the greatest velocity, and under which the least velocity, and why? (Assume that the bullet was initially moving horizontally.)

(a) The bullet passes through the block.

(b) The bullet gets stuck in the block.

(c) The bullet bounces backward off the block.

12. Why does the football player in Figure 5-8 tend to lose his balance and fall when the ball he intended to kick is unexpectedly removed?

6 Gravitational Interaction

Several fundamental laws have been discovered that describe how objects or particles of matter interact with one another. One of these is Newton's law of gravitation, which describes the forces that arise because of the masses of objects. Through these gravitational forces, each particle of matter attracts every other particle of matter. The laws of motion which we have presented in the last three chapters allow us to predict the motion of an object when the forces acting on it are known. The law of gravitation describes some of these forces. Gravitational forces hold celestial objects such as the planets in their orbits and also influence objects thrown or shot by people on the surface of the earth. The behavior of these objects can be predicted with great accuracy by using the law of gravitation together with the laws of motion.

In this chapter we shall discuss the law of gravitational interaction. We shall also discuss circular motion at constant speed, which is approximately the motion of the planets in their orbits around the sun and the moon in its orbit around the earth. This leads us to Newton's "system of the world."

☜ **FIGURE 6-1** The M74 Spiral Galaxy. The same forces that keep us bound to the earth keep the stars in the galaxies, and indeed control the paths of all celestial objects. Courtesy of Mount Wilson and Palomar Observatories.

The Law of Gravitation

According to Newton's law for the gravitational interaction, every material object attracts every other material object. This means that every particle of matter pulls and is pulled by every other particle of matter. In Figure 6-2 we have diagramed two objects, each pulling on the other with gravitational force. We emphasize that the magnitudes of these forces are exactly the same for both objects. The sun pulls on the earth, and the earth pulls back on the sun with a force of equal magnitude. The earth pulls down on rocks and men, and rocks and men pull up on the earth. You might not think that a pebble pulls as hard on the earth as the earth pulls on the pebble, but it does. Of course, the force on the pebble has a greater influence on its motion than the force on the earth has on *its* motion because the mass of the pebble is so much smaller than the mass of the earth. The force on each object in Figure 6-2 points directly toward the other object. In fact, the two forces, one on one object and one on the other, are an action-reaction force pair and together make up the gravitational interaction between the two objects.

Several factors affect the strength of the gravitational forces. Among these factors are the masses of the two objects and the distance between them. If we increase the mass of either object, we *increase* the attractive force on both objects. For example, if we triple one of the masses we triple the force on both objects. Therefore, the force is proportional to either one of the masses. If we increase the distance between the two objects, we *decrease* the attractive force on both objects. In fact, the force between the two objects decreases

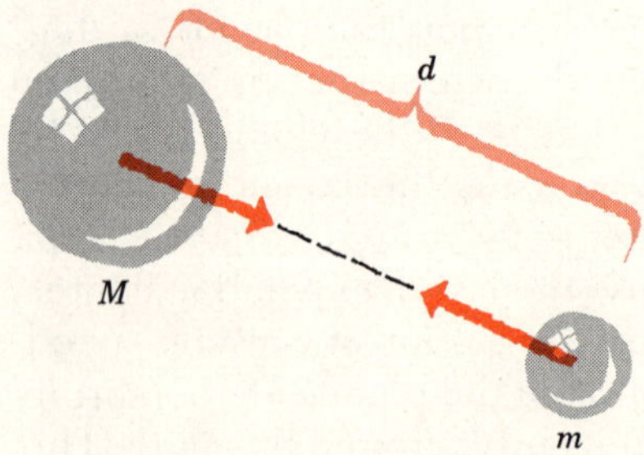

FIGURE 6-2 **The gravitational forces depend on the masses of the objects and the distance between them. The forces on the two objects are equal in magnitude but opposite in direction.**

very rapidly as we increase the distance between them. If, for example, we triple the distance between the two objects we find that the force is *not* 1/3 of what it was before; rather, it is $(1/3)^2$, or 1/9th, of its former value.

These relationships between the gravitational force, F, the two masses, m and M, and the distance, d, between them can be expressed in a single formula:

$$F = \frac{GmM}{d^2}$$

(The quantity G, equal to 6.67×10^{-11} newton meters^2/kilograms^2, will be discussed in the next section.) This equation gives the magnitude of the gravitational force on either object. Using this formula, we can calculate the gravitational force between any two objects when we know the mass of both and the distance between them.

In order to apply this formula to large celestial bodies such as the earth and the sun, Newton had to show that the distance d should be taken as the distance between their centers. This may be done when dealing with spherical objects whose mass is distributed in a sufficiently uniform manner around the center. It cannot be done in the unusual circumstances when one body is inside the other.

The Gravitational Constant G

Unlike the other quantities in the above equation, which may have different values in different situations, the quantity G always has the same value. Quantities that always have the same value are called constants, or constant quantities; hence, G is called the gravitational constant. This quantity is the first of several *fundamental constants* which we will introduce.

The importance of the fundamental constant G can be understood by examining the equation which gives the magnitude of the forces in the gravitational interaction. Suppose, however, the constant G were twice as big as it actually is. Then, according to the equation, the gravitational forces would have twice the magnitude that they actually have. From this we see that, in a given situation, the value of the constant G fixes the magnitude of the forces in the gravitational interaction. We can imagine another universe in which this constant, and therefore all the gravitational forces, might be different. In our universe, however, this quantity is always the same, as far as our experiments tell us. Later, we shall be concerned with other interactions and force laws whose strength is governed by other fundamental constants. We find that the gravitational interaction is the weakest interaction that has been yet observed.

The gravitational force is so weak that for most practical purposes we ignore it, unless at least one of the objects in the interaction has a mass as great as that of the earth, the moon, or some other celestial object. However, by using a very sensitive device, we can measure gravitational forces between objects small enough to hold in your hand. In so doing, we can also measure the value of the constant G.

Such a device, called the Cavendish balance, is shown in Figure 6-3. The masses A and B are fixed to the ends of a thin, stiff rod, suspended by a thin wire, as shown. If two additional masses C and D are placed near the masses A and B, they will pull on these masses because of gravitational forces. This will cause masses A and B to move slightly, twisting the wire. In practice, the masses C and D are first put in one position, which twists the wire one way, and then moved to the other position, shown by the dotted lines in the figure. This twists the wire in the opposite direction. In order to detect the amount of twist, which is very small, a mirror is mounted on the wire. As the wire is twisted, the mirror rotates. A beam of light reflected from the mirror moves along a scale as the mirror turns. This enables us to detect and measure the amount of "twist" in the wire caused by the gravitational force. From this twist we can actually

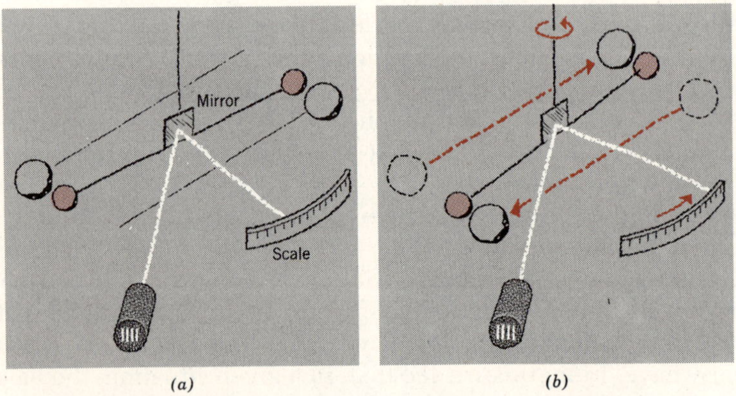

FIGURE 6-3 The Cavendish balance. In (b) the balls have been rolled to a new position, and the "dumbell" with the mirror on it has rotated. The mirror now reflects the light to a new position on the scale.

evaluate the gravitational force, since we know from various prior experiments how much force it takes to twist the wire. Knowing the gravitational force, the masses of the balls, and the distance between them, we can calculate the gravitational constant G, using the equation on page 89.

Weight

We define the weight of an object to be the gravitational force acting on it. For an object on the surface of the earth, the weight is essentially the gravitational pull of the earth on the object. On the moon, however, the weight of the object is essentially the gravitational pull of the moon acting on it. Since these gravitational pulls are not equal, the same object has a different (smaller) weight on the moon than on the earth. Note that the weight of an object is *not* an invariant property of that object. The weight of an object is different in different places in the universe and depends upon its position with respect to other massive objects. In fact, the weight of an object is even slightly different at different places on the surface of the earth (see Fig. 6-4).

The mass of an object is different from its weight. We shall consider the mass of an object to be the same no matter where in the universe the object is located. Therefore, we emphasize that weight and mass are distinctly different quantities. The weight of an object is a force—in fact, the gravitational force on that object. Mass, on the other hand, is a fundamental quantity. The mass of an object determines the acceleration produced by a force, whether it be the force

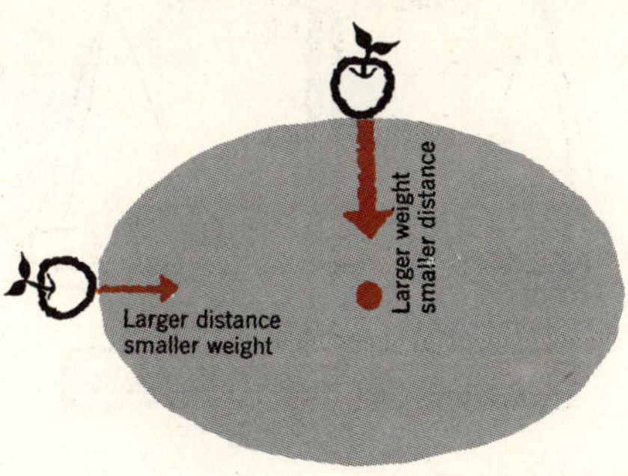

FIGURE 6-4 The apple is closer to the center of the earth at the north pole, and it weighs more there.

 of gravitation or any other force (see Chapter 4, pp. 57–63).

Even though mass and weight are different quantities, they are closely related in practical applications. The most practical (and most accurate) method used to determine the mass of an object is the process known as "weighing." In this process we normally obtain the mass of an object by actually measuring its weight. This process works because, at a particular place, the mass of a given object is proportional to its weight (and vice versa). At the same location, an object with twice the mass (relative to other objects) would have twice the weight, and so on. This relation must hold because the weight of an object is the gravitational force exerted on it and the gravitational force is proportional to either of the masses involved in the interaction.

We shall illustrate the process of weighing by discussing the balance shown in Figure 6-5. Similar balances are used in scientific work to measure masses to an accuracy of one ten-millionth of a kilogram.

The beam with its two pans is balanced on a knife edge. The pans are suspended from the beam at equal distances from the knife edge. In the process of weighing, an unknown mass is placed in one pan and known masses are placed in the other pan until the system is again balanced. This will occur when the gravitational force on the unknown mass is the same as the gravitational force on the known masses. In this process of weighing, therefore, we are comparing the

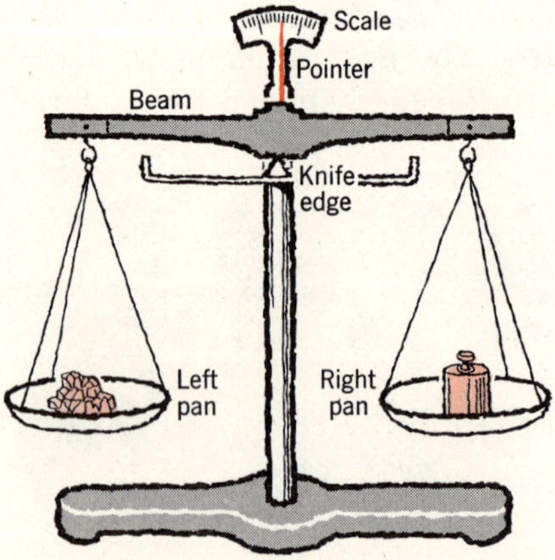

FIGURE 6-5 The mass of an object may be measured by weighing it on a balance, such as the balance illustrated here.

weight (gravitational force) of the known mass with the weight (gravitational force) of the unknown mass. Since, for a given location, the mass is proportional to the weight, the masses will be equal when their weights are equal. (There are other weighing devices and also other ways to use this device that do not require such symmetry in construction.)

The Gravitational Acceleration

The gravitational acceleration of an object is the acceleration it would experience if the gravitational force were the only force acting on it. The fact that the weight of an object is proportional to its mass has this important consequence: all objects (in the same position with respect to the earth) receive the same acceleration from the gravitational pull of the earth, regardless of their mass. It is true that the gravitational force is greater on an object of greater mass. However, it requires this greater gravitational force to give this greater mass the same acceleration. In fact, this principle is true not only for objects falling near the earth, but for objects experiencing a gravitational pull from the moon or any other object. At a given location, the gravitational acceleration does not depend on the mass of the object being accelerated.

Because the gravitational acceleration at the same location is the

FIGURE 6-6 Any two objects would fall together in a vacuum.

same for all objects, a feather and an elephant dropped at the same time from the same height will hit the ground at the same time, in the absence of air friction and other interfering forces (see Fig. 6-6). Of course, if we actually tried this experiment the elephant would hit the ground long before the feather because the retarding effect of the air is more important for the feather than for the elephant. The retarding effect of the air is also important in the time of descent for the two men shown falling in Figure 6-7. When, in the absence of air, gravitation acts alone, skydivers, parachutists, feathers, and elephants all fall together with the same acceleration—the gravitational acceleration. Near the surface of the earth the gravitational acceleration is about 9.8 meters/second². We shall use the symbol g for this quantity:

$$g = 9.8\ \frac{\text{meters}}{\text{second}^2}$$

Near the surface of the moon, the gravitational acceleration would be less.

FIGURE 6-7 **If there were no air, the parachutist would fall with the same acceleration as the sky diver.**

Air resistance is often negligible for such objects as marbles, billiard balls, elephants, and other things much more compact than feathers, falling for short distances at low speeds. These objects can therefore be used to demonstrate that objects of different mass fall with the same acceleration. We can do better than this, however, by using the apparatus shown in Figure 6-8. Nearly all the air can be removed from the large transparent tube by means of an ordinary air pump. Once we have removed the air, we can actually observe the coin and the feather inside the tube falling together side by side. Galileo and others, from their study of physics, realized that this would occur in a vacuum although air pumps had not yet been invented to create one.

The exact relation between mass and weight can be obtained in the following way. A freely falling object is subject to only one force—its weight. Such an object is, as we have seen, falling with the gravitational acceleration. From the second law of motion we know that this force, the weight, must be equal to the mass of the object multiplied by its acceleration. Thus the weight of an object is equal to its mass multiplied by the gravitational acceleration.

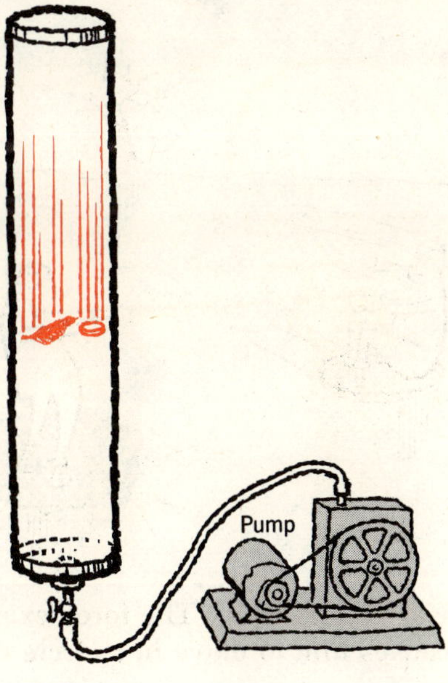

FIGURE 6-8 If the air is removed, the coin and the feather fall together.

The Law of Gravitation and "The System of the World"

After Newton had developed his laws of motion and his law of gravitation, he proceeded to apply them by demonstrating "the frame of the system of the world."[1] Before discussing the motion of the planets in this system, let us consider circular motion.

Circular Motion at Constant Speed

In Chapter 4 we pointed out that we must exert a force on an object if we wish to change its direction of motion. Force is necessary to change the direction even if the speed of the object remains the same. In this section we will discuss a very special case of accelerated motion: motion in a circle with a constant speed.

In Figure 6-9, which illustrates this kind of motion, a man is holding onto one end of the rope and his son, sitting in a pan, is holding the other end. We assume that the ice is so smooth that friction does not slow down the boy's motion. As the boy moves in a circle his

FIGURE 6-9 The force exerted by the rope on the boy causes him to move in a circle rather than a straight line.

[1]Isaac Newton, *The Mathemtaical Principles of Natural Philosophy and His System of the World*, translated by Andrew Motte 1729, rev. by Florian Cajori, Berkeley: University of California Press, 1934, p. 397.

speed remains constant. The direction of his velocity, of course, is changing all the time—that is, as long as both man and boy hang onto the rope. The force causing the change in direction is the force exerted through the rope on the boy. This force is along the direction of the rope and pointed toward the center of the circle. The boy would not move in a circle if this force were not maintained. If, for example, either the boy or his father lets go of the rope, the boy will cease moving in a circle and start moving in a straight line, in the direction in which he was moving at the instant that he or his father let go of the rope.

An object moving in a circle with constant speed is changing its direction of motion at all times, and a force acting on it at all times is necessary to maintain this kind of accelerated motion. This force must act toward the center of the circle. Whenever a force causes this kind of motion, it is called a *centripetal* force. The centripetal force is the force actually exerted on an object that is performing circular motion. In the previous example, the centripetal force is the force exerted on the boy. It is *not* the force exerted on his father. As we have stated, the direction of the centripetal force is always toward the center of the circle. This means that it is perpendicular to the velocity. Therefore, the direction of the centripetal force is also always changing as the object moves in a circle (see Fig. 6-10). The magnitude of the centripetal force, however, does not change as long as we are dealing with an object with the same mass traveling in the same circle at the same speed. It *does* depend on the mass m of the

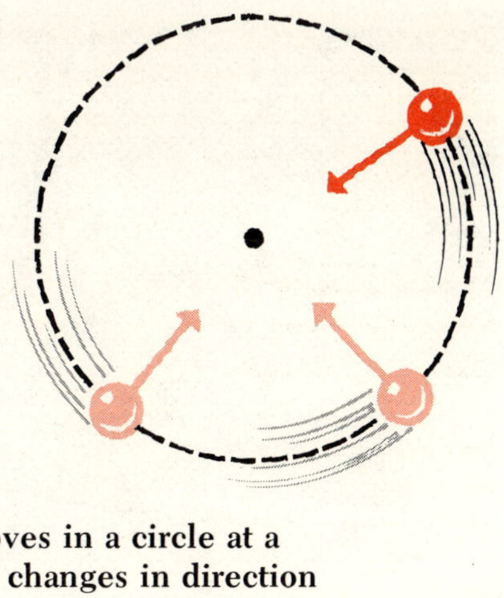

FIGURE 6-10 As the object moves in a circle at a constant speed the centripetal force changes in direction but not in magnitude.

 object, the speed s of the object, and the radius r of the circle. The following equation gives the magnitude of the centripetal force F:

$$F = \frac{ms^2}{r}$$

A force of this magnitude is required for circular motion regardless of how the force arises, whether from gravitation, a man pulling on a rope, or any other source.

The Planetary Model

In Newton's day, people had begun to believe that the earth and the other planets moved in orbits around the sun. The orbits of these planets, as we have noted, are not perfect circles but ellipses. However, they are so close to perfect circles that we can consider them as such in order to explain the main features of plenetary motion.

Figure 6-11 shows the planets moving in their orbits around the sun in approximately circular motion. As we have explained, a centripetal force is required to maintain this motion. What interaction, acting across millions of miles of space, can supply this force on the planets, pulling them toward the sun? Newton realized that it was the gravitational force that was responsible. The same force that controls the motion of falling objects also controls the motion of the planets and even the stars. There is a gravitational interaction between each planet and the sun. This causes a gravitational force on each

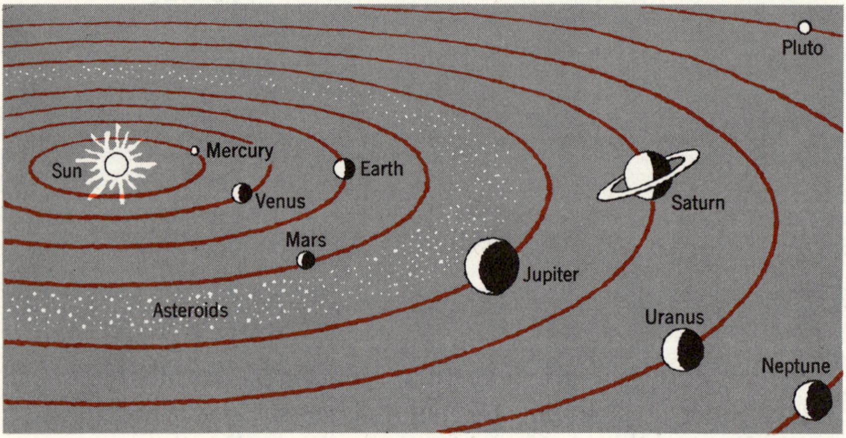

FIGURE 6-11 **The sun and the planets.**

planet that acts in the direction of the sun. The gravitational force is the centripetal force that maintains the nearly circular motion of the planets.

The planets, of course, also exert gravitational forces on the sun, which have the same magnitude as the forces exerted on them by the sun. The sun's mass is so large, however, that the effect of these forces on the motion of the sun is small.

The moon, pulled by the gravitational attraction of the earth, moves in a nearly circular orbit around the earth, just as the planets do around the sun. Other planets also have moons, or natural satellites, orbiting about them. On a still greater scale, the sun and billions of other stars form a system, called a galaxy. The stars move around the center of the galaxy. The sun is about two thirds of the way out from this center, a distance of about 35,000 light years. (A light year is the distance that light travels in one year, at a speed of 300 million meters per second – 186,000 miles per second.) Galaxies themselves occur in groups. The nearest galaxy in our group is supposed to be about two million light years away. The modern picture of the universe is thus one of systems of systems of systems, held together by gravitational forces, moving through the "immensity of space."

Newton, in his law of gravitation, made no attempt to explain why it is that the earth or any other mass can exert a gravitational force. Nevertheless, using this law we can predict, with beautiful precision, the accelerations of the moon and the planets. For example, the planet Uranus showed variations in its path that could not be accounted for by the gravitational forces of the other known planets. Therefore, the gravitational law predicted the existence of an unknown mass, a planet even more distant from the sun than Uranus, that was responsible for the disturbance. Faith in the prediction led Leverrier to compute the required orbit of this new planet and to direct the astronomer Galle to look at the precise spot in the heavens where the new planet should be located. The planet was found as predicted and named Neptune. Much later, the planet Pluto was discovered in a similar manner. Again, the bright star Sirius was observed to have variations in its motion with respect to the other stars in the heavens. Bessel's faith in the law of gravitation led him to predict an invisible star, which he called "Little Dog." Later the star, though very faint, was found at the predicted position.

The list of successful predictions resulting from Newton's laws of motion and his law of gravitation is impressive. The time finally came, however, when a planet was predicted and not found. A planet named Vulcan was predicted to explain very small variations in the orbit of the planet Mercury, but Vulcan was never observed. Today we use a new theory of gravitation devised by Albert Einstein (see Chapter 15) to account for these small variations.

Artificial Satellites and Weightlessness

In the course of his discussion of planetary orbits, Newton showed the possibility of artificial satellites.[2] In doing this he used a diagram similar to Figure 6-12. We imagine a mountain so high that the air near its top is too thin to cause much resistance to the motion of objects. An object thrown horizontally from this tall mountain will fall to the earth in a curved path. The faster it is thrown, the further it will travel before hitting the ground. In fact, given sufficient speed, it will travel clear around the earth in a nearly circular path. (It *falls* clear around the earth.) In order for this to happen, the object must attain a speed of about 7,700 meters/second (about 5 miles/second). Still greater speeds will cause the object to travel in an elliptical path around the earth. If the object is given a speed greater than about 11,000 meters per second (about 7 miles per second), it will leave the earth and not return.

The technique used in putting a satellite in orbit, then, is to first send it high enough for the air resistance to be very small, then get it pointed in a nearly horizontal direction, and finally increase its speed to at least five miles per second. For artificial satellies, just as for the moon, it is the gravitational force of the earth that acts as a centripetal force holding the satellite in nearly circular motion around the earth, once it has attained the proper speed in a nearly horizontal direction.

It will seem to an astronaut as though he and the other objects in his capsule are "weightless." The astronaut will observe no tendency for any of these objects to "fall" or move away from him, since both he and they are moving at essentially the same speed in essentially the same orbit. Actually, of course, these objects are not weightless, since it is their weight (the force of gravitation) that keeps them in orbit. Their weight supplies the centripetal force that holds them in circular motion. A person or object in orbit is already in free fall, whether the orbit is circular or elliptical, because gravitation is the only force acting on him (see again Fig. 6-12). Thus he cannot drop a pencil or a cup, since these objects are already falling

[2]*Ibid.*, p. 551. Figure after Newton.

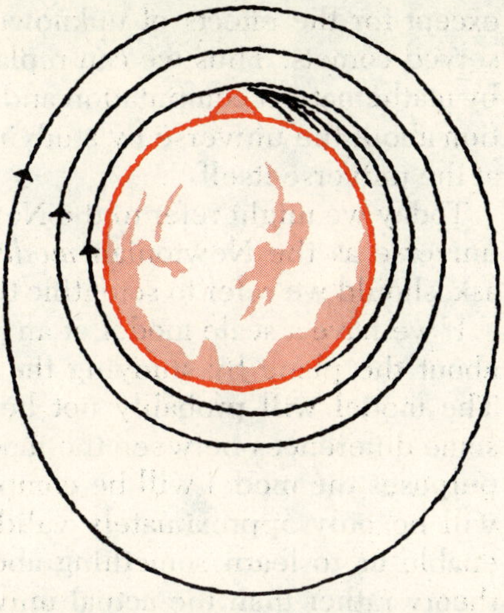

FIGURE 6-12 **Diagram used by Newton which suggests the possibility of artificial satellites.**

with him. The effect is essentially the same as if a man falling from a building tried to drop a rock. If he let go of the rock, it would still fall along with him.

The Newtonian Model

We have now completed our discussion of the first comprehensive mathematical theory of the universe. It is made up of the following principles and their consequences.

I. The three laws of motion.
II. The law of gravitational interaction.

If we know the positions and velocities of the various objects of the universe at a given time, and if the gravitational interaction supplies the only force acting on them, we can, in principle, accurately predict the future (and past) behavior of the objects. We do this by solving equations that express the above laws. (The solutions may be extremely difficult to obtain.) Since the objects in the heavens—the earth, the stars, the sun, and the planets—interact mainly through gravitation, we can predict with amazing accuracy how they behave. Knowing them as they are now, we can reconstruct the appearance of the heavens many thousands of years in the past and in the future, except for the effects of unknown objects such as previously unobserved comets. Thus we can replace observation, in many instances, by mathematical computation and prediction. We can obtain information about the universe by studying our equations instead of looking at the universe itself.

Today we might refer to the Newtonian mathematical theory of the universe as the Newtonian *model* of the universe. Why, we might ask, should we refer to scientific theories as models?

If we have a scale model of an airplane, we can obtain information about the plane by studying the model instead of the plane itself. The model will probably not be exact or complete—there will be some differences between the model and the real airplane. For some purposes the model will be completely satisfactory, and for others it will be only approximately valid. Similarly, our scientific theories enable us to learn something about the universe by looking at the theory rather than the actual universe. From the mathematical laws of motion and gravitation that comprise the Newtonian model of the universe, we can learn many things about the universe. Like most models, however, this is not perfect. There are some inadequacies in it, as we shall see later. Nevertheless, it is so accurate that for a long time most scientists thought that it was precisely true in every detail. This mathematical model has been one of the most successful intellectual constructions of mankind. It is still satisfactory for most celestial motions and for normal motions on earth which do not approach the speed of light.

Rotation

Angular Momentum

In discussing the planets and other objects in nature, we often describe the motion of rotating or revolving bodies. The earth spins (rotates) on its axis as it revolves around the sun. On earth, we often deal with the spinning wheels of automobiles and other machines.

There is a conserved quantity which is closely related to the rotations and revolutions of objects. This quantity is called *angular momentum*. We may consider the conservation of angular momentum very roughly as a conservation of the spin or rotation of objects and systems of objects. A top, once set spinning, tends to continue rotating forever unless it is slowed down by friction. The earth continues to spin, the planets continue to revolve around the sun, etc.

Let us consider the behavior of a rotating object, such as the spinning bicycle wheel shown in Figure 6-13. A rod forms the fixed axis around which the bicycle wheel rotates. Most of the mass of this particular wheel is located on the rim, since the spokes have relatively little mass. For this discussion, therefore, we consider all the mass to be on the rim, at a distance r from the axis. There are three quantities which are essential in considering the rotation of this wheel: its mass m, the speed of the rotating rim, s, and the radius r. The derived quantity, angular momentum, has a magnitude which is defined as follows:

$$\text{Magnitude of angular momentum} = msr.$$

Angular momentum is a vector quantity, and indeed can be defined generally about any given axis even for objects which are not rotating or revolving about that axis in the normal sense of the terms. Here, however, we are mainly concerned with the rotation of objects about a fixed axis. In such cases we do not need to worry about the vector nature of angular momentum except to remember that if rotation in one sense about the axis is taken as positive, then rotation in the other sense about the axis must be taken as negative. For example, if counterclockwise rotation is taken as positive, clockwise rotation must be taken as negative.

Although we can take the angular momentum about any axis, there

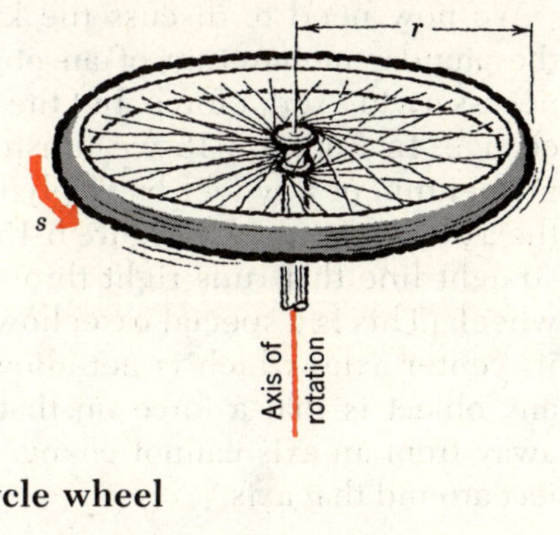

FIGURE 6-13 A spinning bicycle wheel

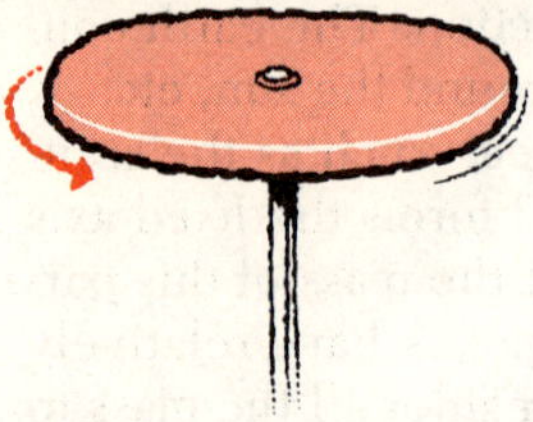

FIGURE 6-14 **Any rotating object has angular momentum about its axis of rotation.**

will often be some particular axis that is most convenient in a given situation. For the bicycle wheel considered above, and for similar objects, it is most convenient to choose the axis through the center.

To find the *total angular momentum* of a group or system of objects around a certain axis, we simply find the angular momentum of each object around that same axis and add these values together. A rotating object such as the disk in Figure 6-14 does not have all its mass at the same distance from the axis. Also, different parts of the disk will be moving at different speeds. To find the angular momentum of the disk we must add up the contributions, msr, of each small part of the disk. This, of course, explains why we chose the bicycle wheel at the beginning of our discussion—nearly all the mass is at the same distance from the axis and is moving at nearly the same speed. The total angular momentum will be *conserved* for any system that has no interaction with objects outside the system. (Actually, we can make a stronger statement than this, since forces can be applied to a system in such a way that they do not change its angular momentum.)

We now need to discuss the kind of interaction that will change the angular momentum of an object about an axis. Of course, it is necessary to exert a force, but the force alone does not determine the change. In Figure 6-15 we illustrate attempts to change the angular momentum of a wheel by applying force at different distances from the axis of rotation. In Figure 6-15*a* and *b*, we apply the force along a straight line that runs right through the axis. This fails to rotate the wheel. (This is a special case, however, because the wheel is held by its center axle, which is not allowed to move.) The general rule for any object is that a force on that object directed exactly toward or away from an axis cannot change the angular momentum of the object around that axis.

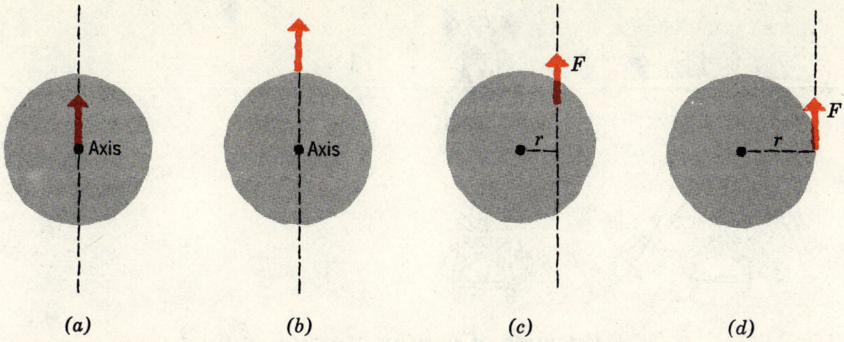

FIGURE 6-15 **When the line of force is directed right through the axis, as in (*a*) and (*b*), it can have no direct effect on the object's rotation around that axis. The force in part (*c*) is more effective in changing rotation than the equal force in (*d*) because it is along a line farther from the axis.**

In the next parts of Figure 6-15 we apply the same force at different distances from the axis. Now we do indeed cause the object to begin to rotate. Furthermore, the greater the perpendicular distance *r* from the axis to the line along which the force acts, the greater the effectiveness of the force in causing the wheel to start to rotate. Thus, the force in Figure 6-15*d* is more effective than the force in Figure 6-15*c*.

Examples of Conservation of Angular Momentum

The law of conservation of angular momentum is easily illustrated. In Figure 6-16 we see a man sitting on a stool. The stool is so designed that the seat can rotate freely around a vertical axis. When the seat of the stool is rotating, there will, of course, be some friction, causing a twist, which tends to slow it down. We will assume that this friction is so small that it can be ignored.

In our first experiment a man is sitting on the seat, rotating with it. He holds a large mass in each of his outstretched hands. What happens when he pulls in his hands? We find that his rotation speeds up. The law of conservation of angular momentum explains why.

The angular momentum *msr* must be conserved. When the man pulls in his hands, the *r* (or distance out) for the mass in his hands is reduced. In order for the quantity *msr* to remain the same, something

FIGURE 6-16 **Because of conservation of angular momentum, his rotation increases as he pulls the weights in.**

else in the formula must increase. Since the mass m is constant, the speed s must increase. Thus the speed of rotation of the whole system must increase. (Actually, it is the angular momentum of the whole rotating body that must be conserved, not the angular momentum of the masses alone. The net effect is that the number of rotations per second increases.)

Now consider a second demonstration. The man is standing stationary on the same stool, holding the bicycle wheel above his head like an umbrella (see Fig. 6-17). At first neither the wheel nor the man on the stool is rotating. Then the man reaches up and begins to spin the wheel. What happens? Since the angular momentum of the system must remain the same, the increased rotation of the wheel must be balanced by an opposite rotation somewhere else in the system. As the man spins the wheel, he and the seat of the stool begin to rotate in the opposite direction.

The Newtonian Synthesis

Newton the Person

About 1684 the English astronomer Edmund Halley asked Newton if Kepler's first law would follow from what we call Newton's law of gravitation. Newton said that indeed it would, but that he had lost the calculation. As C. C. Gillispie has so aptly put it, "While others

FIGURE 6-17 **Because of conservation of angular momentum, he begins to rotate clockwise as he spins the wheel counter-clockwise.**

were looking for the law of gravity, Newton had lost it."[3] Evidently he had mentioned the discovery to no one. In fact, Newton appeared to be more interested in his private researches than in risking controversy by publishing their results. He threatened to suppress part of his great work, the *Principia*, because of his controversy with another English scientist, Robert Hooke.

Newton also did important research in the nature and behavior of light. In addition he is responsible for one of the great advances in mathematics—the differential and integral calculus. This is the kind of mathematics needed to deal with quantities which change in a nonuniform way. (The calculus was also discovered independently by the German philosopher-mathematician G. W. von Leibnitz (1646–1716), and a dispute over priority resulted.) Newton's powers of concentration were evidently very great; when asked how he made his discoveries, he replied, "By always thinking on them."[4]

In 1696 Newton left Cambridge University to accept the post of Warden of the Mint (later Master of the Mint). According to Derek de Solla Price, this post was intended as a sinecure, but Newton took it "so conscientiously and seriously that one must credit quite a lot of the later economic strength and security of England to [his] efficient

[3]C. C. Gillispie, *The Edge of Objectivity. An Essay in the History of Scientific Ideas*, Princeton University Press, Princeton, N.J., 1960, Fourth Printing, 1967, p. 137. Reprinted by permission of Princeton University Press and Oxford University Press.
[4]See Gillispie, *op. cit.*, p. 117.

 reforms and administration."[5] Even at Cambridge, Newton was absorbed by nonscientific studies; indeed, his work in physics and mathematics seems to be the result of discoveries which he made very early, in the years 1665–1666. He was fascinated by the esoteric lore of alchemy. He was evidently considered a very able theologian, and wrote a commentary on the Apocalypse of John and the prophecies of Daniel. Apparently he disagreed with the standard doctrine of the Trinity but had to keep it secret.

About Newton's nonscientific interests the economist John Maynard Keynes commented:

> In the eighteenth century and since, Newton came to be thought of as the first and greatest of the modern age of scientists, a rationalist, one who taught us to think on the lines of cold and untinctured reason.
>
> I do not see him in this light. I do not think that anyone who has poured over the contents of that box which he packed up when he finally left Cambridge in 1696 and which, though partly dispersed, have come down to us, can see him like that. Newton was not the first of the age of reason. He was the last of the magicians, the last of the Babylonians and Sumerians, the last great mind which looked out on the visible and intellectual world with the same eyes as those who began to build our intellectual inheritance rather less than 10,000 years ago. Isaac Newton, a posthumous child born with no father on Christmas Day, 1642, was the last wonder-child to whom the Magi could do sincere and appropriate homage.[6]

We may see Newton as one of the greatest mathematicians of the world, or perhaps the greatest physicist of all time; or even as a magician, "Copernicus and Faustus in one,"[6] as Keynes saw him. He himself said:

> I do not know what I may appear to the world; but to myself I seem to have been only like a boy, playing on the seashore, and diverting myself, in now and then finding a smoother pebble or a prettier shell than ordinary, while the great ocean of truth lay all undiscovered before me.[7]

[5]Derek J. de Solla Price, "Newton as an Elder Statesman," *Science*, **158** (December 8, 1967), p. 1298, a book review of J. F. Scott, ed., *The Correspondence of Isaac Newton, Vol. 4, 1694–1709*, published for the Royal Society by Cambridge University Press, 1967.

[6]John Maynard Keynes, "Newton the Man," in James R. Newman, *The World of Mathematics*, vol. 1, New York: Simon and Schuster, 1956, pp. 277, 285.

[7]E. N. Da C. Andrade, "Isaac Newton," in Newman, *op. cit.*, p. 271.

Some of that "ocean of truth" left undiscovered by Newton will concern us in later chapters.

The Impact of the Newtonian Synthesis

Newton knew that the laws of Kepler for planetary motion were experimentally correct. He also knew that the laws of Galileo for falling bodies near the earth were experimentally correct. He adopted his own three laws of motion as assumptions, and then proceeded to obtain his law of gravitation by reasoning from these laws and experimental results. He showed that the observed behavior of falling objects, and also the behavior of comets, planets, and other celestial objects, could be precisely predicted using his three laws of motion and the law of gravitational force. In this way he combined his results with those of his predecessors into a simple, comprehensive system—the famous "Newtonian synthesis." It was and is considered one of the most important achievements of the human mind. From simple assumptions Newton derived the properties and behavior of matter in motion. The result was a successful mathematical treatment of mechanics—the behavior of matter in motion under the influence of forces.

The men of the eighteenth-century Enlightenment had reason to be impressed with the new physics. When had the world known such an achievement? The universe, both the heavens and the earth, operated according to the same precise mathematical laws. All one had to know was the position and velocity of the particles of matter and the forces between them. Then one could, in principle at least, predict exactly what would happen to the particles at any time in the future—or, for that matter, determine what had happened to them in the past. Thus it is that our modern planetariums can reproduce the face of the sky, the positions of the planets and the stars, for any day in history, past or future. Of course, there is always some experimental error in determining the starting position and velocity of the particles, and there are too many particles in the universe to keep track of them all. In principle, however, there should be an exact position and velocity for every particle, and the laws of motion and force should exactly determine the behavior of the particles.

This type of physics almost inevitably suggested a mechanistic philosophy, in which everything was reduced to particles of matter in predictable motion according to exact and irrevocable mathematical law. It also suggested a deterministic type of philosophy, in which the future is in principle predictable and all "behavior is caused." In this philosophy, because of the invariable and predictable operations of natural law, conditions in the future were irrevocably determined by conditions in the present. Thus the universe appeared to be a "world machine," in which each piece behaved in a perfectly predictable and determined way.

Such a philosophy inevitably had ramifications and effects outside of physics: it clearly contained the foundations of a challenge to free will and other spiritual values. Such a world view would influence the other sciences, the humanities, religion, and the general culture.

There is a temptation to blame the French Revolution on the philosophers of the Enlightenment: "It is the fault of Rousseau," "It is the fault of Voltaire." Could Newton have had anything to do with it? Voltaire was an avid fan of Newton; he even considered it worthwhile to write a book about Newton's philosophy. In the Age of Reason, Newton was evidently considered an apostle of Reason, and therefore the influence of his ideas was important among philosophers. Woodrow Wilson suggested an influence of Newtonian ideas on the American Revolution:

> The government of the United States was constructed upon the Whig theory of political dynamics, which was a sort of unconscious copy of the Newtonian theory of the universe. In our own day, whenever we discuss the structure or development of anything, whether in nature or in society, we consciously or unconsciously follow Mr. Darwin; but before Mr. Darwin they followed Newton. . . .[8]

We certainly do not contend that the ethical and political principles that arose in the Enlightenment are dependent upon Newton's mechanics. It is interesting, however, to see the wide influence of Newtonian ideas.

In the "Romantic Reaction" which followed the "Age of Reason," and at other times, Newton and the mechanistic world view came in for some strenuous criticism. From a scientific point of view, however, the nineteenth century saw an increase in mechanistic views and what appeared to be a near-triumph for exact mathematical law. Lord Kelvin, one of the ranking scientists of the age, wrote in 1884:

> I am never content until I have constructed a mechanical model of

[8]Woodrow Wilson, *Constitutional Government in the United States*, New York: Columbia University Press, 1921, pp. 54–55. On this general subject we must acknowledge our debt to the following authors without holding them responsible for our opinions: J. Bronowski and Bruce Mazlish, *The Western Intellectual Tradition from Leonardo to Hegel*, New York: Harper Torchbooks, 1962; and Gillispie, *op cit.*

the object I am studying; If I succeed in making one I understand, otherwise I do not.[9]

Another example of mechanistic ideas was the theory of the evolution of living things proposed by Charles Darwin (1809–1882) and Alfred Wallace (1823–1913). These men suggested a mechanism by which, in succeeding generations, animals and plants could change to produce new species.

We cannot discuss all the mechanistic, materialistic, or deterministic philosophies of the nineteenth and twentieth centuries, nor would we contend that they all depended upon Newtonian physics. However, the unprecedented success of mechanistic and deterministic physics through the nineteenth century, has, it appears, helped to spread and strengthen these philosophies.

Mechanistic ideas are probably still among the most important and influential effects of physics on the general culture, even though they are no longer dominant in physics itself. There are, for example, less mechanistic theories of evolution than Darwin's, but they are not generally regarded as quite scientific. It is doubtful that extrasensory perception will be considered wholly respectable until some mechanism is suggested to account for its alleged effects. Freudian psychology might also be reproached for its lack of a material mechanism—behavioristic psychology seems more "scientific" to some. It is hard for some scientists to look at an animal, for example, which appears to be made up of material particles, and not believe that its properties and behavior can be explained in terms of those particles. The science of molecular biology attempts to do this as far as possible, and has been amazingly successful in uncovering the chemical mechanisms by which molecular particles determine the nature and heredity of the living cell.[10]

The strict determinism of Newton's laws is gone from physics today. Strict prediction is also gone, although prediction is even more important in physics than ever. Today, however, physicists make *approximate* predictions, forecasting a probability rather than a certainty. Knowledge of the future is of great practical importance, since it is only the future that we can change, not the past or the present. The models of the universe incorporated in our scientific theories can help us to do this.

[9]William Thomson, Lord Kelvin, as quoted by Julien Ripley, *The Elements and Structure of the Physical Sciences*, New York: John Wiley & Sons, 1964, p. 370.

[10]See, for example, J. D. Watson, *Molecular Biology of the Gene*, New York: Benjamin, 2nd ed., 1970.

 To close this section, we quote from a modern poet, Robert Frost.

Let me ask you to watch a metaphor[11] breaking down here before you. Somebody said to me a little while ago, "It is easy enough for me to think of the universe as a machine, as a mechanism."

I said, "You mean the universe is like a machine?"

He said, "No, I think it is one. . . Well, it is like . . ."

"I think you mean the universe is like a machine."

"All right. Let it go at that."

I asked him, "Did you ever see a machine without a pedal for the foot, or a lever for the hand or a button for the finger?"

He said, "No – no."

I said, "All right. Is the universe like that?"

And he said, "No. I mean it is like a machine, only . . ."

". . . It is different from a machine," I said.

He wanted to go just that far with that metaphor and no further. And so do we all. All metaphor breaks down somewhere. That is the beauty of it. It is touch and go with metaphor, and until you have lived with it long enough you don't know where it is going. You don't know how much you can get out of it and when it will cease to yield. It is a very living thing. It is as life itself.[12]

FUNDAMENTAL IDEAS

I. FUNDAMENTAL LAWS OR PRINCIPLES

  A. THE LAW OF GRAVITATIONAL INTERACTION. Every two objects attract each other with a force of magnitude $F = GmM/d^2$.

  B. CONSERVATION OF ANGULAR MOMENTUM

II. QUANTITIES

  A. DERIVED QUANTITIES

    1. Angular momentum.

  B. FUNDAMENTAL CONSTANTS

    1. The gravitational constant $G = 6.67 \times 10^{-11} \dfrac{\text{newton meter}^2}{\text{kilogram}^2}$

[11]A scientist might prefer the word *model* to *metaphor*.

[12]From "Education by Poetry" by Robert Frost, from *Selected Prose of Robert Frost* edited by Hyde Cox and Edward Connery Lathem.

III. APPLICATIONS
   A. WEIGHT
   B. ACCELERATION OF GRAVITY. The same for all objects at the same location.
   C. CIRCULAR MOTION AND CENTRIPETAL FORCE
   D. "SYSTEM OF THE WORLD"
   E. SATELLITES AND "WEIGHTLESSNESS"

IV. METHODS OF SCIENCE
   A. MODELS—THE NEWTONIAN MODEL
   B. THE NEWTONIAN SYNTHESIS

QUESTIONS

Gravitation

1. Does an apple exert a gravitational force on the earth? How does this force compare with the force exerted on the apple by the earth?
2. A cannonball and a postage stamp are allowed to fall in a vacuum. They are released at the same time from the same height. Describe the motion of each. What is the acceleration of each object? Are the forces on the two objects the same as they are falling? Why is there no contradiction between the answers to the last two questions?
3. If you were suddenly transported to another planet, how would you determine its mass?
4. What are the units of weight?
5. What is the difference between weight and mass?
6. Explain why mass can be accurately measured by weighing.
7. Would the process of measuring mass by weighing give the same result on the moon as it does on the earth?
8. Is the acceleration of gravitation always and everywhere the same? Explain.
9. The air resistance to a moving object increases with speed. How could this account for the fact that falling objects generally reach a maximum velocity if they fall far enough?
10. What sort of effect would you expect if the gravitational constant were ten times its actual value?
11. What is the velocity of a ball thrown straight up when it is at its highest point?

12. What is the acceleration of the ball in Question 11 at its highest point?
13. What is the force on the ball in Question 11 when it is at its highest point?
14. In Figure 6-18 the monkey and the object have the same mass. What happens if the monkey starts to climb the rope? (Ignore friction in the pulley.).
15. In Figure 6-18, we now allow the monkey to be less massive than the object. Is there any way in which the monkey can get the object to rise by climbing his end of the rope?
16. Will the system in Figure 6-19 move if there is no friction but M is a much greater mass than m?

Circular Motion

17. What else is constant when an object moves in a circle at constant speed? What is not constant?
18. Suppose that each of the following objects is traveling in circular motion: (a) a rock in a sling, (b) water in a spinning bucket, (c) an airplane in a horizontal circle, and (d) an automobile on a circular race track. How is the centripetal force applied in each case?
19. Can we exert a single force on an object without changing its speed? Give an example where this occurs (approximately). Does this example disprove the second law of motion?
20. Why will a flywheel fly apart if it is spun too rapidly?

FIGURE 6-18 What happens when the monkey starts to climb? See questions 14 and 15.

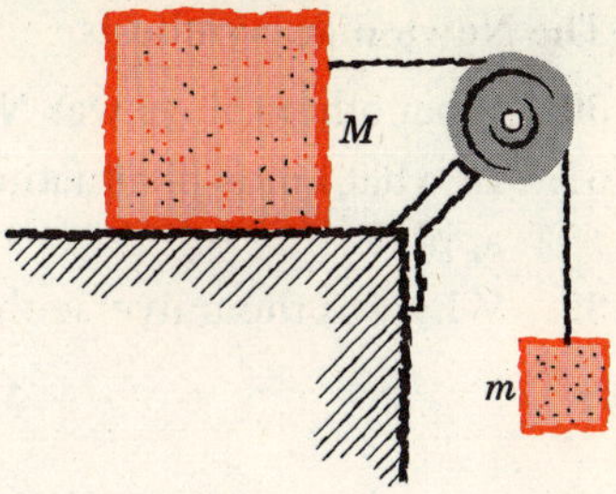

FIGURE 6-19 **Will it start to move? See question 16.**

21. Why is it difficult to turn an automobile which is moving along an icy road?
22. A ball is whirled in a horizontal circle on the end of a string. There is a force in the string constantly pulling the ball toward the center of the circle; yet the ball keeps going around at the same speed. How can this fact be reconciled with Newton's second law of motion, which states that force causes an acceleration?
23. We usually think of the sun as stationary and the earth as moving around it. Actually, if the earth were the only planet, the earth and the sun would both revolve about a common center which is inside the sun, close to its center. How do we know that both the sun and the earth revolve to some extent? (Assume that Newton's laws of motion and his law of gravitation are correct.).
24. Did Newton know how to orbit a space capsule?
25. Describe the motion of an astronaut who leaves his capsule for a space "walk." Is he massless? Is he weightless? Why does he not fall to the ground?
26. In what sense can Newton's system of the world be regarded as a model?

Angular Momentum

27. Suppose that a rotating ball of hot gas cools down and contracts. What happens to its rate of rotation? Its angular momentum?
28. If the man in Figure 6-9 pulls in or lets out on the rope, thus changing the distance at which his son swings around him, is angular momentum conserved in the process?
29. In Figure 6-17, after the man has reached up and started the bicycle wheel spinning, he turns the wheel upside down, without stopping it from spinning. What happens to his own rotation?

30. What other things was Newton involved in besides physics?

31. In what kinds of situations do Newton's laws of motion and gravitation apply?

32. Why was the universe thought to be like a machine?

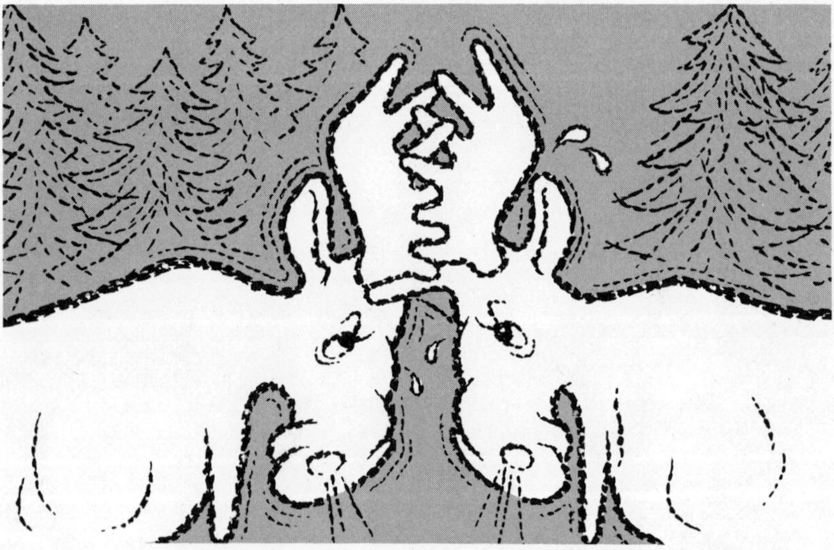

7 Further Applications of the Laws of Force and Motion

Forces on Objects

The Resultant Force Responsible for Motion

Frequently we are interested in the motion of an object that is subject to more than one force. In order to apply the laws of motion to such an object and to find its acceleration, it is necessary to add up all of these forces as vectors. It is this vector sum, or resultant force, that determines the acceleration of the object. To obtain the resultant we must be sure that we take into account all of the forces acting on the object.

In order to find all the forces on an object, we first determine all the interactions in which the object participates. All of the interactions we have discussed up to now have been the result of the direct contact of an object with other objects (through rods, strings, touching, etc) or of the gravitational interaction. Later we will discuss other kinds of interaction. Each of the interactions the object has with another object will produce *one force* on the object. (Each interaction will also produce one force on the other object, but since this force does not affect the motion of the object with which we are concerned, we can ig-

☞ **FIGURE 7-1**

nore it here.) Different parts of the same object can have interactions with one another, but these interactions may be ignored when we are studying the motion of the object as a whole.

To make sure that all of the forces are considered, it is helpful to draw a diagram of the object alone, separate from its surroundings. The next step in the procedure is to represent each of the forces acting on the object by an arrow drawn on the diagram. The vector resultant of these forces is then obtained by the method of vector addition described in Chapter 4. This resultant force is responsible for the acceleration of the object.

Before considering an example, let us outline the procedure for finding the resultant force on an object, the force responsible for its acceleration. The steps are as follows.

1. Select the object in whose motion you are interested.
2. List each interaction that this object has with other objects or bodies.
3. Draw a diagram that isolates the object, and draw an arrow on the diagram for each force acting on the object.
4. Obtain the resultant force by vector addition.

As an example of this procedure, we shall find the resultant force on a boat that is being pulled upstream by two boys (Fig. 7-2). In

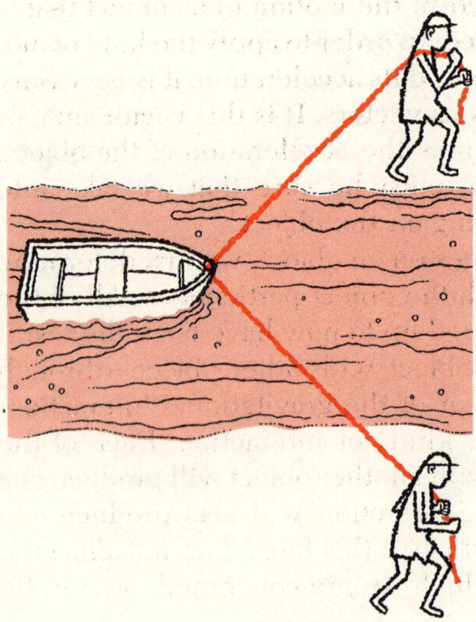

FIGURE 7-2 Two boys pulling a boat upstream.

accordance with the first step above, we have selected the boat as the object in whose motion we are interested. The second step is to list all of the interactions the boat has with other objects or bodies. Since we know that the acceleration of the boat is in a horizontal direction, we know that the resultant force must also be in a horizontal direction. Therefore, we ignore all vertical forces caused by the interaction of the boat with its surroundings. For example, we ignore the gravitational force of the earth pulling down on the boat. We will show later that it is all right to ignore these vertical forces when we are dealing with motion in a horizontal plane. We find that there are just three interactions that produce horizontal forces on the boat. We list these interactions as follows.

(a) The interaction with the rope being pulled by the boy on the right bank.
(b) The interaction with the rope being pulled by the boy on the left bank.
(c) The interaction with the water due to the retarding friction as the water passes by the boat.

In Step 3 we draw a diagram of the isolated object together with arrows representing the forces acting on it. Since there are three interactions, there will be three forces acting on the boat. This step is illustrated in Figure 7-3*a*. We are assuming that the magnitudes and directions of the forces are known so that the vectors can be drawn to scale. Step 4, obtaining the resultant force, is illustrated in Figure 7-3*b*. (We are using the "head to tail" rule for adding vectors, discussed on pp. 66–69.)

In the example we have chosen, the resultant force turned out to

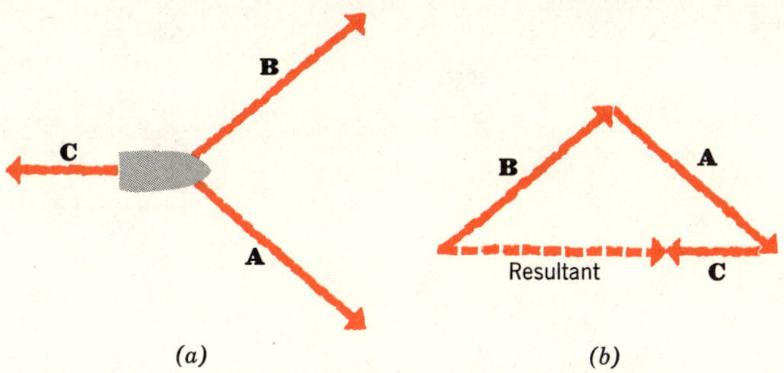

FIGURE 7-3 **Obtaining the resultant force on the boat in Figure 7-2.**

be upstream. If the three forces exerted by the boys and the water were different, however, the resultant force could turn out to be downstream, to the side, or the resultant force could turn out to be downstream, to the side, or even zero. These possibilities are illustrated in Figure 7-4.

Equilibrium

An object is in *equilibrium* whenever all of the forces acting on it add up to zero. When this resultant force is zero, the acceleration is also zero. Conversely, any object which is not accelerating is in equi-

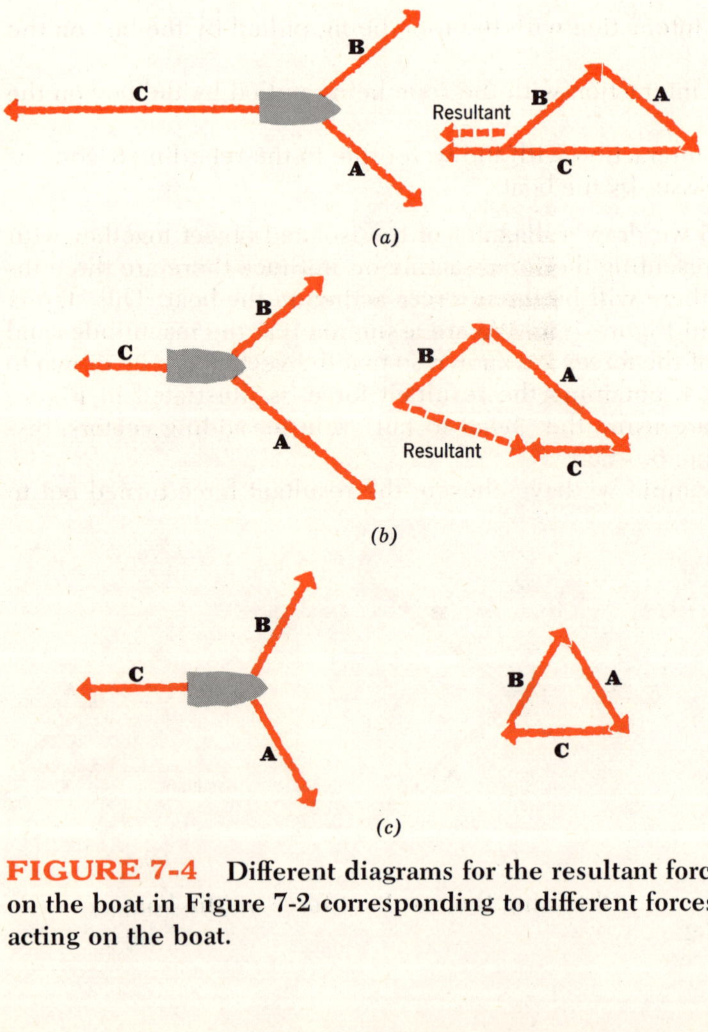

FIGURE 7-4 **Different diagrams for the resultant force on the boat in Figure 7-2 corresponding to different forces acting on the boat.**

librium, and the forces acting on it must add up to zero. Since an object can be moving without accelerating an object in equilibrium does not have to be at rest. It may either be at rest or in motion with a constant velocity. When the vector forces on an object in equilibrium are added together by the "head to tail" procedure, the head of the last vector comes right up to the tail of the first vector, forming a closed figure (see Fig. 7-4 *c*).

A simple example of an object at rest is the motionless moose shown in Figure 7-5. Following the steps outlined above, we first identify the object we are interested in—namely, the moose. Then we list the interactions between the moose and other objects.

(a) The gravitational interaction between the earth and the moose.
(b) The interaction of direct contact between the moose's body and the ground.

This second interaction consists of a pair of forces. One of these forces is exerted by the moose pressing down on the ground, and the other is exerted by the ground pressing up on the moose. The force of the ground pushing up on the moose is responsible for the difference in the motion of the moose in Figure 7-5 and its motion in Figure 7-6, where the ground has been removed. In Figure 7-6 the moose is involved in only one interaction, that of gravitation, and he is therefore accelerating. The moose in Figure 7-5 is at rest; the force of the ground pushing up on him is required to keep him from falling.

We next draw a diagram in which we isolate the moose and represent the two forces acting on him by means of arrows (see Fig 7-7). Since the moose remains at rest, he is in equilibrium, and the resultant of the two forces acting on him must be zero. The force of the ground pushing up must have the same magnitude as the gravitational force of the earth pulling him down. This same analysis can be

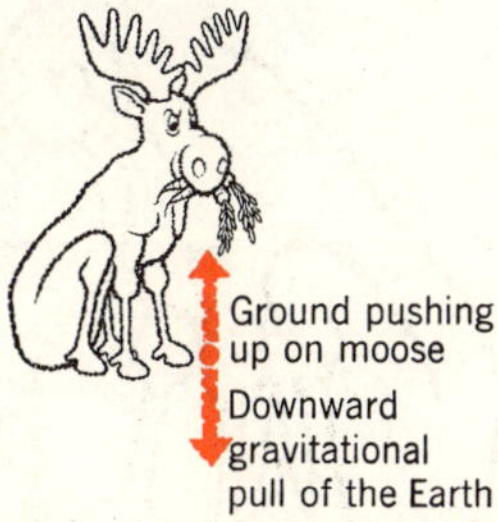

FIGURE 7-5 **Moose in equilibrium.**

FIGURE 7-6 Moose not in equilibrium.

applied to any object on the earth that has no vertical acceleration. This explains why we were able to eliminate the discussion of all vertical forces in the example shown in Figure 7-2. The force of gravitation pulling down on the boat is balanced by the force of the water pushing up on it.

The Moose-Log Paradox

We now hook the moose to a log and ask him to pull it for us (Fig.

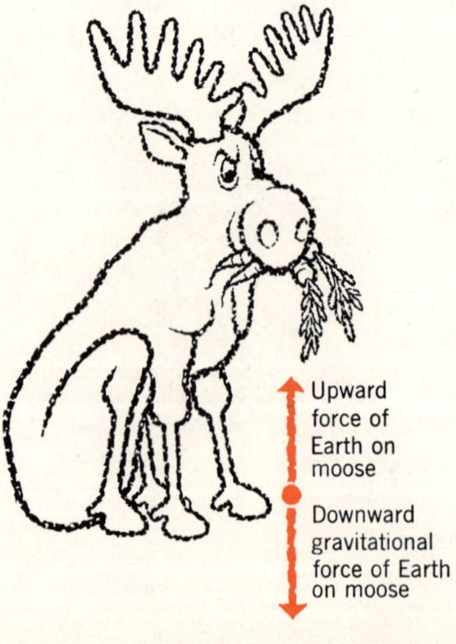

FIGURE 7-7 Isolated moose with forces applied.

FIGURE 7-8 We ask the moose to pull a log.

7-8). The learned moose objects, however, quoting from Isaac Newton as follows:

> If a horse draws a stone tied to a rope, the horse (if I may so say) will be equally drawn back towards the stone; for the distended rope, by the same endeavor to relax or unbend itself, will draw the horse as much towards the stone as it does the stone towards the horse, and will obstruct the progress of the one as much as it advances that of the other.[1]

The moose correctly interprets this Newtonian wisdom to mean that however hard he pulls on the log, it will pull back on him just as hard. "Therefore," says he, "it is useless for me to attempt to move this log (or anything else, for that matter), since any force I exert on the log to move it will be exactly balanced by the force with which the log pulls back."

In order to expose the fallacy propounded by this lazy beast, let us analyze the forces on the log as the moose begins to pull it. We have selected the log as the object of our interest. Since the motion of the log is horizontal, we shall again ignore all vertical forces. Thus there are only two interactions that we need to deal with: the interaction between the log and the moose, and the interaction between the log and the ground. The second interaction is made up of the frictional forces produced as the log is dragged across the ground. Thus, when we isolate the log in Figure 7-9, we have only two forces to consider. They are the force exerted on the log by the moose to the right, and

[1]Isaac Newton, *The Mathematical Principles of Natural Philosophy and His System of the World*, translated by Andrew Motte, 1729, rev. by Florian Cajori, Berkeley: University of California Press, 1934, p. 14. Originally published by the University of California Press; reprinted by permission of The Regents of the University of California.

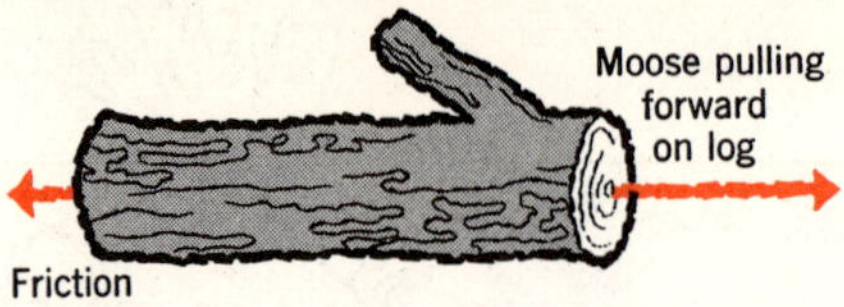

FIGURE 7-9 Isolated log with forces applied.

the frictional retarding force acting on the log to the left. To start the log moving, all that is required is that the moose exert a force on the log greater than the retarding force of friction. In this condition there will be a resultant force to the right, and the log will accelerate in the direction of the moose. If, after the log is in motion, the moose wants to pull it along with a constant velocity, he must exert a force on the log equal in magnitude to the force of friction on the log. In this condition, the resultant force on the log will be zero, and therefore the log will be in equilibrium. It will continue to move, in equilibrium, with a constant velocity as long as this resultant is zero.

Notice that in this analysis we have not mentioned the force of the log on the moose. This force is not acting on the log; although it does influence the motion of the moose, it does not affect the motion of the log. The fallacy of the moose's statement arises from the fact that one of the forces he was talking about does not even act on the object he was supposed to move. The interaction between the moose and the log does involve the two forces he described, but only one of them acts on the log.

Projectile Motion

Let us consider the motion of a sky diver who has just been dropped from an airplane. We assume that the airplane is flying along horizontally and uniformly, and not engaging in aerobatics and that we may ignore air friction in dealing with the motion of the sky diver (see Fig. 7-10). One way of analyzing the motion of the sky diver is to use the principle of relativity. (See again Chapter 3.) From the uniformly moving reference frame attached to the airplane, he falls straight down and remains under the plane. In the reference frame attached to the earth, however, the sky diver continues to move horizontally (to the right in our figure), with the same velocity at which he started. At the same time he is gaining a downward ve-

FIGURE 7-10 **Except for air friction, the sky diver remains under the airplane as he falls.**

locity because of his acceleration by the gravitational force. His vector velocity, therefore, is the resultant of two vector velocities—a vector velocity to the right (which remains the same all the time) and a vector velocity straight down (which is continually increased by the pull of the earth). His downward motion is the same as if he were not moving horizontally at all.

The situation can also be analyzed by using some of the principles discussed in this chapter. The gravitational force on the sky diver is vertical and therefore has no effect on his horizontal motion. Since there are no other forces on him, his horizontal velocity cannot change while he is falling. Figure 7-11 shows what happens to the vector velocity for two sky divers, one dropped from an airplane and one dropped from a stationary balloon. Notice that the velocity to the right never changes for either during the time of fall. It is zero for the sky diver dropped from the stationary balloon, and equal to the velocity of the airplane for the sky diver dropped from the airplane. As the speed of an object falling through the air increases, the forces exerted by the air on the object become more and more important and the above discussion which ignores the effects of the air becomes inadequate. Real sky divers, for example, use the forces of the air to help them control their motion even before they open their parachutes.

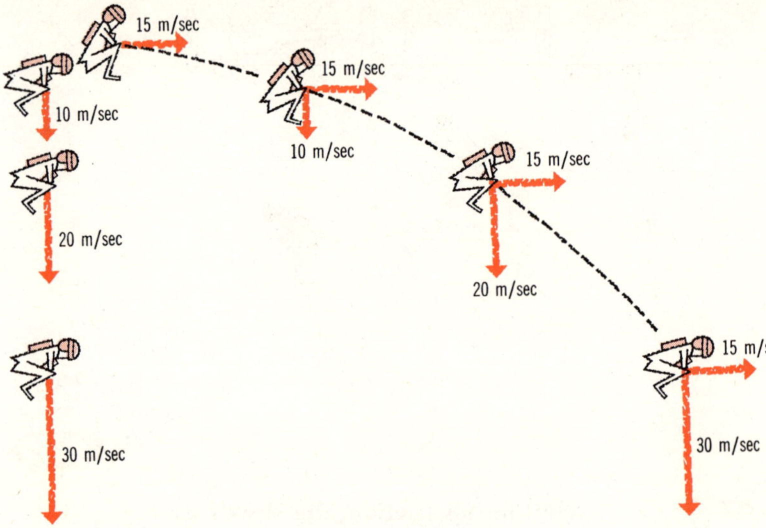

FIGURE 7-11 The horizontal velocity does not change for either sky diver (neglecting air resistance).

Forces in Fluids

Fluids at Rest

A fluid is a substance that can flow. Examples of fluids are liquids (such as water) that can be poured and gases (such as air) that tend to fill their containers completely. Two derived quantities, density and pressure, are expecially useful in describing the properties of fluids.

Density. The *density* of a fluid (or, indeed, of any body of matter) is equal to the amount of mass it has per unit of volume. Water, for example, has a greater density than air because there is more mass in a given volume of water than there is in the same volume of air—a bucket of water has more mass in it than the same bucket filled with air. Since the density is the mass per unit volume of a substance, we can express it in a formula as follows:

$$D = \frac{m}{V}$$

V is the volume of a certain amount of matter, m is the mass of this matter, and D is its density. For example, suppose that we wish to find the density of water. We happen to know that three cubic meters of water has a mass of 3000 kilograms. Since there are 3000 kilograms in three cubic meters there must be 1000 kilograms in one cubic meter. Therefore, the density of water is 1000 kilograms per cubic meter.

Pressure. When a force exerted on a surface is distributed over the surface (as is often the case in fluids) and not just applied at a single point, it is convenient to use a new derived quantity, *pressure*. The pressure on any region of a surface is equal to the total force on the region divided by the area of the region. That is to say, the pressure P is equal to the force F divided by the area A:

$$P = \frac{F}{A}$$

To show that pressure, rather than force, is the important factor in some physical effects, consider the dancer in Figure 7-12. If she weighs 506 newtons (115 pounds weight), she exerts a total force of 506 newtons on the floor, whether she is standing on the floor or sitting on it. The pressure, on the other hand, is much different in these two situations. When she is sitting down, her force is spread over a much more extensive area than it is when she rocks back on a single heel while dancing. If the bottom of her heel has a very small area, she might even dent the floor. The 506 newtons of force exert a much larger pressure when they are distributed over a smaller area.

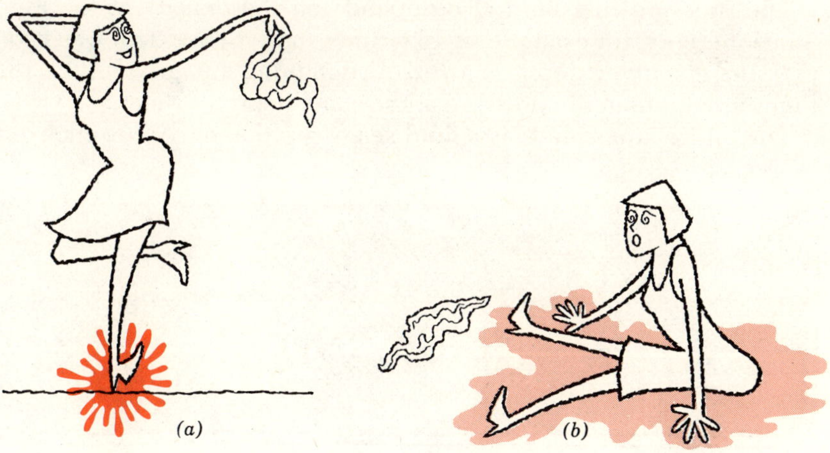

FIGURE 7-12 **The same force distributed over a greater area results in a smaller pressure.**

Pressure in Fluids. The philosopher and mathematician Blaise Pascal (1623–1662) studied fluids at rest, and as a result he obtained several principles that describe pressure in fluids. He found, for example, that the pressure in a fluid is perpendicular to any surface in contact with the fluid. In fact, one may consider that the pressure is perpendicular to any chosen surface in the fluid itself, regardless of its position or orientation. Furthermore, at any given point in the fluid, the magnitude of the pressure is the same in all directions. Figure 7-13 illustrates these facts. The piston in these illustrations is subjected to the same pressure in each of its orientations. In each case the pressure is perpendicular to the surface of the piston.

Pascal also found that the pressure in a fluid increases with depth and does not depend on the total volume of the fluid. For example, in the connected vessels shown in Figure 7-14, the pressure is the same at any point on the same level, as, for example, at points P_1, P_2, P_3, P_4, and P_5. This is true even though the vessels themselves differ greatly in size and shape. Because the pressure is the same at any point on the same level, the level of the surface is the same in all three vessels. (For small tubes, etc, there are exceptions to this rule.) If the pressure were not the same at the same level in all three vessels, water would flow from one vessel to another until the pressure was equalized. The pressure at any given level in the connected vessels is, or course, greater than the pressure at higher levels, since each level must help to support the weight of the water in all the levels above.

The pressure in a fluid also depends on the density of the fluid. Since mercury, for example, is 13.6 times more dense than water, the pressure at a given depth in a vessel containing mercury will be 13.6 times greater than the pressure at the same depth in water.

The entire atmosphere is a fluid sea of gas that covers us all, exert-

FIGURE 7-13 The pressure in a fluid at a given depth is the same in all directions.

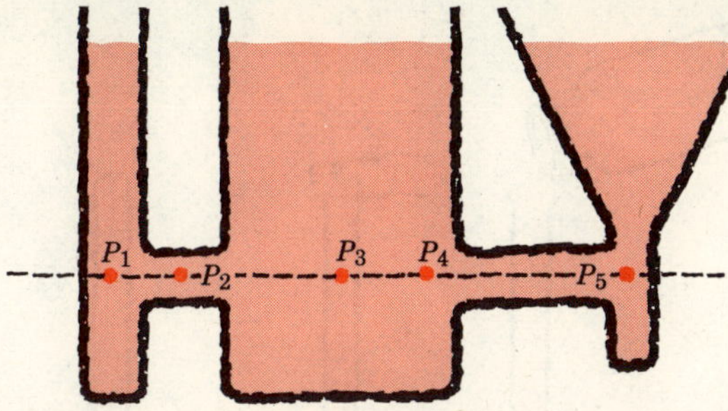

FIGURE 7-14 **The pressure in a stationary fluid depends on the depth, and not on the size or shape of the container.**

ing a pressure of 100,000 newtons per square meter on objects on the surface of the earth. This atmospheric pressure arises from the weight of the air, from the surface to the "top" of the atmosphere. Since the earth's atmosphere merges with the atmosphere of the sun, its "top" is not easy to define. However, the density in the upper regions is so small that the weight of the air above 5×10^4 meters contributes less than one percent to the total weight of the atmosphere.

We measure atmospheric pressure by using an instrument known as a barometer. Figure 7-15 shows how to make one kind of barometer—a mercury barometer. A glass tube, about a meter long, is filled with mercury. Then the open end of the tube is covered (to prevent loss of mercury) while it is immersed in a vessel containing mercury. In this process no air is allowed to enter the tube. When the open end is again uncovered, some of the mercury flows out of the tube into the dish, leaving a column of mercury in the tube with a near vacuum above it at the top (Fig. 7-15*d*). The pressure is essentially zero in the vacuum. Therefore, the pressure at the point P is due only to the weight of the mercury column above it. The pressure at the point Q at the surface of the mercury in the dish outside the tube is due only to the atmospheric pressure. Since these two points are connected by fluid mercury and are at the same level, the pressure at the two points must be equal; that is, the pressure represented by the height of the mercury column is the same as the pressure of the atmosphere outside. If the atmospheric pressure increases, the mercury will rise higher in the tube. If the atmospheric pressure decreases, the mercury will fall lower in the tube. Thus the mercury column can be used to measure air pressure. Changes in atmospheric pressure are used to help predict the weather. Since the atmospheric

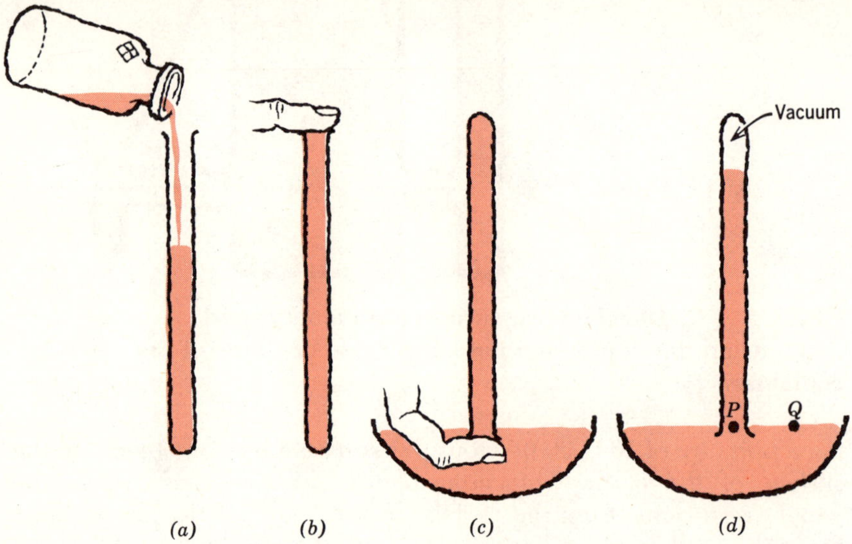

FIGURE 7-15 The construction of a mercury barometer.

pressure also changes with altitude, the barometer can be used to measure altitude.

Air pressure also helps us to breathe, as illustrated in the simple model of the human breathing apparatus shown in Figure 7-16. To begin with, everything is in equilibrium and all pressures are equal to atmospheric pressure. The breathing muscles pull on the diaphragm in order to expand the volume inside the chest cavity. This reduces the pressure on the outside of the lungs, represented as two balloons in the figure. The pressure inside the lungs is still the atmospheric pressure, which therefore expands the lungs, bringing air into them. When the diaphragm reduces the volume enclosing the lungs, the pressure in the chest cavity exceeds atmospheric pressure, causing the lungs to collapse and thus forcing air out.

Buoyancy. If you have ever tried to remove a large rock from a pool of water, you may have discovered that although the rock can be lifted easily while submerged, it is not so easily lifted out of the water and onto the bank. Somehow a fluid is able to exert a force which helps us to lift rocks or other submerged objects. This force is called a *buoyant* force.

Since the most important interaction between the water and the rock arises in the direct contact between the fluid and the rock, the

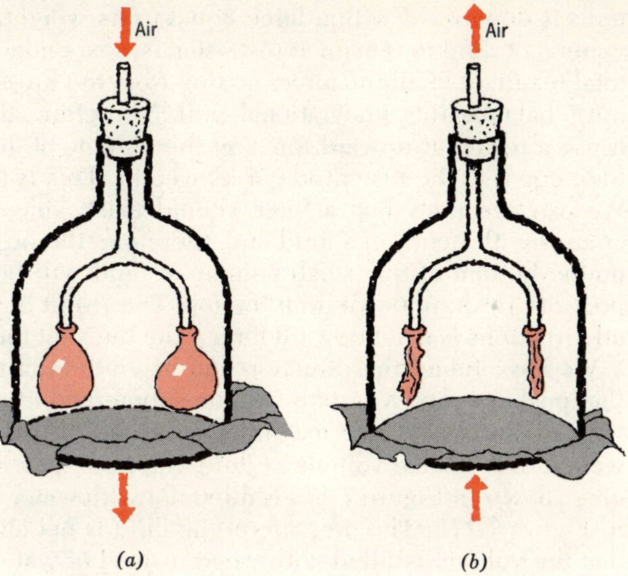

FIGURE 7-16 Diagram illustrating the operation of the lungs.

buoyant force must come from the pressure exerted on the rock by the water. To see how the pressure in a fluid can produce a buoyant force, we first consider the forces on a small part of the fluid itself.

Figure 7-17*a* indicates the forces acting on a small volume of fluid which is submerged in a vessel of the same fluid. The pressure acts on the surface of this volume, perpendicular to it. The only other force on the fluid inside the volume is the gravitational force, which

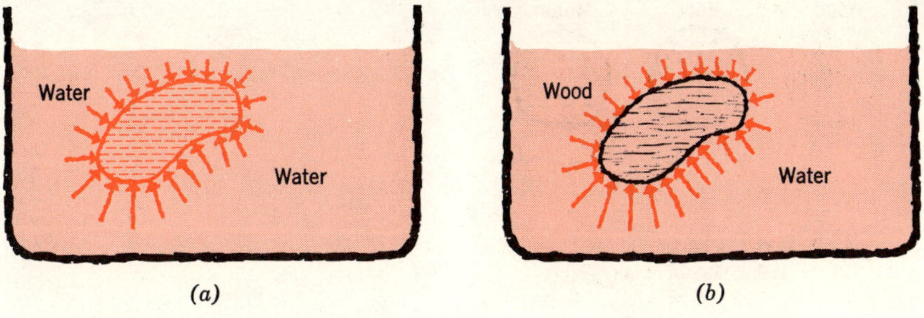

FIGURE 7-17 The bouyant force on an object depends on the volume of the object and not what it is made of.

pulls it downward with a force equal to its weight. In order for this volume of fluid to remain at rest—that is, suspended in the fluid—the total resultant of all the forces arising from the pressure on its surface must balance this gravitational pull. Therefore, the pressure must cause a resultant upward force on the volume of fluid with a magnitude equal to the magnitude of its weight. This is the buoyant force. We can see that such a force should exist; since the pressure increases with depth in a fluid and, therefore, the pressure pushing up on the bottom of this small volume of fluid will be greater than the pressure pushing down from the top. The result of the pressure from all directions is a net upward force—the buoyant force.

We have found the direction and magnitude of the buoyant force that pushes up on a certain volume submerged in a fluid. The direction is upward and the magnitude is equal to the magnitude of the weight of the same volume of fluid. Suppose now that this same volume shown in Figure 7-17*a* is filled not with water but with wood, as in Figure 7-17*b*. The pressure of the fluid is not changed by the fact that the volume is filled with wood instead of water, since the depth to each point of the volume's surface is still the same. Therefore, the buoyant force, caused by the pressure of the surrounding fluid, will still be the same. The buoyant force depends on the volume of the object and not on its nature.

We shall now explain, using the above principles, why some objects sink and some objects float. Figure 7-18 represents three objects submerged in water. One of them is wooden, one is a rock, and one is a small volume of water itself. All of these objects have the same volume and, therefore, they will all have the same buoyant force ex-

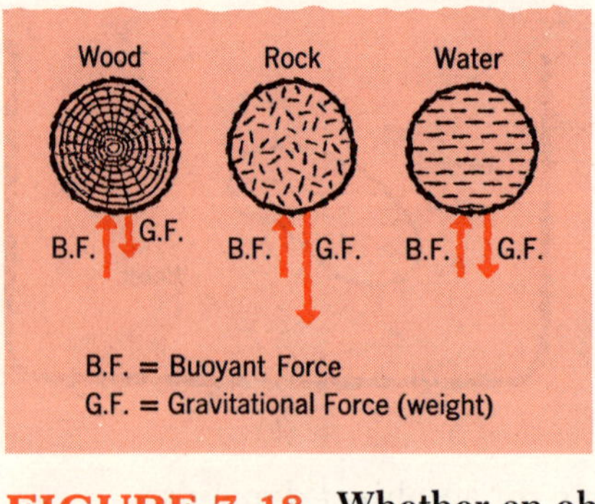

FIGURE 7-18 **Whether an object immersed in a fluid will rise or sink depends on whether the buoyant force is greater or less than its weight.**

erted on them by the surrounding water. They do not all have the same weight, however. The wood has less weight than the same volume of water, and the rock has more weight than the same volume of water. The water, needless to say, has the same weight as the same volume of water. The buoyant force on our small volume of water exactly balances the weight of the water inside; therefore, this small volume of water is in equilibrium. However, the weight of the wood is less than the buoyant force, and the resultant of these two forces pushes the wood up. The weight of the rock is greater than the buoyant force, and therefore the resultant pulls the rock down and it sinks.

Instead of comparing the weights of the wood and the rock with the weight of water, however, we could have compared their masses, since their masses are proportional to their weights. If we do this, it is the mass for a given volume that determines whether an object will rise or sink. Since it is the mass for a given volume and not just the mass alone that is involved, it is really the density that determines whether an object will rise or sink. Objects that are more dense than water will sink in water. Objects that are less dense than water will rise in water. The forces of weight and buoyancy also account for the equilibrium of floating bodies. A floating body sinks partly into the water and therefore occupies a certain volume below the surface level of the fluid (see Fig. 7-19). The weight of this same volume of fluid is equal to the weight of the floating object. This is true because the buoyant forces around this volume must be sufficient to support an equal volume of fluid.

The theory of buoyant forces was first developed by Archimedes of Syracuse (287–212 B.C.), one of the greatest mathematicians of all

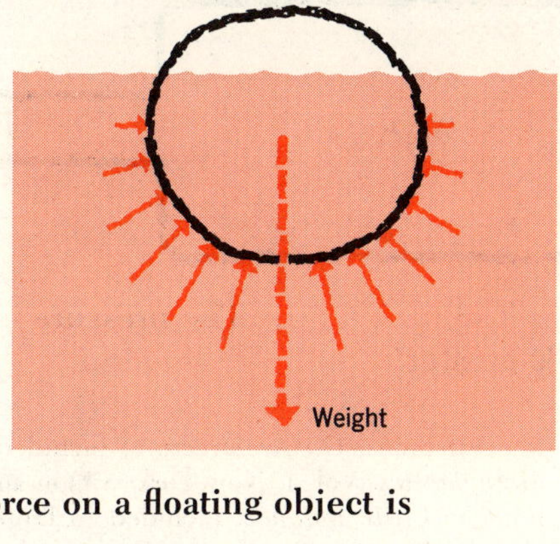

FIGURE 7-19 **The buoyant force on a floating object is equal to the weight of the object.**

 time, whose studies of equilibrium inspired Galileo. The story is that Archimedes was soaking in a public bath, while contemplating a problem set him by King Hiero of Syracuse. The king suspected that the gold in a crown had been diluted with a less dense metal. The mass of the crown could be obtained very easily by weighing it, but to check the density would require a very accurate measurement of the volume as well. One could find this by melting the crown, of course, but this was hardly a satisfactory solution. Archimedes realized that his own loss of weight in the water of his bath depended on his volume, and in this he saw the solution to the case of the king's crown. Elated with this discovery (so it is told), Archimedes jumped from his bath and ran home, shouting "Eureka!" (I have found it!).[2]

Pressure in Moving Fluids

In this section we shall discuss an effect first discovered by Daniel Bernoulli (1700–1782), a member of a famous Swiss family of scientists. According to the Bernoulli effect, in a moving fluid the pressure in the low-speed regions is higher than the pressure in the high-speed regions. Within the stream, therefore, the higher the speed, the lower the pressure. This may seem rather paradoxical, but it is correct. (Pressure effects resulting from increased depth and other matters may obscure this effect.)

Figure 7-20 shows water flowing smoothly through a tube. The tube is narrower in the center section than it is at either end; and therefore, the fluid must flow faster through this region. (If the same amount of fluid must flow through a smaller channel without being

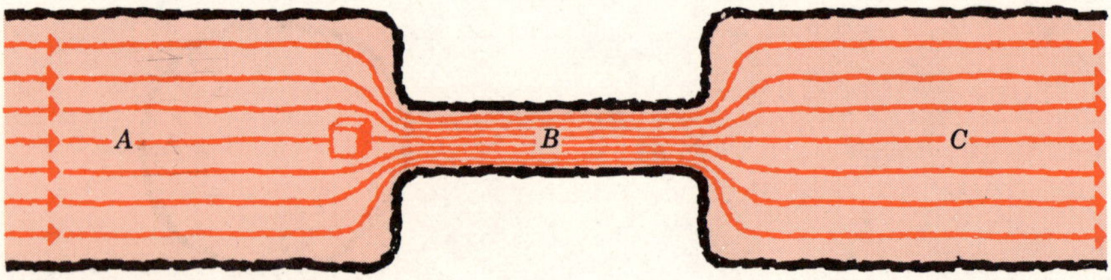

FIGURE 7-20 **The pressure is lower where the speed is greater.**

[2]See Vitruvius, *On Architecture*, included in James R. Newman, ed., *The World of Mathematics*, Vol. I, New York: Simon and Schuster, 1956, p. 185; and Galileo Galilei, *The Little Balance*, included in Laura Fermi and Gilberto Bernardini, *Galileo and the Scientific Revolution*, Greenwich, Conn.: Fawcett Publications, 1965, p. 114.

compressed, it must move faster.) Bernoulli's effect states that the pressure in this section must be less than it is on either side. We can actually show that this must be true by using the second law of motion. Suppose that we have a small volume of fluid just entering the narrow region of the tube. Since it is entering a high-speed region, it must accelerate. The only forces on this volume of fluid arise from the pressure of the surrounding fluid. The pressure must be greater behind it than in front of it, or there will not be a resultant force on it to accelerate it. Where the tube becomes wider, the situation is reversed.

Another example of this effect is illustrated in Figure 7-21, which shows a ping-pong ball suspended in an air jet. We are not concerned with what keeps the ball up (the force of the air pushing against it from below balances its weight). The real question is why it does not fall out of the air jet to the side even when the air jet is moved around or tilted. Bernoulli's effect helps us to explain why the ball stays in the jet. Suppose the ball were at the edge of the jet, in danger of moving out of the stream of air. The air of the jet, on one side of the ball, is moving faster than the air outside the jet, on the other side of the ball. The faster air has the lower pressure; therefore, the ball is pushed back into the jet.

Still another example is illustrated in Figure 7-22. One cannot make the cardboard fall away from the bottom of the spool by blowing smoothly through the tube. According to Bernoulli's effect, air flowing between the bottom of the spool and the cardboard must have a lower pressure than the stationary air on the other side of the cardboard. This difference in pressure balances the weight of the cardboard.

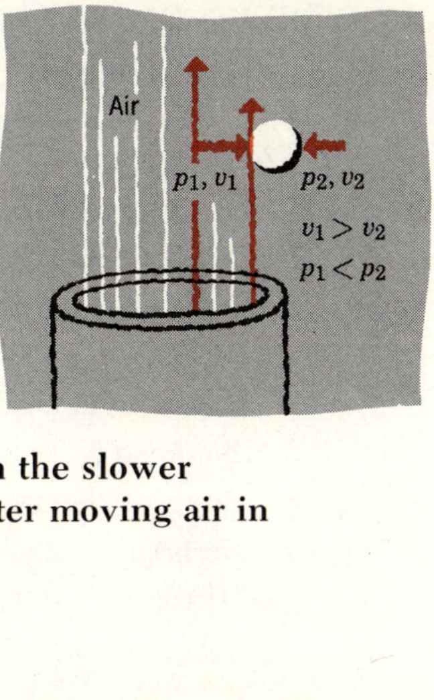

FIGURE 7-21 **The greater pressure in the slower moving air pushed the ball toward the faster moving air in the jet.**

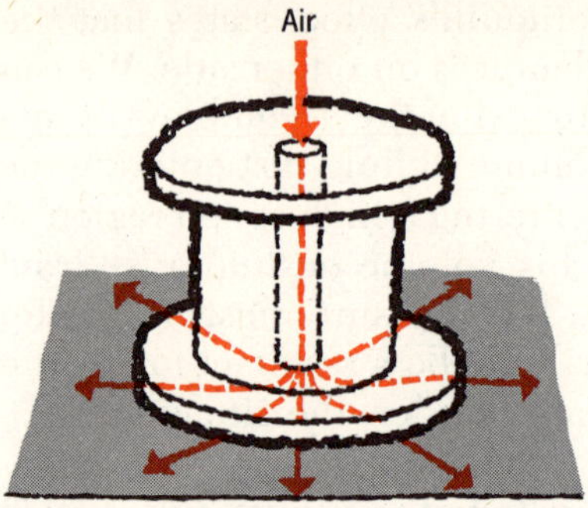

FIGURE 7-22 The pressure in the flowing air between the spool and the cardboard is less than the air pressure on the bottom of the cardboard.

Bernoulli's effect also accounts for the upward force of lift on the wing of an airplane. Figure 7-23 shows an airplane wing in cross section, with air streaming above and below it. The wing is designed so that the airstreams above and below the wing reunite smoothly at the trailing edge of the wing. Under these conditions, the air above the wing travels a greater distance in the same time than the air below the wing; therefore it travels at a higher speed. This higher speed above the wing means that there will be a lower pressure above the wing, and the result is an upward force.

FUNDAMENTAL IDEAS

I. DERIVED QUANTITIES
   - A. DENSITY: a scalar, mass divided by volume.
   - B. PRESSURE: a vector, force divided by area.

II. APPLICATIONS
   - A. FORCES ON OBJECTS, FINDING THE RESULTANT FORCE: for equilibrium and nonequilibrium.
   - B. FORCES IN FLUIDS
     1. Pressure in Fluids: pressure is perpendicular to any surface, is the same in all directions, and increases with depth.
     2. Barometer.
     3. Buoyant Forces.
     4. Bernoulli's effect in moving fluids.
   - C. PROJECTILE MOTION

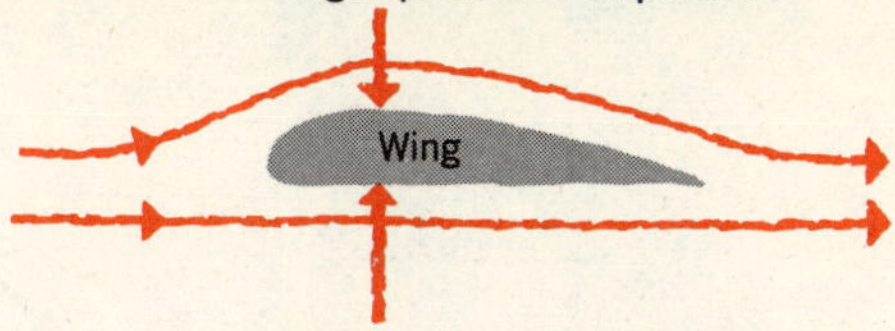

FIGURE 7-23 **The difference between the pressure on the top and the bottom gives the airplane wing its lift.**

QUESTIONS

Forces on Objects: Resultant Force and Equilibrium

1. Why is it necessary to find the resultant force in order to find the acceleration of an object? Are there any exceptions or apparent exceptions to this rule?
2. List the interactions in which the following objects participate. Draw the object isolated from its surroundings, representing the forces on it by arrows, estimating the magnitude and direction of each force. Estimate the resultant force and its direction.
   (a) An airplane in level flight.
   (b) An artificial satellite in orbit.
   (c) An anchor being pulled up through the water.
   (d) A book resting on a table.
   (e) A man walking across the room.
   (f) A tree in a windstorm.
   (g) An automobile slowing down for a stoplight.
   (h) A skier sliding down a hill.
   (i) A shirt hanging on a clothesline.
   (j) A swinging pendulum.
3. In Question 2, estimate the relative magnitude and direction of the forces that the objects listed exert on their surroundings.
4. Figure 7-24 shows a tug-of-war. Identify all the interactions, and then the forces on the object and its surroundings, for the rope and each team in the following situations:
   (a) No motion.
   (b) One team is pulling the other at a constant speed.
   (c) One team is pulling the other faster and faster.

   Which team exerts the greater force on the other in each case?

FIGURE 7-24 See question 4.

5. In Question 4 consider the rope and both teams together as a single object. Identify all the interactions and forces on this object.
6. Can the number of forces acting on an object exceed the number of interactions in which the object is involved?
7. Can an object be in equilibrium if it is involved in only one interaction? In three interactions? Explain.
8. Is a spinning top in equilibrium?
9. Why may we disregard the interactions between different parts of the same object when we are trying to determine the resultant force on the object?
10. We have stated that when the resultant of the forces on an object is zero the object is in equilibrium as far as acceleration of the whole object is concerned. Does this mean that the object cannot move at all? Explain. Does it mean that it cannot accelerate at all?
11. Using symmetry principles, show that the resultant of all the vectors in Figure 7-25 must be zero. Show that this is true for any number of vectors of the same magnitude arranged symmetrically in this way, except of course for one single vector. Why does the argument break down for one vector alone?

Forces on Objects: Projectiles

12. Why is it that an object dropped from a moving airplane does not fall straight down?
13. What is the velocity of a ball thrown directly upward from the ground when the ball is at the top of its path? Is it in equilibrium?

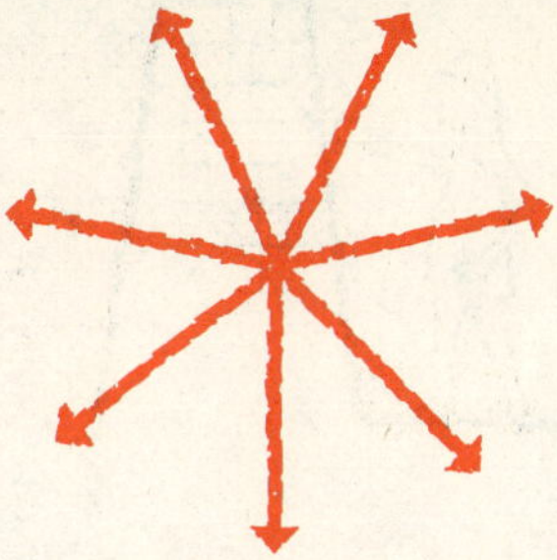

FIGURE 7-25 See question 11.

14. The target in Figure 7-26 is dropped at the same time that the gun is discharged. The barrel of the gun is originally pointing directly at the center of the target. Does the bullet hit the bullseye? Explain?

Forces in Fluids: Fluids at Rest

15. What is meant by the statement that heavy objects sink? Is the statement true, false, or incomplete? Explain.
16. Why do we use mercury in a barometer instead of water?
17. Is it possible for one small boy to hold back the entire Atlantic Ocean by putting his finger in a hole in the dike? Explain. (See Fig. 7-27.)
18. In Figure 7-28, in which case is the force against the dam greater?
19. Does the buoyant force on an object depend on its density? Explain.
20. Does the buoyant force on an object depend on its depth in the fluid? (Consider that the object and the fluid may be compressed by pressure to varying degrees.)

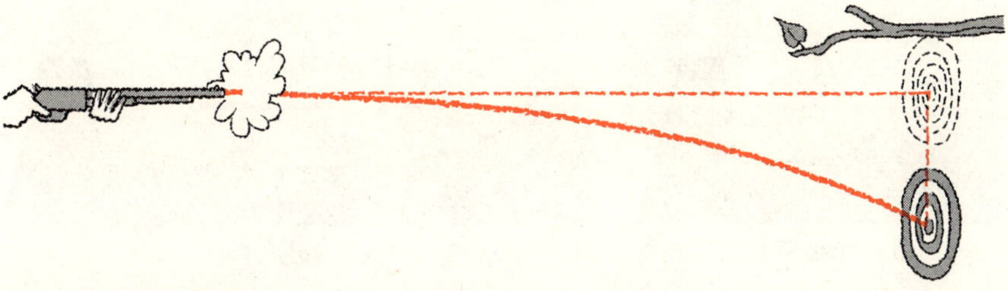

FIGURE 7-26 See question 14.

FIGURE 7-27 See question 17.

21. Why is the atmosphere denser near the ground than at higher altitudes?
22. Why does the pressure in a fluid at a given depth depend on the density of the fluid?
23. What keeps the mercury in a barometer?
24. What determines whether an object will float or sink?
25. How do you know that ice is less dense than water?
26. Is it possible for an object to float on the surface between two fluids in a container, as shown in Figure 7-29? Explain.
27. For which orientation of the pencil shown in Figure 7-30 is the buoyant force greater.
28. The barometer in Figure 7-31 is tilted as indicated by the dotted lines. Which of the two possibilities correctly describes the behavior of the mercury level, and why?
29. How would you measure the height of a building using a barometer?
30. The two vessels in Figure 7-32 have the same area on the bottom and are filled with water to the same depth. Therefore, the total force on the bottom of each vessel must be the same. Yet their weights are not the same, since one contains less fluid than the other. How can the force on the bottom be the same if the weights are different?

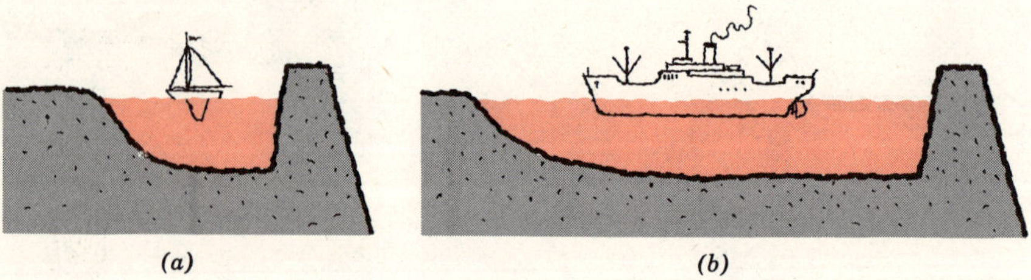

(a) *(b)*

FIGURE 7-28 See question 18.

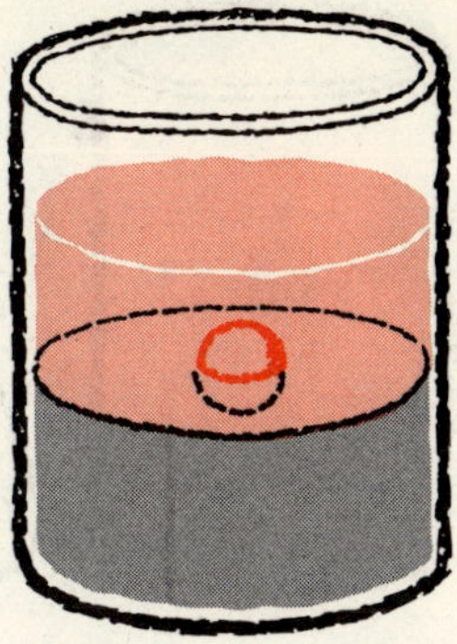

FIGURE 7-29 See question 26.

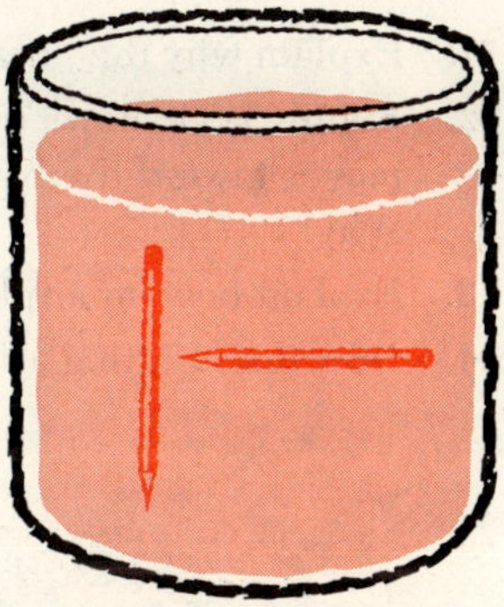

FIGURE 7-30 See question 27.

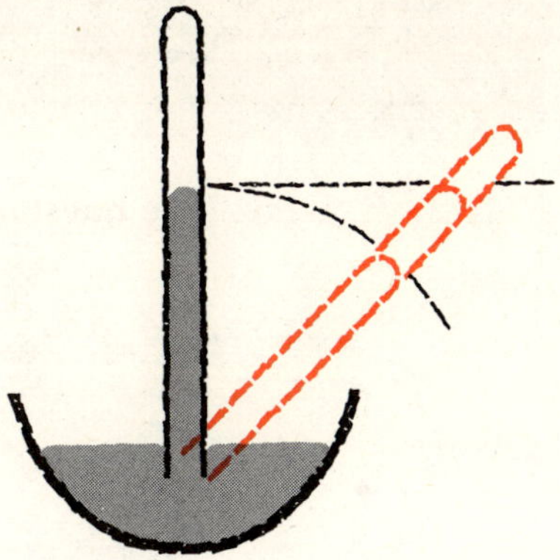

FIGURE 7-31 See question 28.

FIGURE 7-32 See question 30.

Forces in Fluids: Moving Fluids

31. Explain why the ping-pong ball in Figure 7-21 stays in the air jet.
32. Figure 7-33 shows how the air flows around a spinning baseball as it moves toward the batter. Will the ball rise or fall because of this flow? Why?
33. Explain how an airplane wing obtains lift.
34. How does putting a spin on a ping-pong ball affect the game?

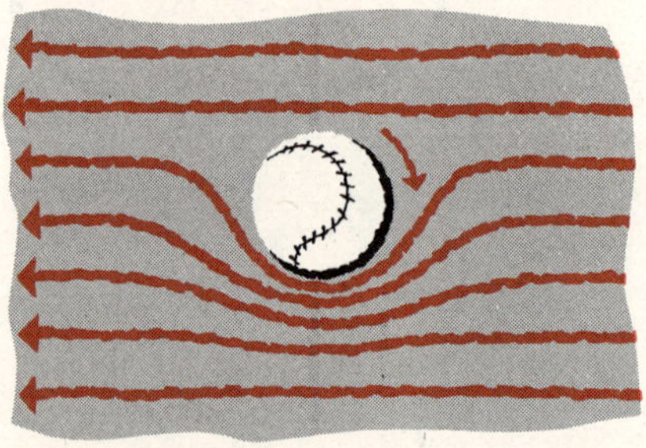

FIGURE 7-33 See question 32.

8 Conservation of Energy

In Chapter 2 we described a thought experiment of Galileo in which he considered balls rolling on inclined planes. We pointed out that a ball placed at a certain level on one plane and allowed to run down that plane and up another would reach a level no higher than the level it started from (see Fig. 8-2).

Galileo knew that he could also show this same effect with a pendulum (see Fig. 8-3). A pendulum bob released from rest at a certain level rises no higher than its starting level, but swings back and forth beneath that level. The experimenter in Figure 8-3 is therefore perfectly safe; if he releases the bob from rest at the position shown, he will not be hit in the face by the returning pendulum bob, provided that he does not move. This is still true even if the motion of the string is obstructed by a peg, as shown in Figure 8-4.

Of course, friction prevents the rolling ball or the swinging pendulum bob from quite reaching its original level. The less friction there is, however, the closer it comes to this level. In the ideal situation, with no friction at all, the pendulum bob would swing forever, returning on every swing to its original level.

FIGURE 8-1 What is it that is "really" wrong with this picture? Humorous design by M. C. Escher. From the collection of C. V. S. Roosevelt, Washington, D.C. Used with permission.

FIGURE 8-2 The ball goes no higher than the level it started from.

Not only is the maximum height of the ball or pendulum bob the same for each swing, but the maximum speed is also the same. The speed of the ball at the bottom of the inclined plane depends on the height through which it has fallen, not on the slope of the plane. The same is true for the pendulum bob. Since the bob falls through the same height in each swing, the speed at the bottom must be the same for each swing. For both the ball on an inclined plane and the pendulum bob, the maximum speed and the maximum height are related. This maximum height and maximum speed keep recurring as time goes on. When the ball or the pendulum bob reaches its maximum height, the speed is zero. When the speed is maximum, the bob or ball is at its lowest point. As the speed increases, the height decreases, and vice versa.

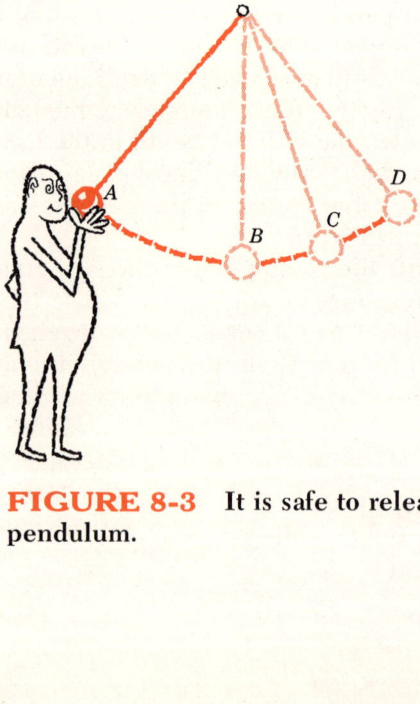

FIGURE 8-3 It is safe to release (not shove) the pendulum.

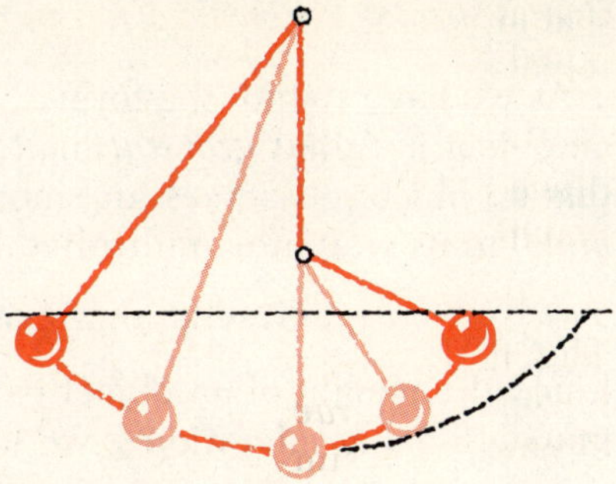

FIGURE 8-4 **Pendulum obstructed by a peg.**

It is very curious that these same heights and speeds keep recurring. One would suspect that there must be some quantity associated with the system which remains the same and is therefore conserved. There is indeed such a quantity. We call it *energy*, and it is a scalar quantity. This conserved quantity, energy, cannot be associated with either the height or the speed alone, since both of these quantities change during each swing. In fact, energy is found in more than one form. One form, called kinetic energy, depends on the speed of an object. Another form, called gravitational potential energy, depends on the height of an object above the surface of the earth. Neither of these two forms of energy alone is conserved in the motion of the ball and bob described above, but the *sum* of the two forms is conserved. This is one reason that energy is important and useful.

In this chapter we shall describe kinetic, gravitational potential, and internal energy. We shall discuss the way in which energy is transferred from one object or system to another.

Conservation of Energy

The forgoing discussion strongly suggests that the quantities speed s and height h should each be associated with a form of a conserved quantity, energy. We now need to know precisely how the forms of energy depend on these two quantities. This information has been obtained by physicists both theoretically and experimentally. Some of the results of their investigations, which established energy as a conserved, scalar quantity, will be discussed in this section.

Gravitational Potential Energy

As we have stated, the form of energy associated with the height of an object is called *gravitational potential energy.* It has been found that an object possesses an amount of gravitational potential energy equal to its weight w multiplied by its height h:

$$\text{Gravitational potential energy} = wh$$

Since the weight of an object is simply its mass m multiplied by the gravitational acceleration g we find that

$$\text{Gravitational potential energy} = wh = mgh$$

The formula mgh is more commonly used. From both formulas, however, we can see that the higher the object, the more of this kind of energy it has.

The *total gravitational potential energy* of a collection of objects is found by simply adding together all the gravitational potential energies of the individual objects.

In this discussion we have talked about the height without specifying any reference frame; therefore, we have as yet no means of determining the height. We need a reference level from which to measure our heights. Actually, it does not matter where we put this reference level, since all of our applications will be dealing with *changes* in the gravitational potential energy, and the changes in height do not depend on the reference level. If, for example, I fall off a 300-meter cliff, I fall 300 meters; it does not matter whether I measure this change in height from a reference level at the top of the cliff, at the bottom, or anywhere in between. The physical effects are determined by the total *change* of height. In each application we are free to choose the most convenient reference level. In our definition, then, the height is simply the distance above or below this reference level. Distances above the reference level are positive; distances below it are negative. The discussion of gravitational potential energy in this section refers to ordinary operations with ordinary objects near the surface of the earth.

Kinetic Energy

The form of energy associated with the speed of an object is called *kinetic energy.* It has been found that an object of mass m and speed s has a kinetic energy of

$$\text{Kinetic energy} = 1/2\ ms^2$$

From this equation we see that as the speed increases the kinetic energy increases. In fact it increases very rapidly because the square of the speed appears in the fomula, and not just the speed itself.

The total kinetic energy of a collection of objects is found by simply adding together all the kinetic energies of the individual objects.

The term kinetic energy was chosen to designate the form of energy associated with speed because the word *kinetic* comes from a Greek word meaning motion. The term potential energy refers to a kind of energy which an object possesses because of its position. The object with potential energy only is not moving, but it has the potential, or possibility, of moving. (If an object is already moving, its potential energy represents the potential it has for increased motion.) Gravitational potential energy refers to the potential energy an object has because of the gravitational forces that are acting on it. This is not the only form of potential energy, as we shall see.

The Unit of Energy

The unit in which energy is measured may be obtained from any of the formulas that express one of the forms of energy. We can use the formula *mgh* (for potential energy) for this purpose, simply replacing the symbol for each quantity by the unit in which it is measured. Thus

$$mgh = (\text{kilogram})(\text{meter/second}^2)(\text{meter}) = \frac{\text{kilogram meter}^2}{\text{second}^2}$$

This derived unit is called a *joule*, after the English physicist James Prescott Joule (1818–1889):

$$\text{joule} = \frac{\text{kilogram meter}^2}{\text{second}^2}$$

Total Energy

The total amount of energy possessed by an object may be found by adding up the amounts of each of the different forms of energy the object possesses. Thus far we have discussed two forms of energy. In situations where these are the only two important forms, we may obtain the total energy by adding the gravitational potential energy and the kinetic energy:

$$E = \text{total energy} = \begin{pmatrix}\text{gravitational}\\ \text{potential energy}\end{pmatrix} + \begin{pmatrix}\text{kinetic}\\ \text{energy}\end{pmatrix}$$

Where these two forms of energy are the only important ones, it is the total energy E, as given above, which is conserved. For a single object of mass m, this will be

$$E = mgh + 1/2\, ms^2$$

This equation applies to the objects illustrated in Figures 8-2, 8-3, and 8-4 if friction can be disregarded. For example, the total energy of the swinging bob of the pendulum does not change with time. Both the kinetic and the potential energy of the bob *do* change with time; in fact, the whole swinging process may be thought of as a continuous changing of one form of energy into the other. As the pendulum bob descends, its gravitational potential energy is converted into kinetic energy. As it rises, kinetic energy is converted back into potential energy. However, the total energy remains the same throughout the process.

Our understanding of many processes in nature is deepened by discovering how energy is changed from one form to another. This approach supplements our understanding of these processes in terms of force and acceleration. Many, if not most, modern physicists consider energy as more "real" than force. In fact, energy, although it is technically a derived quantity, appears to be one of the most fundamental and important "realities" that we study in physics. We have already seen that mass is one of the most fundamental and important "realities" in physics. The picture of mass as the measure of matter and energy as a measure of its motion, real or potential, is very appealing.

Another reason for thinking of mass and energy as important "realities" is that both of these quantities are conserved. Both of them can appear in many forms. With some limitations, they can be transformed from one form to another, but the total amount of each of these quantities remains the same. In a later chapter we shall discuss some surprising relationships between these two quantities.

The concepts of conservation and conserved quantities are considered very important in modern physics, for practical reasons in addition to the aesthetic and philosophical reasons mentioned above. Frequently we are interested in an interaction or a process that either is so complex or happens so quickly that we cannot observe details of the event. We may not even know the exact law that governs the interaction. All we can do is study the conditions that exist before and after the event. Fortunately, we find in nature that there are measurable quantities, such as mass, energy, and momentum, that are unchanged by an interaction. There is as much of each of these quantities after as there was before. This fact gives us information about the conditions after the interaction and allows us to make predictions about them.

Internal Energy

The existence of friction should convince us that there are more forms of energy than we have yet discussed. Let us consider, for example, a real process in which friction is present (see Fig. 8-5). A block sliding along a horizontal slab finally comes to a stop. Since the slab is horizontal, there has been no change in the potential energy of the block. Nevertheless, the block loses kinetic energy as it comes to rest. Does the energy of its motion really disappear entirely, as it appears to do? Where could the energy go if it is not lost? The only obvious possibility is that it somehow disappears into the slab or into the block or into both. Is there any observable change in these objects that would justify the notion that there is an increase of energy inside these objects? Is there reason to believe that there is an increase in what we shall call their *internal energy*? Yes, in fact there is an observable change: as a result of the friction, the two objects get hotter. The kinetic energy, which is lost because of friction, becomes part of the internal energy of the block and the slab. This change in internal energy causes the temperature of the block and the slab to rise. In general, when two objects rub together, most of the energy apparently lost because of friction is transformed into internal energy and usually causes the objects to become hotter. This additional internal energy leaves the objects if they are cooled.

Changes in internal energy may be responsible for other changes besides changes of temperature—for example, melting and boiling. Accurate experiments justify this concept of internal energy. There are other ways in which objects can acquire internal energy. For example, the internal energy of objects is ordinarily increased in collisions.

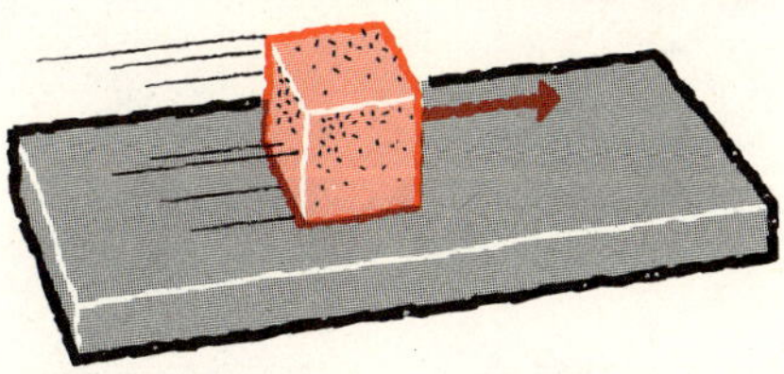

FIGURE 8-5 Kinetic energy is being transformed into internal energy as friction slows the block down.

 If we take into account this new form of energy, internal energy, we can more accurately understand the behavior of objects subject to friction. We may now allow friction to exist in the pendulum and inclined plane in the examples with which we began this chapter. The total energy of these systems must include the internal energy.

$$\text{total energy} = \text{gravitational potential energy} + \text{kinetic energy} + \text{internal energy}$$

It is *total* energy that is conserved, although the amount of each form will generally change. Actually there are still other forms of energy that we have not yet discussed. For example, energy is associated with sound, with light, and with electricity.

The basic conservation-of-energy principle is that the energy of a system before an interaction is equal to the energy after the interaction.

$$\text{Energy before} = \text{Energy after}$$

or, more generally, during a process

$$\text{Energy at one time} = \text{Energy at another time}$$

We must be sure to include all the important forms of energy at both times. Sometimes we shall be concerned with situations in which the only two important forms of energy are kinetic energy and gravitational potential energy. These two forms of energy are usually the only important ones when we are considering objects falling or sliding without friction on smooth stationary surfaces, subject only to the pressure from the surfaces and the gravitational force. If there is friction between the object and the surface, some energy will be transformed into internal energy. Also, if there is an impact, some energy

(*a*) (*b*)

FIGURE 8-6 **Internal energy is generally increased during an impact, such as occurs when the ball in (*a*) bounces.**

may be transformed into internal energy. For example, the ball rolling down the inclined plane in Figure 8-6a has an impact when it reaches the bottom of the plane because the surface forces the ball to change direction suddenly. This impact could cause the ball to lose kinetic energy and gain internal energy. An impact does not occur at the bottom of the plane in Figure 8-6b because the path curves gently, and the ball does not strike a corner. Idealized impacts which do not involve increases of internal energy are called *elastic* impacts.

We must bear in mind the ways in which the energy of a system of objects can be changed. A force arising from an interaction with an object outside the system can change the energy of the system. Objects sliding without friction along stationary surfaces are subject to a force exerted by the surface, but this force does not change their energy, since it is perpendicular to the direction of motion. A frictional force, however, will retard the object and change its energy. Whenever the energy of a system changes, it is because energy has been transferred into or out of the system.

Energy Transformations

In this section we shall discuss two ways in which energy can be *transferred* from one object to another or *transformed* from one form

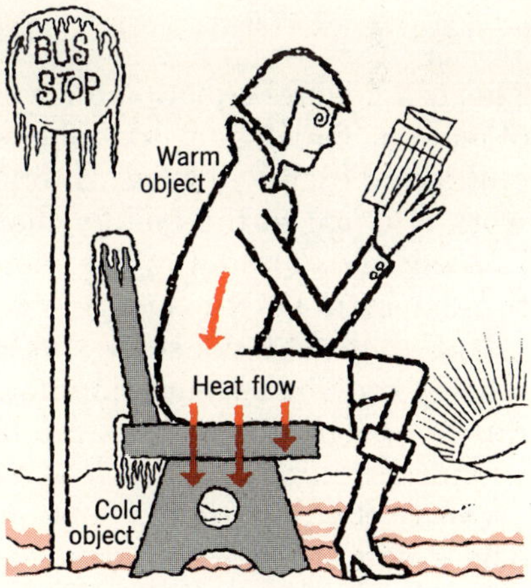

FIGURE 8-7 **Energy is transferred by heat flow from a warm object to a cold one.**

 to another (It is possible for energy to be transferred and transformed in the same process.)

Heat Flow

If we put a warm object in contact with a cold one, we find that the cold object gets warmer and the warm object gets colder (see Fig. 8-7). This means that some of the internal energy of the warmer object is flowing into the colder object. An energy transfer of this kind is called a *heat flow*. We reserve the term "heat" to refer to this kind of energy transfer. We do not say that an object "contains heat"; we prefer to say that it "contains internal energy."

Work

Energy may be transferred or changed in form if a force is exerted on (or by) an object as it moves through a distance. This kind of energy transfer or transformation is called "doing *work*." Work requires both a force and a distance. In this section we shall assume that the force is either in the direction of motion or in the opposite direction.

Figure 8-8*a* illustrates how energy is supplied to an object by doing work on it. The block is first at rest on a frictionless surface; then it is pushed by a force of magnitude F. This force keeps acting on the block until it has gone a distance d. We define the amount of *work* done by a force to be

$$\text{Work} = Fd$$

This is a general equation for work when the force is in the direction of motion. Thus, the work done is equal to the magnitude of the force multiplied by the distance through which it acts. In this example the work is equal to the energy that has been transferred to the block. The energy supplied to the block takes the form of kinetic energy: the change of kinetic energy is equal to the work done.

In Figure 8-8*b* we see a similar example, except that the block is already moving in one direction and the force acts in the opposite direction. This force slows the block down, and therefore decreases its kinetic energy. In a case like this, when the force on the object is opposite the direction of motion, energy is taken out of the object. The energy taken out is just equal to the work. That is, in such a case, the work is equal to the energy transferred *out*. In this case, again, the force multiplied by the distance gives us the work, and this work is equal to the energy transferred.

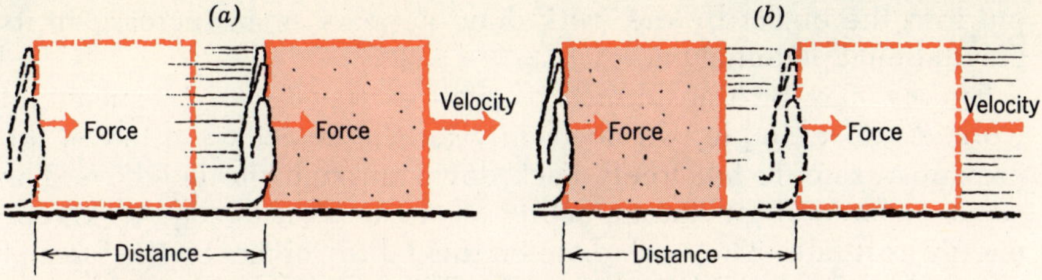

FIGURE 8-8 **Energy is transferred to an object by using a force applied through a distance to do work on it.**

Another example of putting energy into an object by doing work is shown in Figure 8-9. In order to lift an object we must exert a force on it equal to its weight. (It will, of course, require a slightly larger force to start it rising; but once it is rising at a constant speed, a force equal to its weight is sufficient to keep it going upward.) If we lift it through a total height h, we have done some work on it since we have exerted a force equal to its weight w through a distance h. The total work done is

$$\text{Work} = wh = \text{energy transferred}$$

This quantity is equal to the gravitational potential energy the object gains by being raised to that height. Thus, in this example the energy

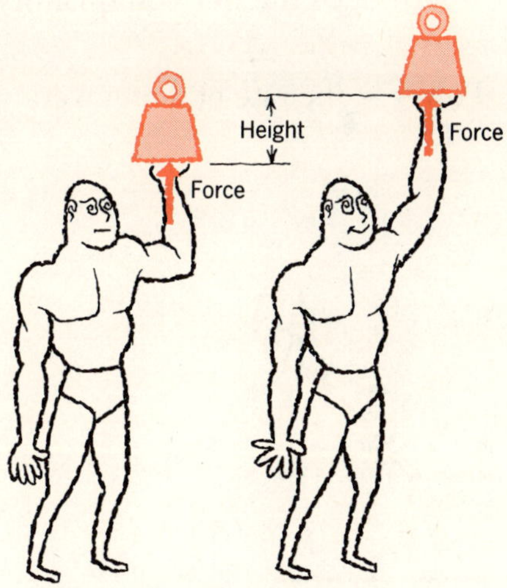

FIGURE 8-9 **The gravitational potential energy of an object is increased when work is done to lift it.**

 put into the object by the work done appears as an increase in its gravitational potential energy.

Energy may be transformed as well as transferred by means of work. As an example, we need only let the object we lifted in the previous example fall freely back down through the height h. The constant force of gravitation acting on it causes it to acquire kinetic energy as it falls. The work done on this falling object by the force of gravitation transforms the gravitational potential energy into kinetic energy.

Even if the applied force is not exactly in the direction of the object's motion, it may still do work on the object. For example, in Figure 8-10, the force does do work on the block, although not as much as if the force were directed horizontally. Only if the force is exactly perpendicular to the direction of motion of the object will it fail to do work on the object. For example, suppose the man in Figure 8-9 does not lift the object but holds it at the same height, meanwhile walking along level ground at a constant speed. He is exerting an upward force on the object but he is not doing any work because the upward force is perpendicular to the horizontal motion of the object.

Power

In practical applications, engineers are interested not only in the total amount of work done but also in the rate at which it is being done. Therefore, we introduce the derived quantity *power*, which is defined as

$$\text{Power} = \text{the rate of doing work}$$

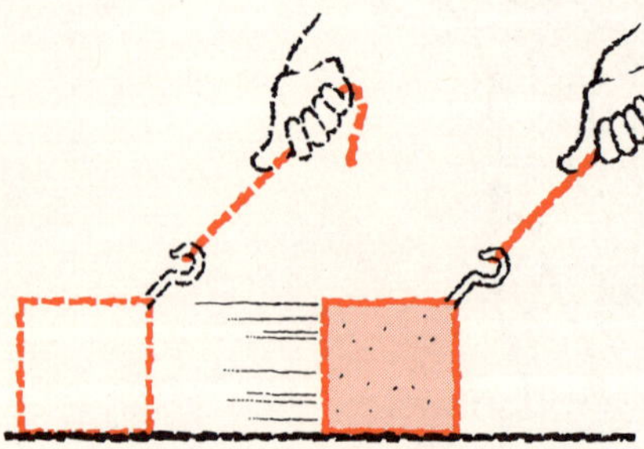

FIGURE 8-10 Work can be done on an object even when the force is not in the direction of its motion.

If the work is being done at a constant rate, then the power can be found easily by dividing the work done by the time required to do it:

$$\text{Power} = \frac{\text{work}}{\text{time}}$$

Sometimes the word *power* is used to refer to the rate of energy transfer or transformation when heat flow as well as work is involved. We have here used a more limited definition of the quantity power, restricting it to work alone.

The unit of power in the system of units we are using is called the *watt*, after James Watt, who helped to develop the stream engine. The watt can be expressed in terms of the joule and the second by inserting the units of power, work, and time into the last equation. The unit of work is the same as the unit of energy (the joule) and therefore

$$\text{watt} = \frac{\text{joule}}{\text{second}}$$

The common English unit of power, the *horsepower*, is equal to about 746 watts. The common electrical unit of power, the *kilowatt*, is equal to 1000 watts.

Applications

Binding Energy

Two objects held together by a force may be separated by a force. Work will be done in separating the two objects, since the force must be exerted through a distance. This means that energy must be supplied to the objects in order to separate them. The amount of energy needed to completely separate two objects is called their *binding energy*. The objects may actually have more energy when they are separated than when they are bound together because of the binding energy.

As an example of binding energy, consider the rock in Figure 8-11, which is held to the surface of the earth by gravitational forces. In order to remove this rock from the earth, we would have to do work on the rock to lift it so far from the earth that the remaining gravitational pull would be negligible. The energy that must be supplied in doing this work is equal to the binding energy. This energy used in

FIGURE 8-11 **The energy of the system is greater after the rock has been separated from the earth because of the binding energy supplied.**

lifting the rock increases the gravitational potential energy, so that the system actually has more energy when separated than before.

In Chapter 6, we mentioned that an object with sufficient speed could escape from the earth. We can now understand this "escape speed." To remove an object from the earth, we must supply energy equal to the binding energy. An object moving as fast as or faster than this escape speed already has the required extra energy in the form of kinetic energy, and therefore can escape from the earth.

Perpetual Motion

It is often said that perpetual-motion machines are impossible. Here we shall discuss one kind of perpetual-motion machine—a device that would be capable of putting out energy without taking any in (see Fig. 8-1). Many such devices have been designed, some of them, to judge from published diagrams, are marvels of engineering skill as far as workmanship and complexity of operation are concerned. Unfortunately, none of them has been demonstrated to work, despite the obvious ingenuity and effort that have evidently gone into their "discovery."

What principle would prevent a device from putting forth energy forever without ever taking any in? The principle, of course, is the law of conservation of energy. Energy cannot be created (or destroyed) by any machine or device. It can be stored in a device, of course, but the stored energy will eventually be used up unless it is replenished by energy from outside the device. Thus, according to the law of conservation of energy, devices which supply an inexhaustible flow of heat or do work forever without ever taking in any energy are impossible. If a successful perpetual-motion machine were actually devised, we would have to abandon the law of conservation of energy. However, since this law has been verified in such a wide variety of phenomena and since all of the perpetual-motion machines produced so far have failed, physicists are very skeptical when a new perpetual-motion machine is proposed. We sometimes express the law of conservation of energy by saying that perpetual-motion machines of the kind discussed here are impossible.

Simple Machines—The Lever

A machine is a device that we put energy into and get energy out of. The input and output energies may be of the same or different forms. Figure 8-12 illustrates the action of one simple machine, a lever. To operate the lever, we push down on the input end, putting energy into the machine by doing work. The other end does work in lifting the rock, thus putting energy out of the machine into the rock. Energy flows from us into the rock by way of the machine. Suppose that the rock has a weight w; how much force must we use to operate the machine and lift the rock? (We will disregard friction in the machine, which would ordinarily divert some of the energy from useful work into internal energy.)

We first push down on the lever with a force F and move the lever through a small distance H. Thus we have supplied an input energy equal to the work FH. The other end of the lever will then lift the rock a small distance h. (The distances H and h are approximately vertical.) Thus there is an output energy equal to the work wh. Ignoring friction, the input energy must be equal to the output energy:

$$FH = wh$$

Dividing both sides of this equation by H we obtain

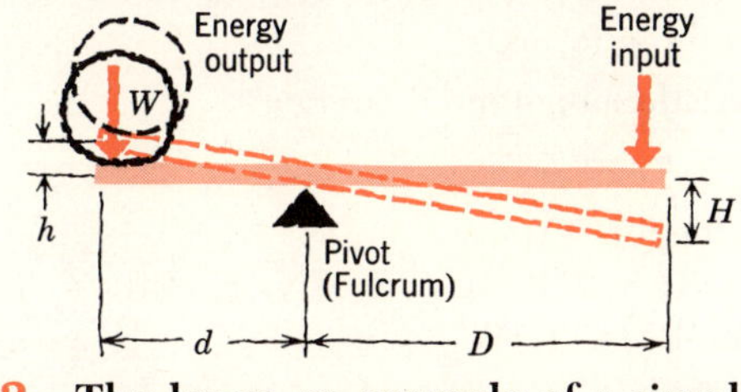

FIGURE 8-12 **The lever, an example of a simple machine.**

$$F = w\,(h/H)$$

The greater the distance of either end of the lever from the pivot, the farther that end will move when the lever turns about the pivot. In fact, the distance each end of the lever moves is proportional to the distance from the pivot to that end. We may write this proportionality as follows, letting d and D represent the distance of the two ends from the pivot:

$$(h/H) = (d/D)$$

Therefore, the force we must exert to lift the rock is

$$F = w(h/H) = W(d/D)$$

If the length of the output side of the lever d is smaller than the length of the input side D, then we can use the lever to lift the rock by exerting a force smaller than the weight of the rock. This shows one use of machines—to increase the force that we can apply to an object. A small force on the input end is transformed by the machine into a large force on the output end. We had to exert our input force, however, over a larger distance than the rock itself moved. Therefore, we had to do the same amount of work as if we had moved the rock ourselves without the use of the machine. Machines cannot create energy, but they enable us to use it more effectively.

FUNDAMENTAL IDEAS

I. FUNDAMENTAL LAWS AND PRINCIPLES

- **A.** CONSERVATION OF ENERGY

II. QUANTITIES

- **A.** ENERGY: a derived scalar quantity. The unit of energy is the joule, which is equal to one kilogram meter^2/second^2. We have introduced three forms.
  1. Gravitational potential energy.
  2. Kinetic energy.
  3. Internal energy.
- **B.** QUANTITIES INVOLVED IN ENERGY TRANSFER AND TRANSFORMATION
  1. Heat flow.
  2. Work = Force times distance.

III. APPLICATIONS

A. ENERGY RELATIONS; INCLINED PLANES AND PENDULUMS

B. BINDING ENERGY

C. PERPETUAL MOTION

D. THE LEVER, A SIMPLE MACHINE

QUESTIONS

Kinetic and Potential Energy

1. Does the kinetic energy of an object depend on the direction of its velocity?
2. When a spring is stretched we say that it possesses potential energy, not gravitational potential energy, but another kind. Why do we say this?
3. Does a stretched rubber band contain energy? Explain.

Internal Energy

4. What makes a swinging pendulum finally come to rest? Where does the energy go in this case?
5. Why does an automobile have to burn fuel continually just to keep going at a constant speed? What happens to the energy supplied by the burning fuel?
6. Suppose you were assigned the task of proving that "internal energy" really exists. What sort of experiment would you devise?
7. What happens to the kinetic energy of a falling object when it hits the ground?
8. Would it be possible to project a snowball against a wall at such speed that it would entirely melt on impact? Explain.
9. Artificial satellites often stay up for a long period of time and then eventually fall to earth. Why do they finally come down?
10. Why do the brakes on a car heat up when the car is stopping?
11. Would a hard-boiled egg roll farther than a raw egg if they both started rolling at the same speed on a hard surface? Explain.[1]

[1]This question arises from a comment by John Perry, *Spinning Tops and Gyroscopic Motion*. New York: Dover, 1957, p. 54.

Energy Transformation

12. What is "heat flow"?
13. What is "work"?
14. How do we know that energy can be transformed from one form to another by the process of work?
15. How do we know that energy can be transferred from one object to another by the process of work?
16. How do we know that energy can be transferred from one object to another by the process of heat flow?
17. A professor performed the pendulum demonstration as illustrated in Figure 8-3. In his enthusiasm to show it to his students, he gave the bob a shove as it left his nose. What happened and why?
18. Figure 8-13 shows two kinds of water wheel. Where does the energy come from to run each kind?
19. It requires work to stir a pan of soup. What do you think happens to the energy thus transferred?
20. Why does a knife get hot when it is sharpened on a grinding wheel?
21. Why is it possible to obtain fire by friction?

Binding Energy

22. How can the energy of two objects held together by a force be greater when they are separated than when they are together?
23. Do the earth and the moon have more or less energy at present than they would have if the moon were lying on the surface of the earth but things were otherwise the same? Explain.

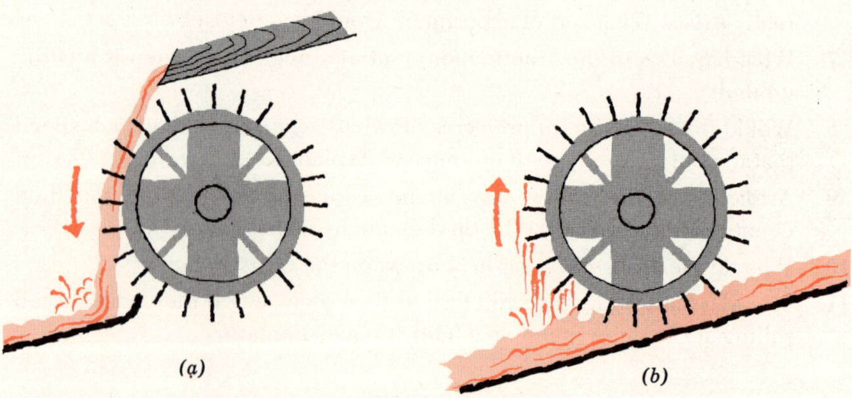

FIGURE 8-13 See question 18.

Power, Perpetual Motion, and the Lever

24. Does a more powerful automobile engine necessarily exert more force than a less powerful engine?

25. Figure 8-14 shows a proposed device for running a grinding wheel. The electric motor runs an electric generator, which generates enough electricity to run the motor and the grinding wheel. Without knowing much about the operation of electric generators and motors, why would you conclude that this device cannot work?

26. Design a machine that gives an output motion greater than the input motion. What must be sacrificed in order to obtain this result?

General

27. Why do we believe that there is more than one form of energy?

28. What kinds of energy transformation would you expect where friction occurs?

29. Why is energy an important quantity?

30. Give examples in which the various forms of energy are transformed into other forms or obtained from other forms.

31. Figure 8-15 shows a device turned by escaping steam. Where does the energy come from? Follow the transformations of energy illustrated in this process.

32. What is meant by the term "conservation"?

33. Can you offer an explanation for the slowing down of the rotation of the earth?

34. What would convince you that sound involves a form of energy?

35. If I drop a certain weight into a bucket of molasses from a height h, the temperature of the molasses rises by one degree. What happens if I drop the weight from a height of $2h$? What assumption are you making about energy, temperature, and molasses?

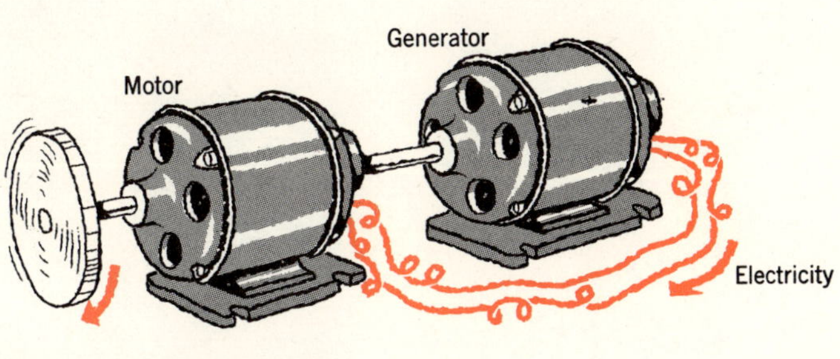

FIGURE 8-14 See question 25.

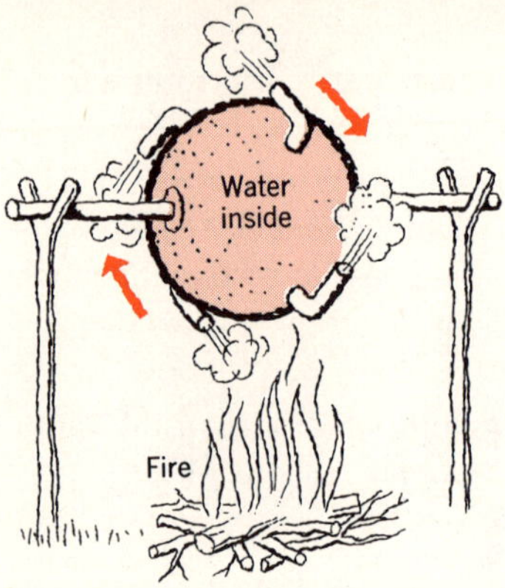

FIGURE 8-15 See question 31.

36. If I throw an object into a bucket of molasses at a speed of v, the temperature rises by two degrees. What happens if I throw it at twice the speed?

37. When a gun is fired, the momenta of the gun and the bullet are approximately equal in magnitude. Why, then, is it not just as dangerous to be hit by the gun as by the bullet?

9 Molecular Order

What is everything made of? In order to answer this question we enter the realm of the invisible. Matter is made up of invisible particles called molecules, in various states of motion. Some of the largest of these molecules can be easily seen today in the electron microscope (see Fig. 9-1). Scientists were convinced that these small particles existed, however, long before they could see any of them, because they could very accurately account for the behavior of matter in terms of the behavior of molecules. For example, the kinetic theory of gases is a theory which accurately predicts the behavior of gases by assuming that they are made up of molecules in motion.

Molecules themselves are combinations of other invisible particles. In studying the realm of invisible particles, you might expect to find different laws than the ones we have already considered. In fact you do, and we shall consider some of these new laws in other chapters. However, a great many phenomena may be explained by assuming that molecules themselves obey Newton's laws of motion. In this chapter and the next we shall use Newton's laws of motion and the molecular theory of matter to explain some of these phenomena.

FIGURE 9-1 **A crystal formed by tobacco necrosis virus molecules; magnification about 80,000.**
Courtesy of Dr. Ralph W. G. Wyekoff.

Molecules and Atoms

The particles of matter in which we are interested in this chapter are called *molecules*. Molecules themselves are made up of one or more particles called *atoms*. Each molecule of the gas hydrogen, for example, is made up of two atoms of hydrogen (see Fig. 9-2). Each molecule of the gas helium, however, consists of a single atom of helium. Each molecule of water consists of two atoms of hydrogen and one atom of oxygen. The various kinds of atom correspond to the various chemical elements. There are laws of chemical combination which explain how atoms combine to form molecules, but we shall not discuss them here.

Although molecules occur in various shapes and sizes, they are extremely small, so small, in fact, that it is difficult to develop an appreciation for their size. To be told, for example, that a molecule of helium has a diameter of about 10^{-10} meters means very little if we have had no experience with anything else that small. The fact that a thimbleful of air near sea level contains 10^{19} molecules is also difficult for us to comprehend—this quantity is more than the number of all the people in the world multiplied by a billion. Later, in other chapters, we will discuss particles still smaller than atoms.

Solids, Liquids, and Gases

We are familiar with matter in three common forms: solid, liquid, and gas. Since gases are ordinarily easily compressed, we think of the molecules of a gas as being far apart so that they can be pushed closer together (see Fig. 9-3). Since liquids are not easily compressed, we may think of them as being made up of molecules which are almost in contact with one another but are free to slide over one another so that the liquid can flow. Solids are more rigid, so we think

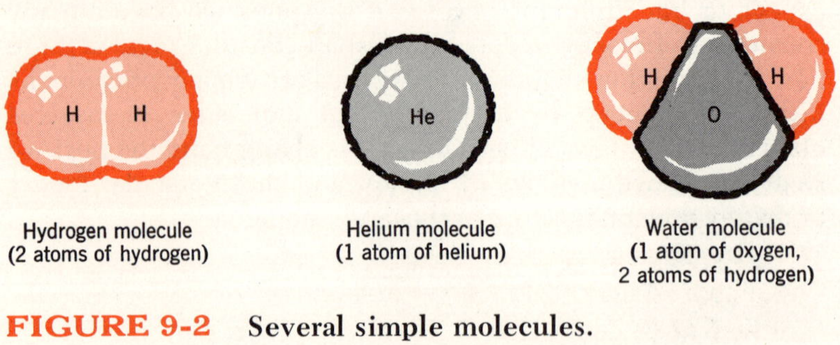

FIGURE 9-2 Several simple molecules.

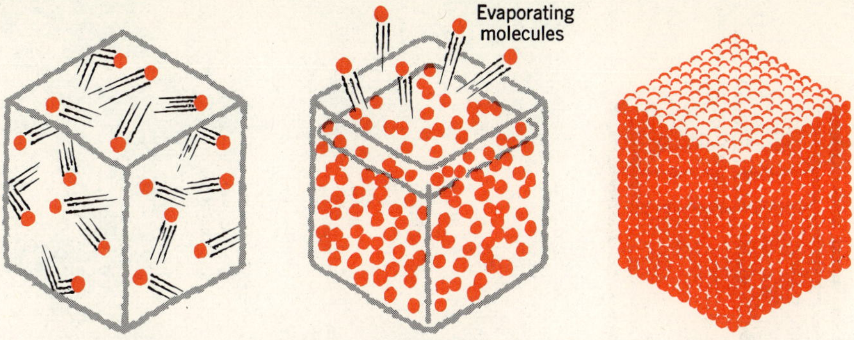

FIGURE 9-3 Molecules in a gas, a liquid, and a crystalline solid.

of them as being made up of molecules that are held more firmly to one another by some force. Although these molecules can sometimes slip past each other, they are more commonly found vibrating in place. Since crystalline solids exhibit a regular shape, we consider that the molecules of these substances are arranged in an orderly fashion, stacked, for example, like oranges in a grocery-store display.

The Explanation of Gases in Terms of Moving Molecules

Many observable properties of matter are explained in terms of molecules. To illustrate, we will discuss the internal energy, temperature, and pressure of a gas, and a few of the simpler aspects of the behavior of liquids and solids.

The basic picture of a gas, according to the kinetic theory of gases, is one of molecules in motion, striking and rebounding from each other and from the walls of their container. These frequent collisions continually alter the speed and direction of individual molecules. As a result of these collisions, the molecules have a wide range of speeds, and there are about as many traveling in one direction as in another.

Pressure. The kinetic theory enables us to account for the pressure that a gas exerts on the walls of its container (see Fig. 9-4). When a moving molecule strikes the wall of its container, a force is exerted on the wall during the impact. The continual rain of moving particles striking the walls continually produces these impact forces all over the walls of the container, accounting for the pressure of the gas.

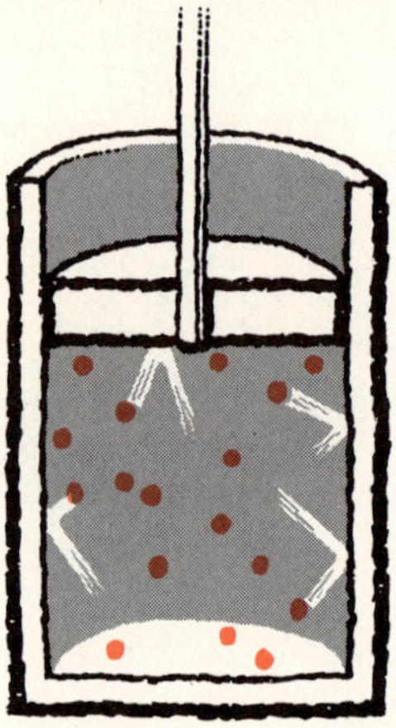

FIGURE 9-4 **Molecules bouncing off the walls of the container are responsible for the pressure exerted by the gas.**

Internal Energy. The internal energy of a sample of a gas is the total energy of all of the molecules within it. In the simplest case, this is approximately equal to the sum of the kinetic energies of each of the molecules. For solids and liquids, and to a lesser extent for gases, the molecules also have nongravitational potential energy that arises from forces between molecules. In these cases, the internal energy is equal to the sum of total kinetic and potential energy of all the molecules. When a moving object slows down because of friction, the energy of the object is not lost but can be accounted for in the total energy of the individual molecules of the object and its surroundings.

Temperature. We all have an intuitive understanding of temperature, derived from our sense of touch. Our sense of touch, however, is insufficient even to distinguish between very hot and very cold temperatures, as is illustrated by a well-known practical joke (see Fig. 9-5). The victim first sees a pot of water boiling on the stove and is then blindfolded. While he is blindfolded, his hand is plunged into a pan of ice water. His great distress at this event is due to his inability to discriminate between the boiling and freezing temperatures of water.

Of course, we can use an ordinary mercury thermometer to measure temperature far more accurately than with our hands. As the temperature is increased, the mercury inside the thermometer ex-

FIGURE 9-5 **Cold seems hot in this experiment.**

pands and therefore rises in the glass tube. In fact, most substances expand when heated. However, we are interested in what temperature means in terms of the motion of the molecules.

According to the kinetic theory, the temperature is related to the *average* kinetic energy of the molecules. In a very simple gas, the temperature is proportional to this average kinetic energy. When we heat a gas, the internal energy increases. The molecules start to move with greater speeds. Their average kinetic energy increases, and we observe a higher temperature. If the gas is not confined in a rigid container, it will expand as its temperature is increased. The faster-moving molecules collide, on the average, more frequently and with greater impact with each other and with the walls of the container. This results in a greater pressure on the walls, which can cause expansion.

The kinetic theory helps us to understand how heat flow occurs. Figure 9-6 shows a cold thermometer in hot air. Since the air is hot, its molecules will have high kinetic energies. Some of these molecules collide with the slower-moving molecules in the cold glass of the thermometer. In time, the molecules in the glass begin to move faster as a result of these collisions, and the temperature of the glass increases. These faster-moving molecules in the glass then speed up the mercury molecules as a result of collisions, and in this manner the temperature of the mercury is also increased. Finally, both glass and mercury are at the temperature of the air. There has been a heat flow into the thermometer. Thus the kinetic theory helps us to explain heat flow as a result of molecular collisions.

As we have said, most substances expand when their temperature rises. This is true for both the mercury and the glass of the thermometer. However, the mercury expands more than the glass, and therefore the mercury rises in the thermometer tube.

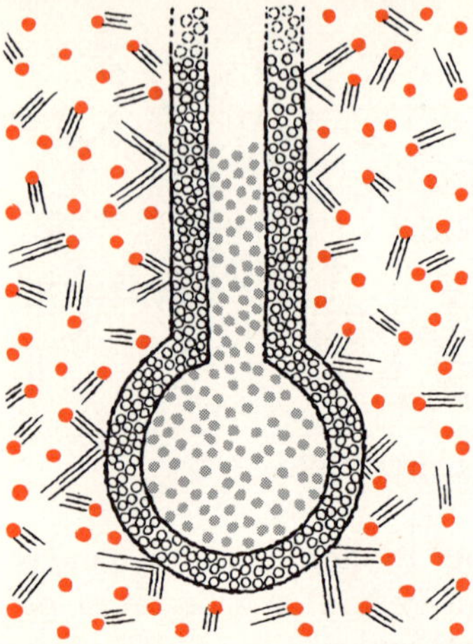

FIGURE 9-6 **Heat flow from the air into the thermometer is a result of molecular collisions.**

Absolute Zero of Temperature. If the temperature is closely related to the average kinetic energy of the molecules, then there should be some lowest possible temperature that would be attained when the molecules lose all their kinetic energy. There appears to be just such a temperature, called the *absolute zero*. Actually, even at absolute zero there is still a small amount of kinetic energy in the molecules, the so-called zero point energy. (The presence of this small zero-point energy can be explained by modern quantum theory.) We shall not explain the zero point energy here but shall simply point out that the absolute zero is the temperature at which the molecules have their minimum kinetic energy.

Most of us are used to measuring temperatures using the ordinary Fahrenheit scale, on which water freezes at a temperature of 32 degrees and boils at 212 degrees. The *absolute zero* of temperature corresponds to a temperature of about 459 degrees *below* zero on the Fahrenheit scale. Most scientists use the Celsius scale of temperature (corresponding to the previously used centigrade scale). On this scale the freezing point of water is zero degrees, and the boiling point is 100 degrees. The absolute zero of temperature corresponds to a temperature of about 273 degrees *below* zero on the Celsius scale. We can see that the absolute zero is far below the normal temperature of our environment. Physicists, however, have recently been able to obtain temperatures within 0.00001 of a degree of absolute zero in very small samples of matter in the laboratory.

Evaporation and Melting

The molecules of a small sample of matter will tend to disperse as moving molecules leave the sample, unless something prevents them from doing so. In a gas, the walls of the container prevent these moving molecules from escaping. In a liquid or a solid, there are forces of attraction (stronger than gravitational forces) that are able to hold the molecules together. However, the molecules of solids and liquids can escape in the process of evaporation.

Let us examine the process of evaporation in liquids—for example, water. (Evaporation in solids is similar.) In the continual jostling of molecules by other molecules that occurs in a sample of water, some of the molecules acquire, by collisions, more energy than others. For some of these molecules, this acquired energy will be greater than the binding energy that holds them to the other molecules. If these high-energy molecules are near the surface and moving toward it, they will escape from the liquid and form or join the gas above (see Fig. 9-7). In this way it is possible for a sample of water to evaporate completely and thus "disappear." Water molecules in the gas above, however, can also reenter the liquid. If there are the same number of molecules leaving and entering, on the average, in the same length of time, then the amount of water in the sample remains the same.

Under the right conditions, evaporation can take place within a liquid, forming bubbles of gas inside it which then rise to the surface. This process is known as boiling. The pressure in the fluid

FIGURE 9-7 **Evaporation occurs when the molecules escape from the surface of the liquid.**

 tends to collapse these bubbles of gas, however, and therefore they will not form unless there are enough evaporating molecules moving fast enough to exert a pressure that can withstand this surrounding pressure. As the temperature of a liquid increases, the pressure that can be exerted by the evaporating molecules increases because there are more high-speed molecules in the liquid. Finally, a temperature is reached at which bubbles can form. This temperature is the boiling temperature.

We can change the boiling temperature by changing the pressure on the liquid. In a pressure cooker, for example, the pressure rises higher than atmospheric pressure. In order for the evaporating molecules to form bubbles (boil) against this higher pressure, the temperature must be higher than the ordinary boiling temperature. This means that the water in the kettle can be raised to a higher temperature without boiling. This higher temperature causes the food to cook more quickly.

As we have seen, evaporating molecules have on the average more energy than the molecules left behind in the liquid. The loss of these high-energy molecules leaves the liquid colder than it was before. Therefore, evaporation cools the liquid. This effect explains why a wet hand feels colder than a dry one. The liquid on the wet hand is being cooled by evaporation. Some evaporated molecules will bounce back into the liquid. If a person blows on his wet hand so that these molecules are carried away and cannot return, evaporation is more rapid, and his hand feels even cooler.

Suppose that we increase the internal energy of some ice by letting heat flow into it. As we put energy into the ice, the temperature increases until it reaches the point at which some of the ice begins to melt. To melt the ice we must supply enough energy to overcome that part of the binding energy that holds the molecules of the solid rigidly together and keeps them from sliding over one another. Even if we continue to supply energy, the temperature will not increase further until all the ice is melted, since all the energy we are supplying is being used to melt the ice. When all the ice has been transformed into water, then the energy that we are supplying raises the temperature of the water until the boiling point is reached. At this temperature the energy supplied is used to overcome the binding energy that keeps the molecules of the liquid together. Again, the temperature does not change until all the liquid has been transformed into a gas even though energy is still being supplied. Then the temperature again rises.

Evidence for the Existence of Molecules

The evidence for the molecular theory of matter is its success in accurately explaining the behavior of matter. We have briefly outlined some of these explanations, and we shall present others in the remainder of this chapter. There are some phenomena that would be especially difficult to explain without assuming the existence of molecules. We shall discuss two of these here—diffusion of gases, and Brownian motion.

The diffusion of gases is illustrated in Figure 9-8, which shows two chambers at the same temperature and pressure separated by a porous wall. This wall has tiny holes in it that will permit molecules to pass through. We start with helium gas on the right and a heavier gas, such as oxygen, on the left. (Oxygen molecules have eight times more mass than helium molecules.) The gases are at the same temperature, and therefore the average kinetic energy of the molecules is the same for both. Since the helium molecules have less mass than the oxygen molecules, they must have greater average speed in order to have as much energy. The faster moving helium molecules move through the small holes in the porous wall into the oxygen chamber more rapidly than the oxygen molecules move into the helium chamber. Consequently, the pressure will rise in the oxygen chamber. After the process is complete, half of the helium will have diffused into the oxygen chamber and half of the oxygen will have diffused into the helium chamber. At this time the pressure becomes and remains the same in the two chambers. In summary, first the pressure rises on the left, and then the two pressures become equal. This effect is experimentally observed and therefore supports the kinetic theory.

Brownian motion, discovered by Robert Brown in 1827, provides

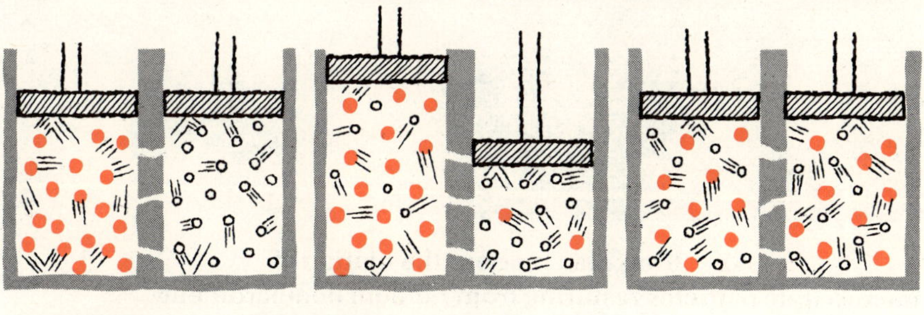

FIGURE 9-8 **The diffusion of gases.**

evidence for the existence of molecules which seems more direct. A particle in a liquid or gas is bombarded by individual molecules. The effect of this molecular bombardment is negligible for large particles: the mass of the particle is so much greater than the mass of the individual molecules that the effect of each collision is small, and the collisions occur on all sides of the particle so frequently that on the average the effects cancel out. However, particles of microscopic size may be seen to bounce around as a result of molecular collisions (see Fig. 9-9). Albert Einstein in 1905 calculated the expected rate of drift for such particles bouncing around in Brownian motion. The calculation agreed so well with the observed drift that it convinced the noted chemist Wilhelm Ostwald that atoms and molecules really existed.

FUNDAMENTAL IDEAS

I. QUANTITIES

- **A.** TEMPERATURE. From the standpoint of the kinetic theory we may regard temperature as a derived quantity. Originally it was thought to be fundamental.
  1. Temperature is related to the average kinetic energy of molecules.
  2. Absolute zero of temperature.

II. MODELS

- **A.** THE MOLECULAR THEORY. Assumes that matter is made up of particles subject to the laws of motion. Used to explain:
  1. Temperature of a gas.
  2. Pressure of a gas.

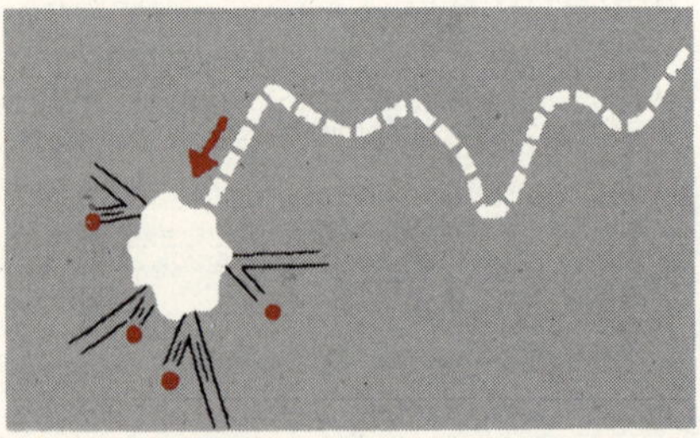

FIGURE 9-9 **Brownian motion: the motion of microscopic particles resulting from random bombardment by molecules.**

3. Internal energy.
4. Heat flow.
5. Evaporation and melting.

B. EVIDENCES FOR THE MOLECULAR THEORY
1. Success of explanations for items 1–5, II-A, above, and others not covered in this book.
2. Diffusion phenomena.
3. Brownian motion.

III. PARTICLES
A. ATOMS
B. MOLECULES

QUESTIONS

General

1. Why does the pressure on a gas enclosed in a rigid container increase as the temperature increases?
2. Why do we believe that matter is made up of small particles in motion?
3. Distinguish between atoms and molecules.
4. How does the kinetic theory account for the temperature and pressure of a gas?
5. How does the kinetic theory account for heat flow?
6. Explain the absolute zero of temperature.
7. What causes Brownian motion?
8. If an astronaut put a conventional thermometer out of the hatch of his orbiting spacecraft, would it record the temperature? Explain.
9. In what sense can it be said that temperature arises from the energy of the molecules and pressure from their momentum?
10. A water molecule has a mass of about 3×10^{-27} kilograms. How many molecules of water would be contained in a thimbleful? (Water has a density of 1000 kilograms per cubic meter.) Estimate how many it would take to fill up the ocean.
11. From your answers to Question 10, estimate how many of the same molecules you would get back again if you emptied your thimbleful into the ocean, mixed thoroughly, and then took out another thimbleful.
12. From the information in Question 10, estimate the size of a molecule of

water. Suppose the information had been given for a gas. Would that enable you to estimate the size of the gas molecule?

Evaporation and Melting

13. Why does evaporation occur?
14. How does evaporation cause cooling?
15. Why do we sometimes hold up a wet finger in trying to find out the direction of the wind?
16. Suppose that our discussion of evaporation and melting (pp. 175–176) had been entitled "condensation and freezing" instead. How would the discussion have been changed?
17. Some water is put into a bottle, air is pumped out of the top, and the bottle is sealed. It is then observed that evaporation takes place inside the bottle for a while, and then apparently stops. Why?
18. Why might an astronaut's blood boil if he stepped out of his capsule without his space suit?
19. Why does it take longer to boil an egg until it is cooked on top of a mountain than at sea level?
20. Why does the temperature of boiling water remain the same as long as the boiling continues?
21. Would it be possible theoretically for water to boil and freeze at the same temperature?

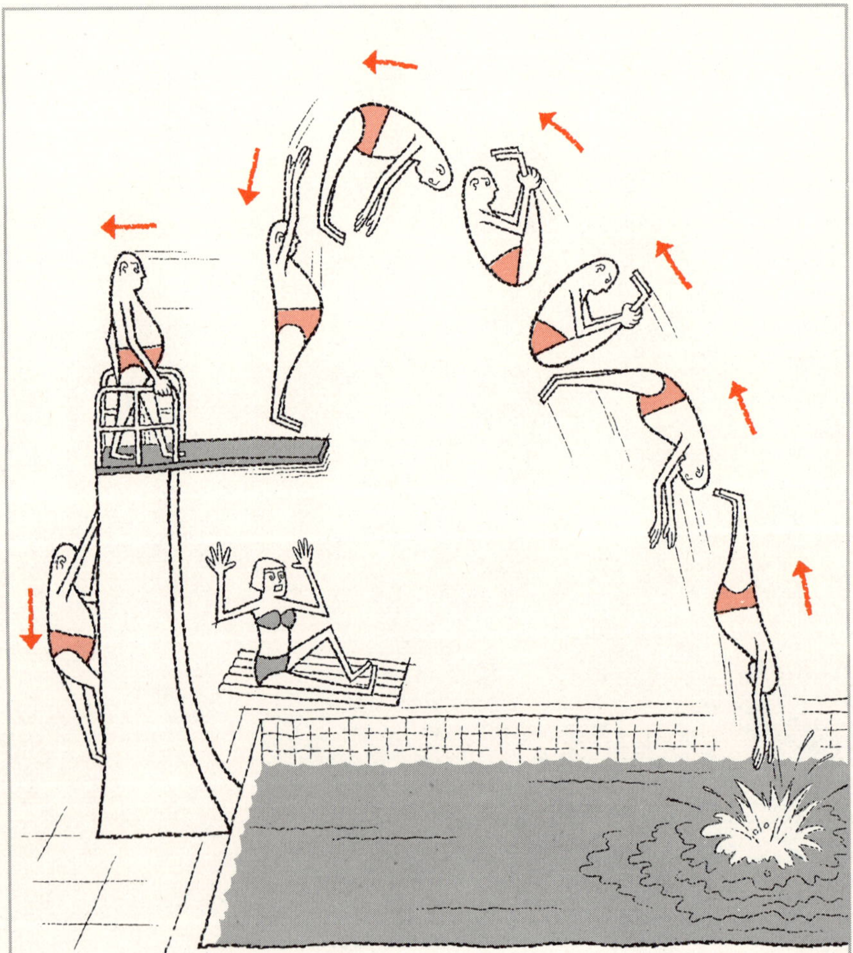

10 Molecular Disorder

Processes which cannot be reversed in time, such as that illustrated in Figure 10-1, present us with a real problem, because apparently the molecules obey only time-reversible laws. In this chapter we shall discuss time-irreversible processes and the problem that arises in trying to explain them in terms of molecules in motion. We shall introduce a scalar quantity, *entropy*, which is conserved in reversible processes, but always increases in irreversible processes.

Time-Reversible Processes

All of the fundamental laws we have presented thus far are the same whether time is running forward or is reversed to run backward—that is, they obey the principle of time-reversibility. The easiest way to understand this principle is to suppose that we take motion pictures of all physical processes. Then we determine the laws of physics from what we see when the film is run backward. If the laws turn out to be the same, then we say that the laws of physics are *time-reversible*.

Newton's laws of motion are time-reversible—they would

FIGURE 10-1 What is "really" wrong with this picture?

operate backward in time in the same way they do forward. For example, a motion picture of a rock thrown into the air would show the rock obeying the same laws whether the film were run backward or forward. Reversed motion is a possible motion for the rock. The principle of time-reversibility has recently been challenged for interactions of certain small particles; in this section, however, we shall assume that the principle of time-reversibility does hold for molecules.

Time-Irreversible Processes

Despite the fact that the behavior of molecules appears to be time-reversible, there are certainly processes in nature which are not time-reversible. We list here several processes that do not occur in reverse.

1. When two objects at different temperatures are pressed together, heat flows from the hot object into the cooler one, and never in the opposite direction. You would not sit on a block of ice to get warm.
2. A ripe peach decays. You have never seen a rotten peach become whole and edible again.
3. When a man dives into a pool, waves are started which spread out and finally disappear. There are no reports of waves arising in a pool by themselves, converging on a swimmer, and tossing him out of the pool onto the diving board.
4. A block of wood on an inclined plane slides, with friction, down the plane and comes to rest at the bottom. The block and the plane are warmed as the result of friction. We have never noticed the block and plane start to cool down by themselves, propelling the block back up the plane.

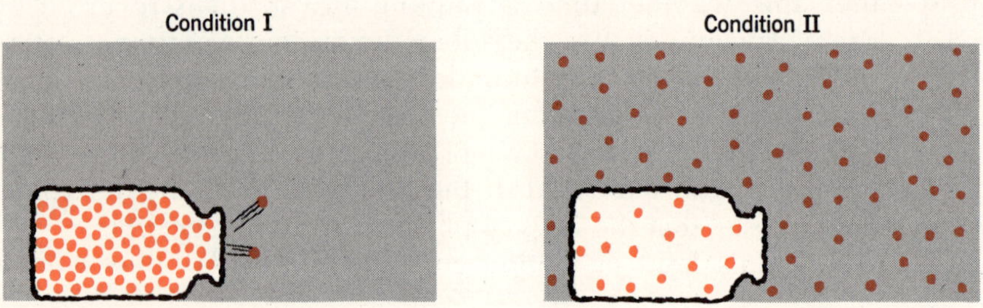

FIGURE 10-2 Gas expanding from a small bottle into a large box.

FIGURE 10-3 An "impossible" process: the grains of sand become unmixed as they are shaken.

5. A gas released from a small container under high pressure expands to fill a large box (see Fig. 10-2). No one has ever seen the gas in a box (as, for example, a room) gather itself together like Aladdin's djinni and flow back into the bottle, building up pressure as it does so.
6. A layer of orange sand is carefully spread over a layer of white sand in a bottle. Vigorous shaking will mix the grains of orange and white sand together. Continued shaking has never been known to cause the orange and white grains of sand to separate themselves into two layers again (see Fig. 10-3).

All these processes are irreversible; each of them establishes a sequence of conditions in time that cannot be reversed. Most commonly observed processes are actually irreversible. Reversible processes (as, for example, motion without friction) are usually idealizations of some actually irreversible process.

The Law of Entropy Increase

The existence of irreversible processes presents us with two problems.

1. Objects are made of molecules in motion, which obey only reversible laws. How, then, can they undergo irreversible changes? This is one of the long-standing puzzles in physics. We shall consider this question again later.
2. What single law or principle can be devised to account for all irreversible changes? In the remainder of this chapter we shall search for such a principle. We shall find a scalar quantity which

usually increases as time goes on and never decreases. This quantity is called *entropy*. Entropy is conserved in reversible processes, but it is *not* conserved in irreversible processes. After an irreversible process there is always more entropy than there was before. The principle that will enable us to account for irreversible processes is, therefore, the *Law of Entropy Increase*.[1]

The total entropy of the objects involved in a process never decreases. In irreversible processes the total entropy increases.

The entropy of a single object or system may well decrease. When we consider *all* the objects and systems involved in a process, however, it turns out that the total entropy always increases or remains the same. In later sections we shall attempt to explain what is meant by the quantity entropy, so that the law of entropy increase may become more meaningful.

Entropy as Disorder

In order to obtain our first clue to the nature of entropy, let us consider the previous example in which white and orange sand were mixed together. What was it that changed in the mixing process? There was order in the arrangement of the sand to start with; the orange and white grains were separated. In the process of mixing, this order was lost. *Disorder*, therefore, increased. Entropy also is supposed to increase. Perhaps this disorder is the entropy we are looking for. Indeed, this seems to be the case. The increase of the quantity entropy corresponds to an increase in disorder among the molecules that make up material objects and substances. We can check this interpretation by examining the other examples of the last section.

The decay of the peach is an obvious example of increasing disorder. It is a breakdown of the remarkable organization of molecules present in living cells. The waves coming from the diver show an ordered motion of the water molecules, which is broken up as the waves rebound against the sides of the pool and are dissipated. In this manner, cooperative motion of the molecules in a wave finally becomes part of the normal random motion of the molecules. In the

[1]The law of entropy increase is generally known to scientists as the second law of thermodynamics. The first law of thermodynamics is the law of conservation of energy.

fourth example, the ordered motion of the entire block moving down the plane is also reduced, by friction, to the increased random motion of the molecules in the block and plane. The molecules of the expanding gas were at first restricted to one small volume; after the expansion they are miscellaneously distributed over the whole volume of the box. Each of these irreversible processes can be thought of as an increase in disorder.

These examples strongly suggest that we might write the law of entropy increase in terms of disorder.

> The total amount of disorder never decreases, and it increases in irreversible processes.

Actually, this statement gives a good idea of the meaning of entropy increase.

The first example in the last section was concerned with the heat flow from a hot object to a cooler one. This process can also be understood as increasing disorder. The temperature of an object is related to the average kinetic energy of its molecules. The molecules of the hot object have, on the average, more kinetic energy than those of the cold object. To begin with, therefore, we have a partial separation between two classes of molecules—those of high kinetic energy and those of lower kinetic energy. After heat has flowed from one object to the other to make the temperatures of the two objects equal, the ordered arrangement of the molecules into these two classes is lost. The high-energy molecules are now mixed in with the low-energy molecules. Thus disorder increases here also. This example enables us to state the law of entropy increase in one of the ways in which it was first expressed.

> No process can occur in which the net effect is the flow of heat from one object into another at a higher temperature than the first.

The most complex and cunningly devised machines cannot make heat flow "uphill" against a higher temperature—unless some other effect also occurs at the same time. For example, a refrigerator can take energy from a cool system and expel it into a region of higher temperature. To do this, however, the refrigerator uses electrical energy or energy supplied by burning gas. It cannot make heat flow uphill without using some other energy.

 Entropy as Probability

We can explain to some extent why disorder increases by introducing the concept of *probability*. Suppose that we tossed a large number of pennies. Would we expect fo find them coming up either all heads or all tails? No; we would expect to find them in a more disordered and nonuniform condition, with about as many heads as there are tails. The most probable condition for these pennies is not the ordered state in which they are all heads or all tails but, rather, a disordered state. Indeed, for most systems a disordered condition is the most probable.

The mixing of the sand provides us with another example. The first condition, in which the orange and white grains of sand are all separated, is certainly possible. It is even possible that shaking the bottle after the sand has been mixed will happen, by chance to separate the grains again into two groups, orange and white (see again Fig. 10-3). This event is so unlikely, however, that you would not believe anyone who said he had seen it happen. A condition in which the orange and white grains are all mixed together is so much more probable that it is the only one we expect to see as a result of random shaking. Thus, if a system of objects is tossed, bumped, jostled, or shuffled, there is a great probability that its order will be lost. The molecules of matter are constantly in motion, jostling, bumping, and disturbing each other. Thus it is not surprising that in irreversible changes their disorder increases. The less probable conditions are replaced by the more probable, disordered conditions. We may, then, state the law of entropy increase as follows.

In an irreversible process the system changes to a more probable condition.

This statement also gives a good idea of what is meant by the law of entropy increase.

We can explain further why the more probable conditions of a system are usually disordered by considering what happens when a number of coins are tossed. If we toss two coins, there are four possible ways in which they can come up. Using T to stand for tails and H to stand for heads, we can list these four possibilities in a table (see Fig. 10-4). There are two conditions that might be called ordered. These are the conditions in which both coins come up heads or both come up tails. The other two conditions, each with one head and one tail, might be called mixed or disordered. If we increase the number

FIGURE 10-4 **The possible results of flipping two coins.**

of coins, we increase the number of possible disordered conditions. Figure 10-5 shows the possibilities for various numbers of coins. If we tossed a large number of coins, the overwhelming majority of possibilities would be mixed or disordered. Comparatively few would be completely ordered (all heads or all tails) or nearly ordered. If we should randomly disturb a collection of coins, we would be far more likely to obtain one of the many disordered conditions than one of the few ordered conditions. Each of the conditions would be considered equally likely, since we have no information that would lead us to choose one rather than another. The disordered conditions are so numerous, however, that we are much more likely to get one of them. On the basis of probability, then, we can explain why we will normally see an ordered condition change into a disordered condition, and not vice versa. Even a thimbleful of gas contains about 10^{19} molecules. With this many molecules involved, the ordered conditions will be extremely rare in comparison with the disordered conditions. In principle, however, it is possible for the law of entropy increase to be violated—one could by chance get a more ordered condition. The probability that this will happen, however, is ridiculously small.

The relation of entropy to probability enables us to understand how apparently irreversible processes can occur when all the molecules involved in these processes obey reversible laws. What we have called irreversible processes are not, strictly speaking, irreversi-

Case	Coin 5	Coin 4	Coin 3	Coin 2	Coin 1	
1	H	H	H	H	H	
2	H	H	H	H	T	1 Coin
3	H	H	H	T	H	
4	H	H	H	T	T	2 Coins
5	H	H	T	H	H	
6	H	H	T	H	T	
7	H	H	T	T	H	
8	H	H	T	T	T	3 Coins
9	H	T	H	H	H	
10	H	T	H	H	T	
11	H	T	H	T	H	
12	H	T	H	T	T	
13	H	T	T	H	H	
14	H	T	T	H	T	
15	H	T	T	T	H	
16	H	T	T	T	T	4 Coins
17	T	H	H	H	H	
18	T	H	H	H	T	
19	T	H	H	T	H	
20	T	H	H	T	T	
21	T	H	T	H	H	
22	T	H	T	H	T	
23	T	H	T	T	H	
24	T	H	T	T	T	
25	T	T	H	H	H	
26	T	T	H	H	T	
27	T	T	H	T	H	
28	T	T	H	T	T	
29	T	T	T	H	H	
30	T	T	T	H	T	
31	T	T	T	T	H	
32	T	T	T	T	T	5 Coins

FIGURE 10-5 The possible results of flipping five coins or less.

ble. The reversed process is not impossible; it is merely extremely improbable. In fact, it is so improbable that for practical purposes, and most theoretical purposes, it can be ignored.

In the above discussion we have not attempted to define order or probability precisely, and we have not discussed the effect of force on disorder. The presence of forces can make it appear that order is increasing, not decreasing, in a process. In fact, where we see order in the behavior of particles, we usually assume that a force is acting, and we seek to discover this force. Also, our use of the term *order* in this discussion is actually rather curious. No matter how the coins fall when we toss them, they will have some kind of order. Similarly, the molecules of a gas, expanding into the box from the bottle in Figure 10-2, have a definite order whether they are cooped up in the bottle or moving around in the box. Why should we call one arrangement (or order) ordered, and another arrangement (or order) disordered? The answer seems to be that the arrangement is considered ordered if it means something to us, or gives us some information. We have more information about the location of the individual molecules of a gas when they are confined to a small volume than when they are allowed to move around in a large volume. Hence we consider the latter condition more disordered. Even in the small bottle, however, the molecules appear to be behaving as randomly as possible under the circumstances. Molecular behavior appears to be chaotic.

Perpetual Motion of the Second Kind

In Chapter 8 we discussed the impossibility of a perpetual-motion machine that would create energy and therefore violate the law of conservation of energy. This type of machine is called a perpetual-motion machine of the first kind. In this section we shall discuss another type of impossible machine—a perpetual-motion machine of the second kind. A perpetual-motion machine of the second kind violates the law of entropy increase. As an example, imagine a machine that can take in energy by heat flow and convert this energy entirely into work done by the machine, leaving the machine in its original condition. This process would violate the law of entropy increase; indeed, one of the early statements of this law stated the impossibility of such a process.

No process can occur whose total overall effect is to take internal energy from a material by heat flow and convert this energy entirely into work done in the process.

We can see why such a process would violate the law of entropy increase by considering again the example of the block sliding with friction down an inclined plane (see p. 184, 187). As the block comes to rest, the energy is converted to internal energy, thus, as we have seen, increasing the disorder of the molecules. If the process occurred in reverse, the disorder of the molecules would have to decrease and would therefore violate the law of entropy increase. But this is just one example of the forbidden process described above: internal energy flowing out of the block and plane and being converted to the work needed to move the block back up the plane, with no other change to the system.

The law of entropy increase can be expressed as a principle of impossibility as follows.

Perpetual-motion machines of the second kind are impossible.

We have already stated the law of conservation of energy as a principle of impossibility—the impossibility of perpetual-motion machines of the first kind. It is often considered elegant to express physical principles as principles of impossibility.

If perpetual-motion machines of the second kind were possible, many convenient devices could be made. For example, it would theoretically be possible to make a machine which could extract internal energy out of sea water, thus lowering the temperature of the water, and use that energy to propel a ship. As long as the law of entropy increase holds, however, such a machine is impossible.

The energy in a heat flow can be partly converted into work, as occurs, for example, in a steam engine. There is, however, a heat flow out of the machine, to surroundings at a lower temperature than the source of the incoming heat flow. The steam engine cannot operate unless the surroundings are at a lower temperature than the boiler (the source of the incoming heat flow). Any engine that converts part of an incoming heat flow into work operates between at least two temperatures and has an outgoing heat flow at the lower temperature.

Free Energy and Heat Death

The fact that energy supplied as heat flow cannot be completely converted into work has some interesting consequences. Energy converted to internal energy in irreversible processes, by friction or any other means, can never be completely recovered and used to do work. Where differences of temperature or the means to produce them exist, we can use *some* of this internal energy to do work, but not all of it. Some of it flows into material at the lower temperature, tending to reduce the temperature difference. We can expect, therefore, that as time goes on, more and more energy will be irreversibly converted into disordered internal energy, and less and less will be available to do work. Although the energy is conserved, we say that the *free energy* is decreased. (*Free energy* may be defined roughly as the energy available to do work.) Another way of stating the law of entropy increase is to say:

> Free energy never increases, and it decreases in irreversible processes.

The continual increase of entropy or decrease of free energy within a system can eventually result in a total absence of free energy in the system. In this condition, unless more free energy is supplied from outside, nothing but random molecular motion can occur in the system. If we consider the whole universe as an isolated system, we can imagine a time in the future when the free energy has totally disappeared and the whole universe has reached the same temperature. In such a condition, often called the "heat death" of the universe, no work can be done.

Entropy as Lack of Information

A more recent point of view considers the law of entropy increase as a law of decreasing information. In the irreversible processes we have been discussing we do not have complete information about the condition of the system in question. For example, consider again the expansion of a gas from a small bottle into a large box. (See Fig. 10-2.) Neither before the expansion nor after it do we know the exact state or conditions of the individual molecules of the gas. To determine these conditions we should have to measure the position and velocity of each molecule—a totally and absurdly impractical task for each of the over a billion trillion molecules involved, even assuming that it were possible to make such measurements without hopelessly disturbing the system. (Furthermore, even if we had somehow obtained this information, it would generally be lost in the disturbances caused by outside molecules randomly striking the system in question.)

Our descriptions of the two conditions of the gas, however, give us even less information about the molecules in the final condition when they are spread out in the box than in the beginning condition when they are still in the bottle. In the box the molecules are free to wander in a larger volume of space than they are when they are in the bottle, and therefore we have less information about their location. In this sense, the irreversible process has resulted in a loss of information. Similar results hold for other irreversible processes, suggesting the following expression of the law of entropy increase:

In an irreversible process, information decreases; is lost.

The relation between the law of entropy increase and information, which we have merely sketched above, has applications which extend beyond physics to such matters as the theory of information transmission in radio and telephone communications. We shall not discuss these applications here.

Entropy Changes Defined by Heat and Temperature

Let us look for a quantity that changes when heat flows from a hot object to a colder one, something we can identify with entropy (see Fig. 10-6). The quantity that might seem most real to us in this flow is the energy that is being transferred. However, the amount of energy that leaves the hot body is equal to the amount of energy that enters the cold body. Since the amount of energy is not changed in this

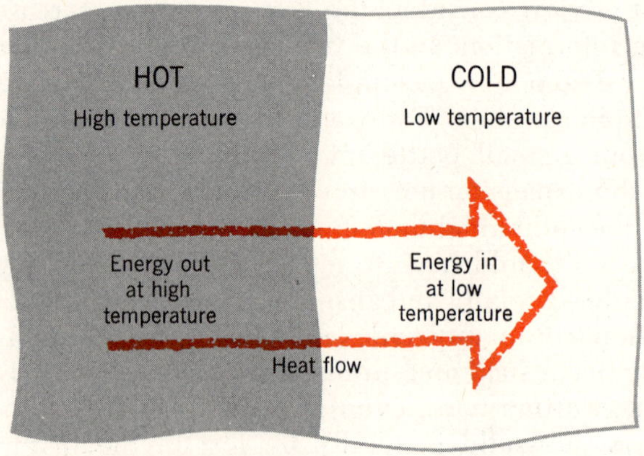

FIGURE 10-6 **Heat flowing from a hot to a cold object results in an increase in entropy.**

process, it cannot be the quantity we are looking for. The *condition* of the energy has changed, however. At first it existed in a high-temperature condition; later it exists in a low-temperature condition. We need to devise a quantity whose change will depend on both the energy and the temperature. We can do this by dividing the amount of energy transferred by heat flow by the temperature. This gives us a quantity that we shall identify as the *change of entropy*:

$$\text{Change of entropy} = \frac{\text{Energy transferred}}{\text{Temperature}}$$

(This is the way in which entropy was defined before its relation to disorder was understood.)

The entropy lost by the hot object is obtained by dividing the energy transferred (out) by the high temperature. The entropy gained by the cold object is obtained by dividing the energy transferred (in) (the same energy) by the smaller low temperature. Since these two temperatures are different, the entropy gained by the cold object is not equal to the entropy lost by the hot object. Dividing the same amount of energy by the smaller temperature results in a greater quantity representing entropy change. Thus the entropy gained by the cold object is greater than the entropy lost by the hot object, and entropy increases for the entire process. This suggests that we have indeed chosen the proper quantity for entropy. More careful analysis confirms this choice of entropy for systems of objects in which heat and temperature are important.

FUNDAMENTAL IDEAS

I. FUNDAMENTAL LAWS AND PRINCIPLES

A. THE LAW OF ENTROPY INCREASE. Entropy never decreases, and it increases in irreversible processes.
or: Heat never flows from an object into a hotter object;
or: Perpetual motion machines of the second kind are impossible.

II. QUANTITIES

A. ENTROPY, a derived scalar quantity.

1. Entropy as disorder.
2. Entropy as probability.
3. Entropy as lack of information.
4. Entropy in terms of heat and temperature.

QUESTIONS

Entropy and Irreversibility

1. Give an example of a time-reversible process.
2. Give examples of time-irreversible processes.
3. How do you know that the processes in Question 2 are not time-reversible?
4. The quantity entropy has been called "times arrow." Can you imagine why?
5. Why are conservation laws not adequate for explaining irreversible processes?
6. How does the kinetic theory of matter seem to contradict the idea of irreversible processes?
7. Give as many reasons as you can to explain why heat passes from a hot body to a cold body instead of vice versa.
8. Can the entropy of a material system decrease? Explain.
9. Can the entropy of an isolated material system decrease? Explain.

Entropy as Disorder

10. Give some examples not mentioned in the text in which irreversible processes cause increased disorder.
11. Give an example of a situation in which the presence of a force makes a system appear to be moving toward a condition of increasing order, in apparent contradiction to the law of increasing entropy.
12. Suppose that a person claims that living processes do not obey the law of increasing entropy because living matter grows, reproduces, and thus increases the amount of ordered material. Comment on this claim.
13. Suppose that another person claims that evolution is impossible, because the appearance of higher forms of life implies an increase in order, which would contradict the law of increasing entropy. Comment on this claim.

Entropy and Probability

14. List all the possible results of an experiment in which four coins are tossed.
15. Why are some hands in a card game more unlikely than others?
16. Explain entropy in terms of order and probability.

Entropy and Information

17. What does information have to do with irreversibility? How can what we know about a process have anything to do with whether or not it is reversible?
18. In Figure 10-2, why do we know more about Conditon I than Condition II?
*19. Do you think we could obtain information about one situation without giving up the possibility of having or getting information about another situation? Explain.

Perpetual Motion and Heat Death

20. Distinguish between the two kinds of perpetual motion.
21. What advantage would there be in a perpetual-motion machine of the second kind?
22. Why can't a machine take in sea water, put out sea water at a lower temperature, and do work with the energy thus obtained?
23. In Question 22, why doesn't the machine, by ejecting the water at a lower temperature, obey the law of increasing entropy?
24. Does a steam engine obey the law of increasing entropy? Explain.
25. What is meant by the "heat death" of the universe, and what would cause it?
26. Express the other symmetry principles and conservation laws as principles of impossibility.

11 Electric Charge and Fields

Benjamin Franklin dramatically dispelled the superstitions of the public when he collected the electricity of lightning through a kite string and showed it to be the same type of stuff that is in the sparks produced by scuffing shoes on a rug. (Incidentally, we do not advise our readers to risk their lives trying to repeat this experiment of Franklin's.) Schneer describes the impact of the experiment as follows:

> The effects of Franklin's discovery were literally electrifying. His identification of the lightning and thunder universally associated with supernatural power as the corporeal, mechanical phenomenon, electricity, was in itself of the character of miraculous revelation. Franklin seemed like a sage from a nobler world, soothing the fears of mankind. To the philosophers of the Age of Reason it was one more blow, as great perhaps as had ever been struck, against the forces of ignorance and superstition which they thought of as an enemy. In Newtonian mechanics, the rationalists found the ordered world-system to fit their ideas, but the *Principia* was after all a remote and forbidding territory of hostile theorems and implacable geometry. Franklin's lightning was immediate to the perception. Franklin had not only rationalized the lightning but with

FIGURE 11-1 An artist's idea of Benjamin Franklin's famous experiment in which he obtained electricity from the atmosphere. Courtesy of Culver Pictures.

 his invention of the lightning rod he had rendered it powerless. To the more literal minded, both in the camp of the clericals who deplored Franklin's impiety and the anticlericals who hailed the backwoods savant, it appeared that Franklin had disarmed the Deity and rendered the awful visitation of divine chastisement an impotent if dramatic display of frustrated rage.[1]

In the years before and after Franklin, scientists discovered many facts and principles about electricity and magnetism. In this chapter, and the next, we will discuss some of these facts and principles. The Scottish physicist James Clerk Maxwell (1831–1879) added to this work and gathered these principles together into a few mathematical laws. Maxwell's laws do for electricity and magnetism what Newton's laws do for motion and gravitation. They give us a few precise laws which describe a wide range of electrical and magnetic phenomena.

Most people are familiar with electrical effects produced by friction in many everyday situations. When you scuff your shoes on a rug and then reach for the door knob, you sometimes see or feel a spark between your fingers and the knob. If you comb or brush your hair vigorously you sometimes hear sparks crackle. Similarly, you may produce sparks if you slide across a plastic car seat and then reach for

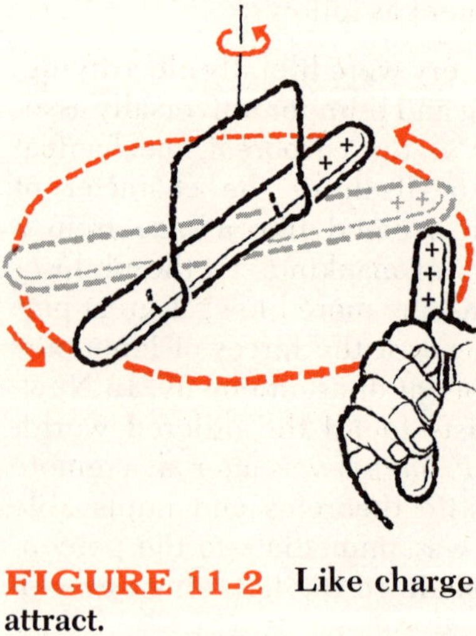

FIGURE 11-2 **Like charges repel; opposite charges attract.**

[1]Cecil J. Schneer, *The Search for Order*, Harper & Brothers (now Harper & Row), New York, 1960, reprinted as *The Evolution of Physical Science*, Grove Press, Inc., New York, pp. 222–223.

the door handle. Electrical effects similar to these have been known for ages.

It has long been known that if a glass rod is rubbed with silk (or plastic) it can be used to illustrate electrical effects. Figure 11-2 shows a rod that has been rubbed with silk and suspended by a thread. When a second rod is also rubbed with silk and brought near the first, the suspended rod is pushed away. A force of repulsion is produced by the electricity of the two rods. Electricity can also produce attractive forces. Suppose that the second rod is made of hard rubber and is rubbed with fur. When this rod is brought near the silk-rubbed glass rod, the two rods are attracted rather than repelled.

Three significant facts are evident in these simple experiments. First, the forces of attraction and repulsion occur even though the rods do not touch one another. Second, the force is extremely large compared to the gravitational force between the same objects. Third, there is something intrinsically different between a rubber rod (rubbed with fur) and a glass rod (rubbed with silk) since they behave differently in the presence of the suspended glass rod (rubbed with silk).

Electric Charge

We can explain the electrical effects discussed above by introducing a new fundamental, scalar quantity: *electric charge.* Rods and other objects can acquire an electric charge by friction when they are rubbed. Because we obtained a different result with the rubber rod than with the glass rod, there must be two kinds of electric charge. The electric charge we obtained on the glass rod is arbitrarily called a positive charge, and the electric charge we obtained on the rubber rod is called a negative charge. In the experiment we have described, the positive charge on the glass rod produced a repulsive force when brought near a similarly charged rod. The negative charge on the rubber rod produced an attractive force when brought near the positively charged rod. These two examples illustrate a general principle:

Like charges repel each other, and unlike charges attract each other.

Two Fundamental Particles

The fact that objects can acquire an electrical charge by friction strongly suggests that matter has an electrical nature. In fact, today we believe that atoms are made up of particles with electrical properties. Two of the *fundamental particles* that make up the atom are the *electron* and the *proton*. For example, every atom of hydrogen is made up of one electron and one proton. The electron has a mass of 9.1×10^{-31} kilograms. The proton has a mass of 1.7×10^{-27} kilograms. These two masses are fundamental constants, since they are the masses of fundamental particles. The proton supplies most of the mass of the hydrogen atom since it is about 1840 times more massive than the electron. Other atoms contain different numbers of fundamental particles.

Although the electron and proton are different in mass, they have exactly the same amount of electric charge. Their charges, however, are not of the same kind. The electron has a negative charge, and the proton has a positive charge. Most objects are electrically neutral: that is, they have the same number of electrons as protons, and therefore have no excess, or net, electric charge. Objects are *charged* when they do not have the same number of electrons and protons. Such an imbalance is ordinarily a result of the gain or loss of electrons by the object. When two objects acquire an electric charge by being rubbed together, the charge is acquired because electrons from one object are transferred to the other.

Conservation of Charge

Charge is a conserved quantity; that is, it cannot be created or destroyed. If an object loses charge, it is because the charge has been transferred somewhere else. For example, we can give the glass rod a positive charge by rubbing it with silk. At the same time, however, the silk acquires a negative charge of the same amount. Some of the electrons in the glass simply move from the glass to the silk, carrying their negative charge with them.

The Unit of Charge

A charge may be measured by the electrical force it exerts on another charge. The unit of charge is called a *coulomb*, after the French physicist Charles Augustus Coulomb (1736–1806). One coulomb of charge is the amount of charge that would exert a force of 8.99×10^9 newtons on an equal charge one meter away. The direction of the force, of course, depends on whether or not the two charges are of like or unlike kinds, but the magnitude does not. Because 8.99×10^9 newtons is a very large force, the coulomb is a very large charge compared to the amounts we normally deal with on charged glass rods, etc.

The Law of Electrical Force for Charges at Rest

Coulomb discovered the exact law that gives the magnitude of the force between any two charged objects at rest. This law is expressed as follows:

$$F = K\frac{qQ}{d^2}$$

In this law, q is the charge on one object, Q is the charge on the other object, d is the distance between the two objects, and F is the magnitude of the electrical force between them. The electrical force depends on the charges of the two objects in the same way that the gravitational force depends on the masses. Both the electrical and gravitational force depend on the distance d in the same way. The quantity which we represent here by the letter K is a fundamental constant that is fixed by the strength of the electrical force. K has a value of 9.0×10^9 newton meter^2/coulomb^2. This value is very large compared to the gravitational constant G. The electrical forces between electrons and protons are much larger than the gravitational forces between them. The direction of the electrical force whose magnitude is given by the equation is along the straight line between the two charges. This force is a force of repulsion for like charges and attraction for unlike charges.

Electric Currents

Some materials, such as most metals, allow electrons to move freely from place to place on and inside them. Such materials are called electrical *conductors*. Other materials do not allow electrons to move freely on or inside them. Such materials are called *insulators*. The glass rod and the rubber rod we discussed above were both insulators. Therefore the electric charge placed on them by friction tended to stay on them, since it could not easily move. If we had used metal rods instead, the electrons would have been able to move out of the rod, into our hands, and perhaps into the ground. (The flow of electrons could also have been in the opposite direction.)

A flow of electric charge is called an electric current. This is usually a flow of electrons through a conductor. The amount of current in a conductor, such as a copper wire, is determined by the rate at which charge is passing through. The unit of current is the *ampere.* A current of one ampere is a current in which one coulomb flows by in one second.

Fields

It may seem strange that a charged object has an electrical force on it because of another nearby charged object, when they are not even touching. Our previous study of gravitation, however, has already given us an example in which one object is influenced by another across a distance without being touched. Thus the falling apple does not stop to ask for directions; it heads for the center of the earth. Often it is convenient to assume that the apple is in a *field* of influence that exists around the earth, and that it falls because it is in that field. We may think of the earth, because of its mass, as setting up a gravitational field of force around it. The apple, because it is *in* this field of force, is accelerated toward the earth *by* this field of force. In this section we shall discuss gravitational fields (which are set up by masses) and electrical fields (which are set up by charges).

Gravitational Fields. We would like to use a picture to represent the force exerted by the gravitational pull of the earth. We want this picture to indicate the force exerted on any object placed in any location with respect to the earth. This picture will represent the gravitational field of the earth, because it will enable us to tell how the presence of the earth affects other objects.

We draw this picture of the gravitational field of the earth by drawing arrows at various points in the space around the earth (see Fig. 11-3*a*). These arrows represent the gravitational force that would be exerted on an object if it were placed at those points. Since objects of different mass would have different forces on them, we agree to use an object of one-kilogram mass. The arrows, then, represent the force that would be exerted on a one-kilogram mass. The one-kilogram mass is pulled harder when it is near the earth than when it is farther away. For this reason the arrows are longer nearer the earth and become shorter as we go farther out into space. The length of the arrows represents the magnitude of the force, and the direction of the arrows gives the direction of the force. The arrows point toward the center of the earth, since gravitational forces pull objects in that direction. Ideally we can think of an arrow drawn at every point in space. This collection of arrows could be used to represent the gravitational field of the earth, since it tells us the gravitational force that would be exerted on one kilogram of mass at any point. If we want the

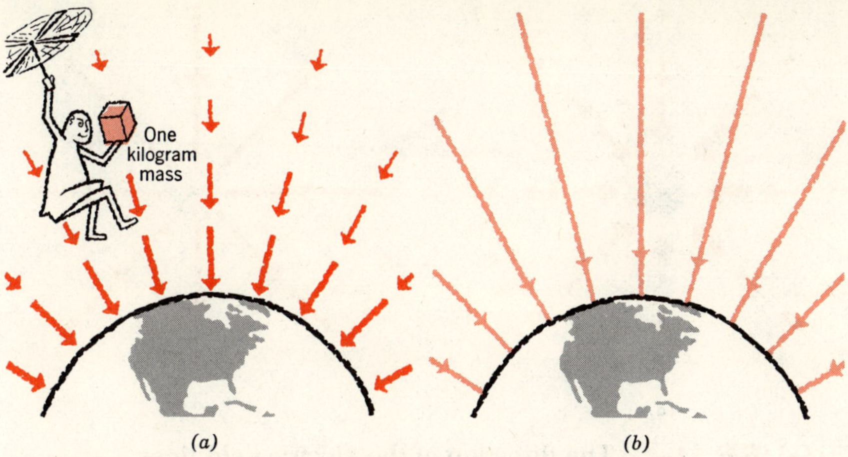

FIGURE 11-3 **The direction of the gravitational field lines is the direction of the gravitational force.**

force that would be exerted on an object of different mass, we simply multiply the mass of the object by the force on the one-kilogram mass. This allows us to find the force that would be exerted on any object at any point in the gravitational field of the earth.

We can represent the gravitational field more conveniently, by lines than by arrows. Figure 11-3*b* shows how these lines look. Each line is drawn by following the arrows from point to point. Therefore, at each point the lines are in the direction of the arrows that would be drawn at that point. Where the force is the strongest, the lines are closest together. Where the force is weak, the lines are far apart. These lines, called field lines, give us the picture that we desired. The lines of the gravitational field do not end except on a mass. (Of course, this is not obvious from what we have said, but like a similar statement to be made later about electric field lines, it is true.)

The Electric Field. We can picture the electric field around a charge just as we have pictured the gravitational field around the earth. Figures 11-4*a* and 11-4*b* show the arrows drawn at various points, and also the field lines. The arrows at a given point represent the force that would be exerted on one coulomb of positive charge if it were placed at that point. (We use a one-coulomb charge here rather than the one-kilogram mass because, while the gravitational force acts on masses, the electric force acts on charges.) The direction of the field lines (or arrows) is the direction of the force that would be exerted on a positive charge. The force on a negative charge would

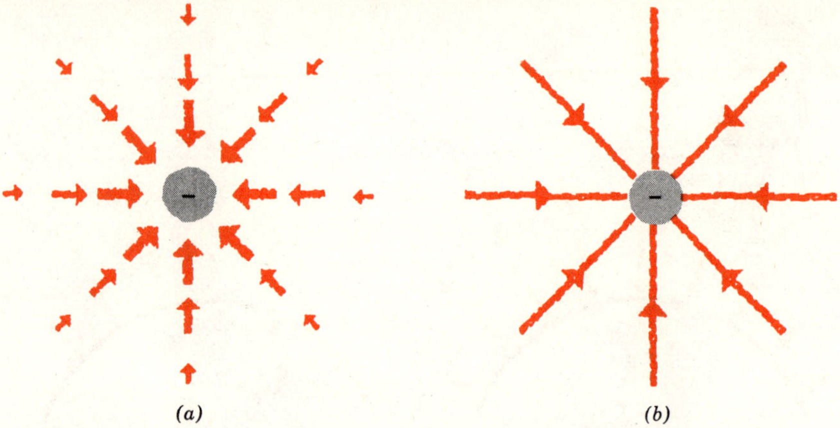

FIGURE 11-4 The direction of the electric field lines is the direction of the electric force on a positive charge.

be in the opposite direction. Once again, the closer the lines of force are to each other, the stronger the force that would be exerted. In Figure 11-5 we show the field lines for several arrangements of charged objects. In each case, a small positive charge would be impelled in the direction of the field lines. A small negative charge, such as an electron, would be impelled in the opposite direction. The field lines of the electric field do not end except on a negative charge, and do not begin except on a positive charge.

It may seem from the above discussion that the field concept is an unnecessary fabrication. In fact, in this discussion, we have simply been looking at the forces between objects in a different way. In-

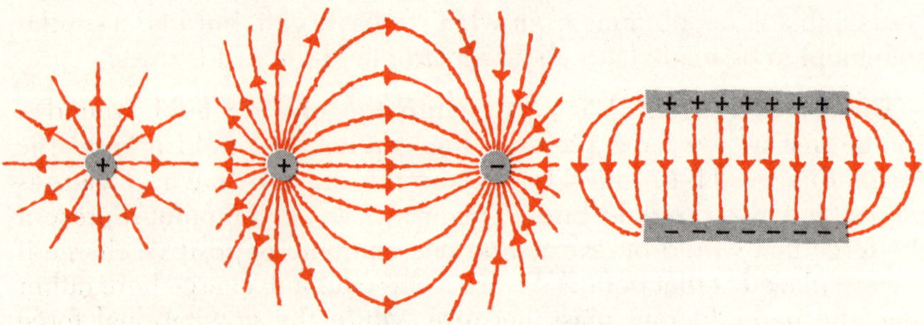

FIGURE 11-5 Electric field lines for several arrangements of charge.

stead of thinking of objects as exerting forces on each other at a distance, we have decided to consider that each one of them has set up a field and that it is this field that exerts the force on the other object. Thus far, there is no reason to choose one approach over the other, except convenience. In Chapter 14, we shall find reason to believe that the field has a real existence.

The English physicist Michael Faraday (1791–1867) used the concept of field lines in his research in electric and magnetic phenomena. Although Maxwell later put these ideas in a more elegant mathematical form, Faraday was able to understand fields sufficiently well to make important discoveries using field lines.

The Charge on the Electron

Although the electron is extremely small, its charge can be measured by a rather direct experiment, conceived and executed by Robert Andrews Millikan (1868–1953) and his students. This experiment is called the "oil-drop experiment" because it measures the charge on small drops of oil. The experimental apparatus is diagramed in Figure 11-6. Droplets of oil from the fine mist sprayed from an atomizer are allowed to fall into the electric field between two charged plates. The droplets are charged as they come out of the atomizer, by friction or some other means. When illuminated by a strong light, they can be seen individually as tiny specks in a microscope.

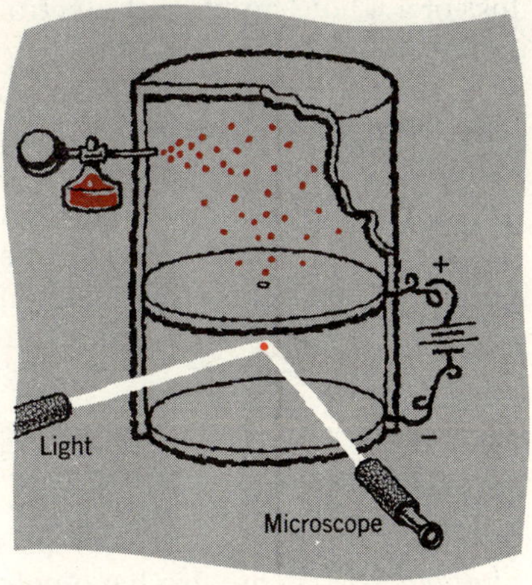

FIGURE 11-6 **The Millikan oil drop experiment.**

As illustrated, there is a positive charge on the upper plate and a negative charge on the lower. This creates an electric field that tends to lift negatively charged objects. It is possible to adjust or reverse the field by changing the charge on the plates, or to eliminate the field altogether by discharging the plates. Suppose, for example, that we have a negatively charged droplet. We can adjust the field so that the electric force pulling the droplet upward is exactly balanced by the weight of the droplet pulling downward. If we know the weight of the droplet we can determine the electric force and thus the charge on it.

Actually, in practice, we find the weight of the droplet by observing how fast it falls against air resistance when no electric field is present. We then compare this speed with its speed as it rises in a known electric field. This method is more convenient than trying to balance the droplet perfectly, but the principle is the same: we compare the gravitational force with the electric force on the same drop. The electric force depends on the charge on the droplet and therefore allows us to measure that charge.

In this way the charge, both positive and negative, on many droplets can be measured. The smallest amount of charge ever observed is 1.60×10^{-19} coulomb. Every other observed charge is some whole number times this smallest charge—either two, three, four, etc., times 1.60×10^{-19} coulomb. Thus charge occurs in small discrete "bundles." Each electron has one such "bundle" of negative charge. A droplet may gain or lose electrons, and thus have a negative or positive charge. The charge, however, always corresponds to the gain or loss of a whole number of electrons.

The amount of charge on a proton is the same as that on an electron, except that the proton's charge is positive rather than negative. Therefore, the charge on a proton is *plus* 1.60×10^{-19} coulomb. Wherever charge is found, it occurs in discrete bundles of this amount; therefore, this quantity is one of the fundamental constants of nature. (Recently there has been speculation and some experimental indications that particles of one third or two thirds this amount of charge may exist, but this is still unverified.)

Electric Potential Energy

We have seen in Chapter 8 that an object may have gravitational potential energy because of its position in a gravitational field. A *charged* object can also have potential energy because of its position in an electric field. This must be true because a charged object in an electric field will experience a force and, if released, will accelerate, acquiring kinetic energy. This energy must come from somewhere. In fact, it comes from the *electric potential energy* the charged object had in its starting position.

Another way to recognize the existence of electric potential energy is to move a charged object in an electric field. We recall that we had to exert a force to lift a mass in a gravitational field. In so doing, we transferred energy to the object by doing work. The gravitational potential energy of the object was increased as we lifted it against the gravitational force. In the same way we increase the electric potential energy of an object when we move it against an electric force. The man shown in Figure 11-7*a* is doing work against the electric force and thus is increasing the electric potential energy of the charged object. The potential energy is transformed into kinetic energy when the charged object is released (Fig. 11-7*b*). It is then accelerated in the electric field by the electric force.

When we discussed gravitational potential energy we found that it was not the potential energy itself that was important but the change in an object's potential energy as it moves from one position to another. Thus we were really interested in the difference between the potential energies of the object at two points in the gravitational field. Similarly, here we are interested in the difference between the potential energies of a charged object at two points in the electric field. We can choose one point to have zero potential energy and use that point as a reference point for the potential energies of other points.

Since the force on a negative charge in an electric field is opposite

FIGURE 11-7 **An electric charge has electric potential energy due to its position in an electric field.**

to that on a positive charge, the two types of charge will have different potential energies at the same point in the same field. The potential energy of a negative charge in an electric field will be the negative of the potential energy that a positive charge of the same amount would have if it were at the same point.

Often in discussing an object in an electric field we speak of its *electric potential* instead of its electrical potential energy. The electric potential of a charged object is simply its electric potential energy divided by its electric charge. The common unit of electric potential is the *volt*, which is defined as follows:

$$\text{volt} = \frac{\text{joule}}{\text{coulomb}}$$

Illustrations of the Effects of the Electric Field and the Electric Potential

Induced Charges. In the experiment illustrated in Figure 11-8, we have suspended a light cotton ball wrapped in tin foil by a thread. The thread acts as an insulator to prevent the "leaking" of electric charge from the ball. The ball is electrically neutral to begin with—that is, it is not charged. We now charge a glass rod by rubbing it with silk. As we bring the glass rod up to the ball, the ball is attracted to the rod, despite the fact that the ball has no net charge (see Fig. 11-8*b*). When the ball touches the rod, it suddenly jumps away. After this, the ball is repelled by the rod (see Fig. 11-8*c*). How are we to explain this behavior?

First we must explain how the ball can be attracted by the positively charged rod although the ball itself has no net charge. The rea-

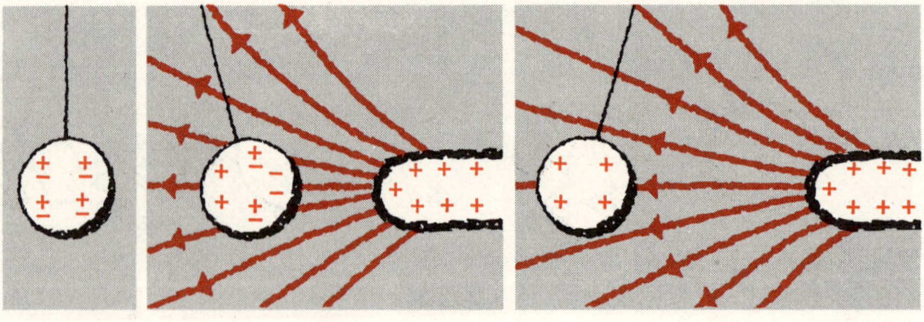

(*a*) (*b*) (*c*)

FIGURE 11-8 A charge separation is induced in the ball by the electric field of the rod.

son is that the positive charge on the rod sets up an electric field. The negative charges in this field are attracted toward the positive rod. This attraction causes many of the electrons in the tinfoil covering the ball to move over onto the side of the ball nearest the glass rod. This sets up a negative charge on that side of the ball and leaves an equal positive charge on the other side. Thus one side of the ball is attracted to the glass rod and the other side is repelled. The two charges on the ball are equal in magnitude, but they are at different distances from the rod. The electric field of the rod is stronger closer to the rod. Therefore, the attraction of the negative charge on the ball overcomes the repulsion of the positive charge, and the ball is pulled toward the rod. We say that the field of the rod *induces* a charge separation in the conducting metal that covers the ball. This causes an induced negative charge on one side of the ball, and an induced positive charge on the other.

When the rod actually touches the ball, some of the electrons flow off onto the glass rod, attracted by the positive charge. Since the rod is an insulator, the electrons from the ball tend to neutralize the positive charge of the rod only in the region where the rod touches the ball. Therefore, the rod still retains part of its positive charge. Now, however, with the loss of these electrons from the ball, the ball *itself* is positively charged. In this condition, the positively charged ball will be repelled by the positively charged rod.

The Electroscope. The behavior of an electroscope may also be understood in terms of induced charges. The electroscope is a device

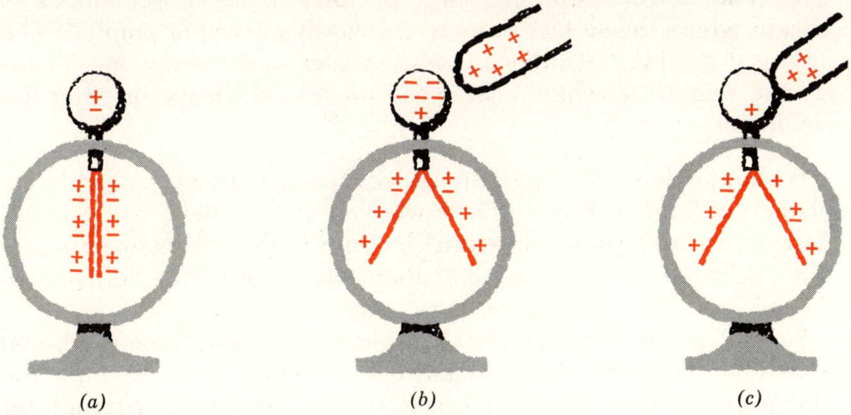

FIGURE 11-9 Charge is induced on the leaves of the electroscope by the electric field of the rod.

used to detect charge (see Fig. 11-9). When there is a charge on the electroscope, there will be a repulsion between the two metal leaves so that they are held apart. Without this force, the leaves fall together.

In Figure 11-9*a*, the leaves of the electroscope are together, indicating that the electroscope is uncharged. If we bring a charged glass rod up to, but not touching, the top of the electroscope (see Fig. 11-9*b*), the positive charge of the glass rod pulls some of the electrons out of the leaves of the electroscope into the ball on the top. The leaves of the electroscope are left with a positive charge. This charge makes the leaves separate because like charges repel each other. When the rod is withdrawn, the electrons return to the leaves and the leaves come together again. If we touch the rod to the top of the electroscope, however, some of the electrons are pulled into the charged glass rod (Fig. 11-9*c*). Then, when the rod is removed, the leaves will remain apart, since the electroscope now has a positive charge. The electroscope appears to behave in the same way if we do the same experiment using a charged rubber rod (negative charge).

Electrical Conditions in Conductors. Here we shall discuss several facts about solid conductors in which the charges are at rest. These facts apply to the solid material of the conductor and not necessarily to holes within it. If currents are flowing, then some of these conditions may not hold perfectly. The facts are as follows.

1. There is no electric field inside a conductor.
2. The excess charge on a conductor is on its surface.
3. The concentration of charge on the surface of a conductor is greater where the surface is more outwardly curved or pointed. The field just outside the surface is also greater in these regions. These are the regions at which charge can most easily leave or enter the conductor.

Without going into detail, the above facts may be very roughly explained as follows. First, if there were an electric field inside a conductor, the field would cause the electrons of the conductor to move. The electrons would rearrange themselves until the field disappeared.

Second, since there is no field inside a conductor, there can be no excess charge there of either kind. Excess charge, as for example the charges in Figure 11-5, would have lines of field issuing from it (or toward it). However, this field cannot exist in a conductor, as we have already noted. Thus the excess charge cannot be inside the

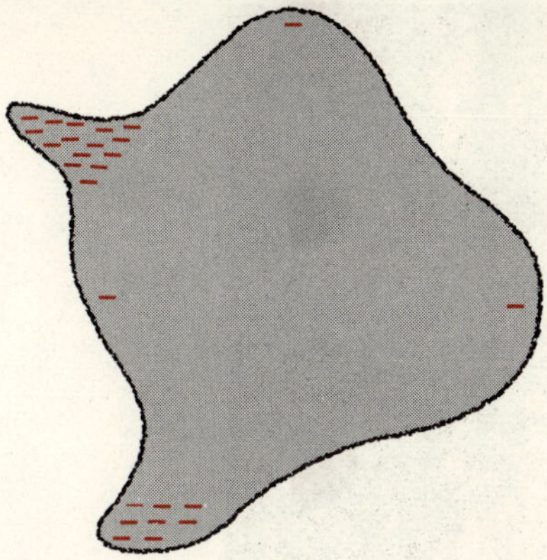

FIGURE 11-10 **Excess charge on the surface of a conductor is more concentrated on the more outward pointed regions.**

material body of the conductor, and therefore must rest on its surface.

Third, the charge which is resting on the surface will repel other like charges. As a result of this repulsion, the charge will tend to distribute itself over the outermost parts of the surface—generally the more outwardly curved or pointed regions—and avoid the inwardly curved or pointed regions (see Fig. 11-10). Where the charge is concentrated, the field (outside the conductor) will also be more concentrated. Because of the high fields, a charge can be more easily pulled on or off the conductor in these outwardly pointed regions.

A lightning rod is a conductor which acts as an outward point on the surface of the conducting earth. The electric charge in lightning tends to enter these high conducting points rather than the nearby structures that the rod is designed to protect (Fig. 11-11).

During the American revolutionary war, King George III of England mistakenly decided that the pointed lightning rods invented by the rebellious Benjamin Franklin were inferior to the blunt rods championed by a loyal British subject. He therefore had the pointed rods replaced on powder magazine and palace. (Of course, this ridiculous, politically influenced decision had no effect on the laws of physics.) The renowned American scientist and statesman took this decision with characteristic good humor:

The king's changing his pointed conductors for blunt ones is a matter of small importance to me. If I had a wish about it, it would be,

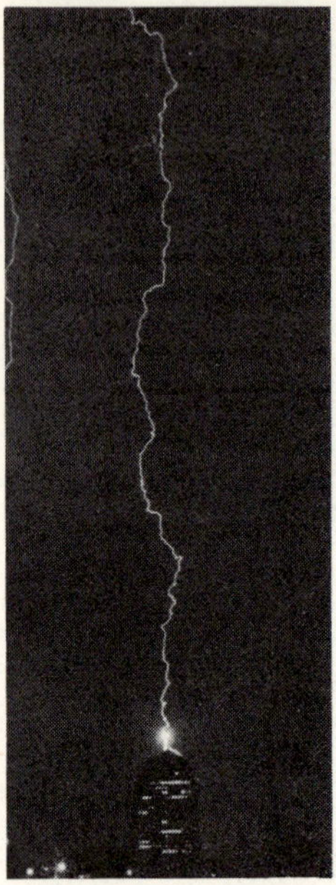

FIGURE 11-11 Lightning striking a lightning rod on a building. Courtesy of NOAA.

that he had rejected them altogether as ineffectual. For it is only since he thought himself and family safe from the thunder of Heaven, that he dared to use his own thunder in destroying his innocent subjects.[2]

The Van de Graaff Generator

The Van de Graaff generator is a laboratory device used to collect charge on a sphere at a very high electric potential energy. This potential energy can be used to accelerate charged particles to high kinetic energies for various experiments. As we have seen, there is no field or excess charge inside a conductor, and the charge stays on the outside. This situation is not changed if part of the material inside is removed to form a hollow conductor, since there is no excess charge in the material removed. There will be no field or charge in-

[2]Benjamin Franklin, as quoted in James Parton, *Life and Times of Benjamin Franklin*, Vol. II, Boston: Houghton, Mifflin and Company and Cambridge: The Riverside Press, Copyright 1864, 1892, p. 217.

side this hollow region unless some charge is brought in and held there. Even if there is a small hole leading into the hollow region from outside the conductor the field inside will be very small.

Part of the Van de Graaff generator (Fig. 11-12*b*) is a large hollow spherical conductor, with a hole in it through which charge can be inserted. Since there is very little field inside this sphere, this charge is easily released onto the inside of the sphere. The charge does not remain there but moves to the outside of the sphere. By bringing in more charge, we are able to build up a large charge on the outside of the sphere. We could not do this by bringing the charge to the outside of the sphere because there is a high electric field building up there as the charge is increased. Charge would flow off the outside of the sphere, rather than onto it.

The charge is brought into the sphere on a belt made of insulating material. It is collected on the sharp points at the top, which are connected to the inside of the conducting sphere. Charge may be put on the belt at the bottom in several ways—for example, by friction.

Electric Circuits. Figure 11-13 represents a simple electric circuit. The battery, by chemical action, maintains an electric potential difference by pulling electrons from one terminal and pushing them onto another. In this case the electric potential difference, or voltage, is twelve volts. Because of the electric potential difference between the terminals, electrons will have more electric potential energy at one terminal than they do at the other. Therefore they travel from one terminal of the battery to the other through the conducting wire and the light.

The wire and the light offer resistance to the flow of electrons which makes up the current. Without the battery to maintain the potential difference, the current would soon cease. The greater the potential difference, the greater the current; in fact in many practical circuits, the current is proportional to the electric potential difference, or voltage. That is, doubling the voltage doubles the current; tripling the voltage triples the current, etc. The current is inversely proportional to the resistance—doubling the resistance cuts the current in half, tripling the resistance decreases the current to one third its value, etc. These facts are summarized in the following relation:

Electric potential difference = Current multiplied by resistance
(voltage)

This relationship is called Ohm's law after the scientist who discovered it. The unit of resistance is called the ohm. A potential dif-

(a)

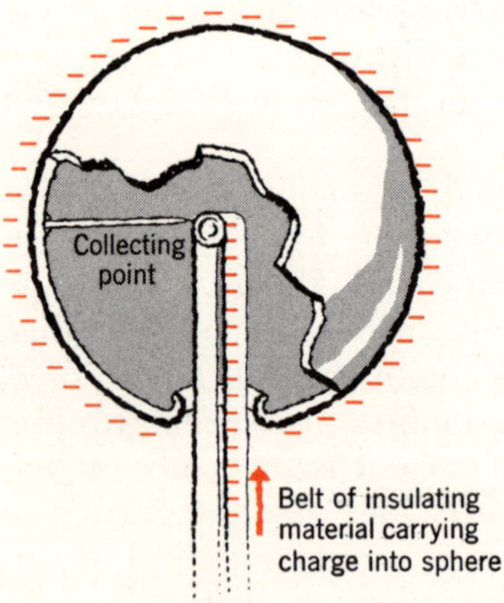

(b)

FIGURE 11-12 **(*a*) A Van de Graaff Generator. Photograph courtesy of the High Voltage Engineering Corporation. (*b*) Diagram showing the operation of a Van de Graaff generator.**

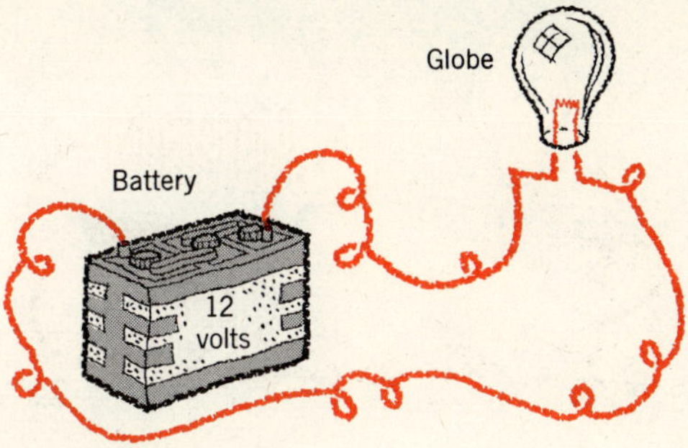

FIGURE 11-13 **An electric circuit.**

ference of one volt will push a current of one ampere through a circuit that has a resistance of one ohm.

For many conductors like the commonly used copper wire, the resistance increases with temperature. For some conductors, however, a very strange thing happens when the temperature becomes very low, say within a few degrees of absolute zero. The resistance apparently suddenly vanishes entirely. This phenomenon is known as superconductivity. Under these conditions the current keeps flowing without any battery or other device to maintain the electric potential difference. Currents can be preserved for weeks at appropriately low temperatures without any observable change.

The battery in Figure 11-13 causes a current to flow in one direction through the circuit. Such a current, flowing in one direction only, is called a *direct current*. The usual household current is generally an *alternating current*, however. Such a current changes its direction—the ordinary household current reverses direction 120 times a second. Since two changes of direction bring it back to the same direction again, there are 60 complete cycles of change per second.

Figure 11-14 shows a simple series circuit with two elements, a light and a toaster. The resistance of the two elements together is the sum of the resistances each would have separately. The full potential difference provided to the circuit through the plug does not act on the light and the toaster separately—rather the potential difference is divided between the two. The voltage experienced by the light plus the voltage experienced by the toaster equals the total voltage supplied to the circuit. If the light bulb should burn out, breaking the circuit, no current would flow through either element.

Figure 11-15 shows a simple parallel circuit with two elements. The resistance of this circuit is less than the resistance of either ele-

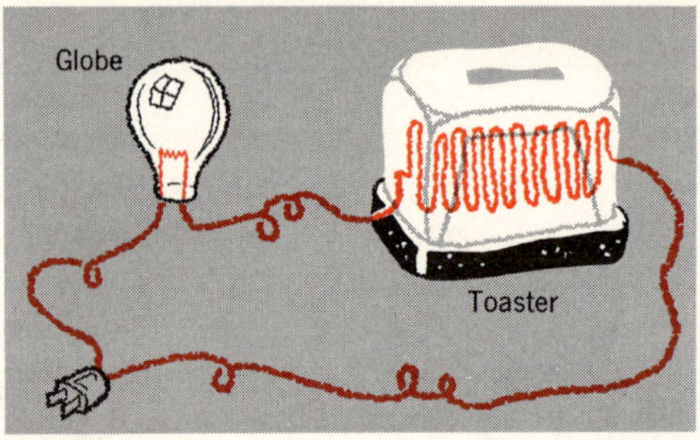

FIGURE 11-14 **The lamp and the toaster are connected in series.**

ment alone. In this case the full voltage supplied to the circuit appears across both elements—they each experience this same voltage. In this circuit, if either element should burn out, breaking its side of the circuit, the other would still function.

We shall give one further useful relation which applies to electric circuits: the power expended in any element of an electric circuit is equal to the current through the element multiplied by the voltage supplied to it.

FUNDAMENTAL IDEAS

I. FUNDAMENTAL LAWS AND PRINCIPLES

A. THE LAW OF ELECTRIC FORCE. The magnitude is given by the formula $F = KqQ/d^2$. The direction is along the line between

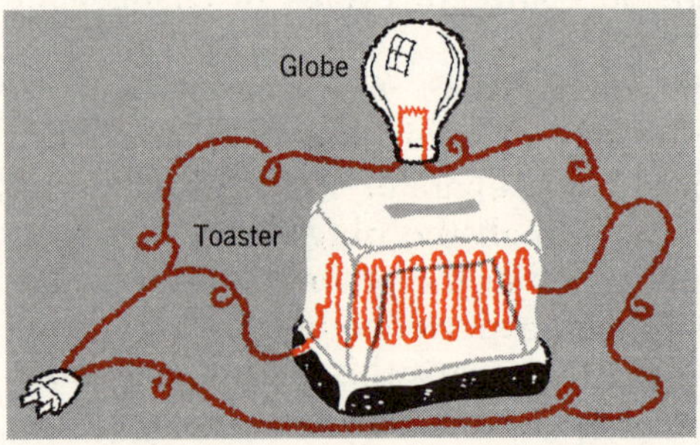

FIGURE 11-15 **The lamp and the toaster are connected in parallel.**

the two charges, and is given by the rule that like charges repel and unlike charges attract.

B. CONSERVATION OF CHARGE

II. QUANTITIES.

A. FUNDAMENTAL QUANTITIES

I. Charge, a scalar. Two kinds, positive and negative. The unit of charge is the coulomb.

B. DERIVED QUANTITIES

1. Current. The amount of charge that flows by per second. The unit is the ampere, which is one coulomb per second.
2. Voltage. Electric potential energy per charge, measured in volts, or joules per coulomb.
3. Resistance, measured in ohms.

III. FUNDAMENTAL PARTICLES

A. ELECTRON. Negatively charged, 1.6×10^{-19} coulomb. Mass 9.1×10^{-31} kilogram.

B. PROTON. Positively charged, 1.6×10^{-19} coulomb. Mass 1.7×10^{-27} kilogram.

IV. FUNDAMENTAL CONSTANTS

A. FUNDAMENTAL CHARGE. the amount of charge on the proton: 1.6×10^{-19} coulomb.

B. MASS OF THE PROTON: 1.7×10^{-27} kilogram.

C. MASS OF THE ELECTRON: 9.1×10^{-31} kilogram.

D. THE ELECTRIC FORCE CONSTANT: $K = 9.0 \times 10^9$ newton meter^2/coulomb^2.

V. FIELDS

A. THE GRAVITATIONAL FIELD

B. THE ELECTRIC FIELD

VI. APPLICATIONS OF THE ELECTRIC FORCE AND FIELD

1. Induced Charges.
2. The Electroscope.
3. The Oil Drop Experiment.
4. Electrical Conditions in Conductors: relating to charge and field distribution in conductors, lightning rods.
5. The Van de Graff Generator.
6. The battery
7. Ohm's law and circuits.

QUESTIONS

Electric Charge

1. What is charge?
2. How do we know that the fur used to rub a hard rubber rod has acquired a positive charge?
3. How do we know there is only one kind of mass but two kinds of electricity?

Electric Fields

4. What is a "field"? How do we know that a field exists?
5. What are the sources of the gravitational and electrical fields?
6. Does a cotton ball wrapped in tinfoil "remember" whether or not it has touched a charged object? Explain.
7. What would happen to an electroscope if the experiment described on pp. 211–212 were repeated using a negatively charged rod? Describe what would happen to the charge on the various parts of the electroscope at each step.
8. How does a Van de Graaff generator work?
9. If I put a piece of fur on top of a Van de Graaff generator while it is being charged, what will happen? Will it be more or less difficult to charge the machine? (*Hint.* Consider the fur as a conductor in these high fields.)

The Charge on the Electron

10. What was measured in the oil-drop experiment? How were these results interpreted?
11. Can a charge of 2.9×10^{-19} coulomb exist?

Electric Circuits (and Electric Current)

12. Is a wire with a current in it necessarily charged? Explain.
13. Explain what is meant by electric potential difference, current, resistance, voltage, volt, ampere, ohm.
14. Explain the difference between direct and alternating currents.
15. How would you connect two light bulbs in the same circuit for (a) maximum resistance and (b) maximum voltage across each element?
16. Does every electric circuit have resistance?

12 Electromagnetic Forces and Fields

Not only is the earth itself a large magnet, but magnetic effects are extremely important throughout our modern technology. Electric motors and generators operate on electromagnetic principles. Magnetic effects, as we shall see, are really electrical in origin.

Magnetic Forces and Fields

Magnetic forces were first studied in connection with magnets. Magnets are generally made of some combination of iron and nickel. They will attract each other or other objects made of iron, nickel, and certain other metals. Most magnets have two regions in which the magnetic forces are especially powerful, called the *poles* of the magnet. The earth itself is a magnet, with two magnetic poles close to the ordinary North and South poles but not quite in the same places. The two poles of a magnet are not alike. If a magnet is suspended by a thread, one of its poles will point to the north and the other to the south. A compass needle is such a magnet. The pole that points

FIGURE 12-1 Representation of the earth's magnetic field.

north is called the north-seeking, or north, pole of the magnet. The other is called the south-seeking, or south, pole. All north magnetic poles repel each other. All south magnetic poles repel each other. All north magnetic poles attract all south magnetic poles, and vice versa. In other words, like poles repel, opposite poles attract.

Cutting a magnet in two will not produce a free (isolated) north or south pole (see Fig. 12-2). If we tried this experiment we would find that another pole automatically appears in each half, so that each piece is a complete magnet with two poles. There are no free magnetic poles—magnetic poles always occur in pairs.

The Magnetic Field Around Magnets and Currents

The concept of the magnetic field helps us to describe the influence of a magnet in the space around it. As in Figure 12-3, we can use a compass needle to map out the field lines. The north pole of the compass needle will point in the direction of the magnetic force on it. The force on the south pole is in the opposite direction. Like the electric field, the magnetic field is strongest where the field lines are closest together and becomes weaker as the lines spread apart.

During a class demonstration, the Danish physicist Hans Christian Oersted (1771–1851) discovered a remarkable connection between electricity and magnetism. Happening to have a compass near a wire, he was surprised to see the compass needle move when an electric current flowed through the wire. Although the compass needle is uncharged, it will turn at right angles to the wire when a current flows, reversing itself when the current flows in the opposite direc-

FIGURE 12-2 **There are no free magnetic poles. When the magnet is broken in two, two more poles appear.**

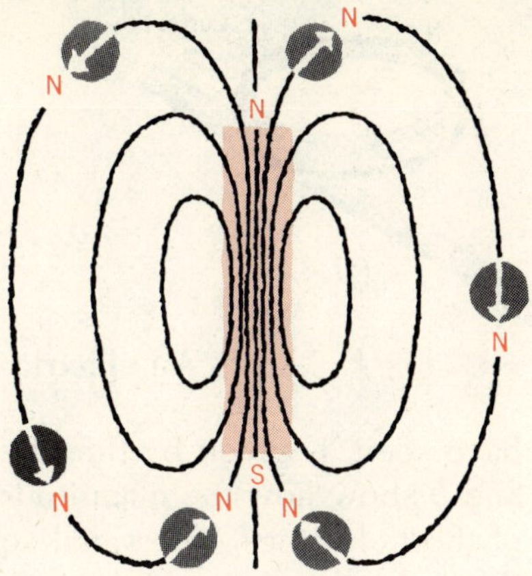

FIGURE 12-3 **The magnetic field of a bar magnet.**

tion (see Fig. 12-4). This means that an electric current, or indeed any moving electric charge, produces a magnetic field. We can use a compass to map out this field just as we did for the permanent magnet. Figure 12-5*a* shows the magnetic field around the current in a straight wire. Figure 12-5*b* shows the magnetic field around an electric current in a loop. The more current flowing in the wire, the stronger the magnetic field it produces. In fact, the field at any point in space is proportional to the current that produces it. The magnetic field lines do not end anywhere. If there were such a thing as a free magnetic pole, similar to a free electric charge, then the magnetic field lines could end on the poles. So far, however, no free magnetic poles have ever been discovered.[1]

The current loop in Figure 12-5*b* has a magnetic field very much like the magnet in Figure 12-3 (see also Fig. 12-6). One face of the loop is a north pole, and the other face is a south pole. If we make the current in the loop flow in the other direction, then the poles will be reversed.

Field lines are imaginary lines drawn in space to indicate the direction and magnitude of the field. The direction of the field, as we

[1]In gravitational, electric, and magnetic fields, there can be points in space at which the field has a magnitude of zero. At such points, field lines can appear either to cross or to come to an end without ending on a mass or a charge. These points are possible exceptions to the general rules that field lines do not cross, that gravitational field lines end only on masses, that electric field lines begin only on positive charges and end only on negative charges, and that magnetic field lines do not begin or end anywhere.

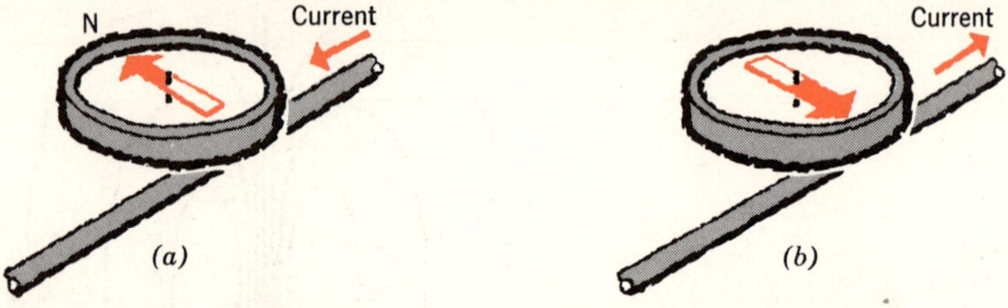

FIGURE 12-4 **An electric current has a magnetic field.**

have seen, is given by the direction of the field lines. Figure 12-7*a* and *b* show how the magnitude of the field is related to the spacing of the field lines. The small square is on a surface perpendicular to the field lines. In a strong field (Fig. 12-7*a*) many field lines go through the square. In a weak field fewer field lines go through it. The strength of the field is proportional to the number of field lines going through a given small area (such as the square) on a surface at right angles to the field lines. (We imagine this area to be small enough so that the strength of the field does not vary much from one part of the area to another.)

Figure 12-7*c* and *d* show what happens to the magnetic field lines when the current through a loop is increased. If the current is doubled, then the number of field lines passing through the coil is also doubled. In fact, the strength of the field caused by the loop is doubled everywhere in space around the wire of the loop.

Scientists believe that the magnetic properties of magnets and

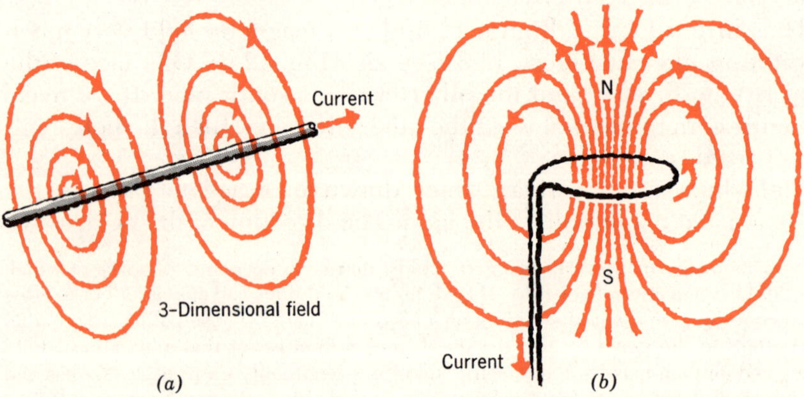

FIGURE 12-5 **Diagram of the magnetic fields due to current in a straight wire and a wire loop.**

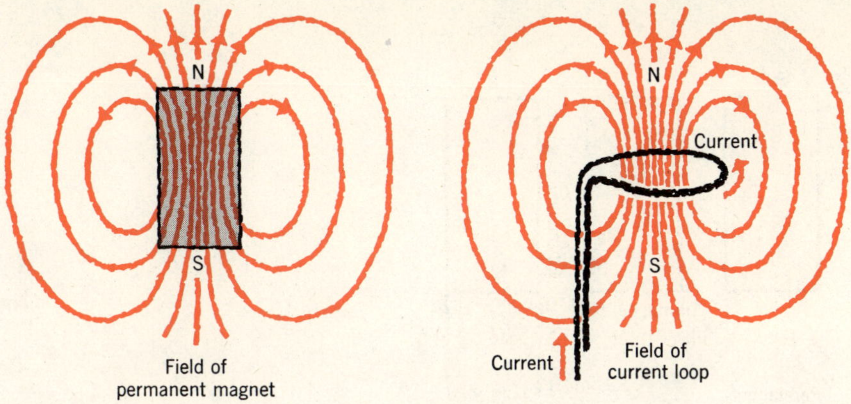

FIGURE 12-6 **Electric currents have the same kind of magnetic field as permanent magnets.**

other types of matter are actually caused by electric currents moving within the atoms. Atoms contain charged electrons moving around in circles or spinning, thus producing tiny current loops which, either individually or in groups, act as small magnets and produce magnetic fields. In some elements, such as iron, the magnetic fields of these tiny current loops added together cause the entire atom to act like a small magnet. In unmagnetized iron, these small magnets are disordered and act in different directions. Therefore, their magnetic forces cancel each other out. In magnetized iron, these small magnets are more nearly lined up, so that they cooperate to produce a large magnetic field.

The small magnets in a bar of iron can be lined up by the influence

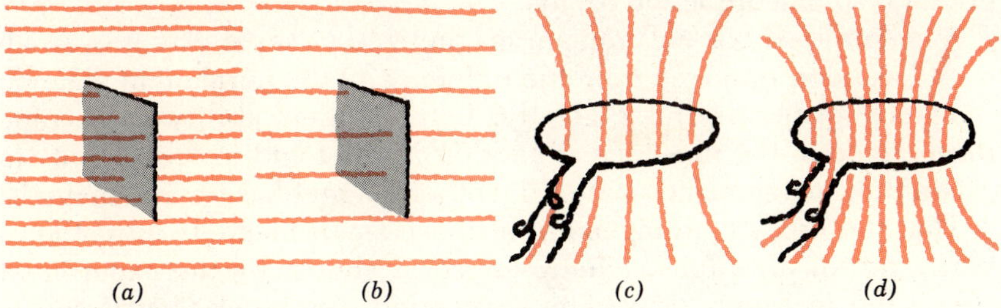

FIGURE 12-7 **A stronger magnetic field is indicated by drawing the field lines closer together. The greater current through the loop in (*d*) causes the stronger magnetic field.**

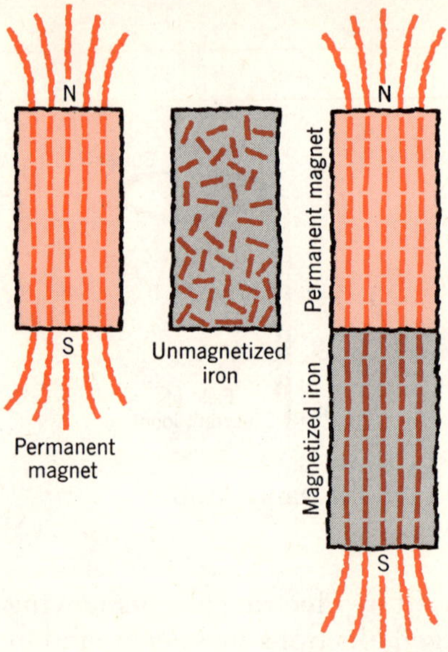

FIGURE 12-8 **Unmagnetized iron can be magnetized by placing it in a magnetic field.**

of the magnetic field from a permanent magnet, just as the compass needles are lined up by the magnetic fields in Figure 12-3. This is known as the process of magnetization. In the influence of this field, the bar itself acts as a magnet and is therefore attracted by the permanent magnet (see Fig. 12-8). This is the process by which magnets are able to attract other objects made of magnetic materials. The object itself becomes an induced magnet, with its north and south poles oriented so that attraction occurs.

A magnetized bar will remain magnetized to some extent even after it has been removed from the influence of the permanent magnet. Magnetic materials differ greatly, both in their ability to become magnetized in the presence of another magnet and in their ability to retain their magnetization after the other magnet has been removed.

That the orderly arrangement of these small magnets in matter is partly responsible for magnetic effects is shown by the experiment illustrated in Figure 12-9. The suspended iron block is attracted to the permanent magnet. When the block is heated over a burner, it eventually loses its attraction to the magnet and swings away. As it cools, it is once more attracted. The heat flow into the block of metal increases its internal energy and, therefore, the random motion of

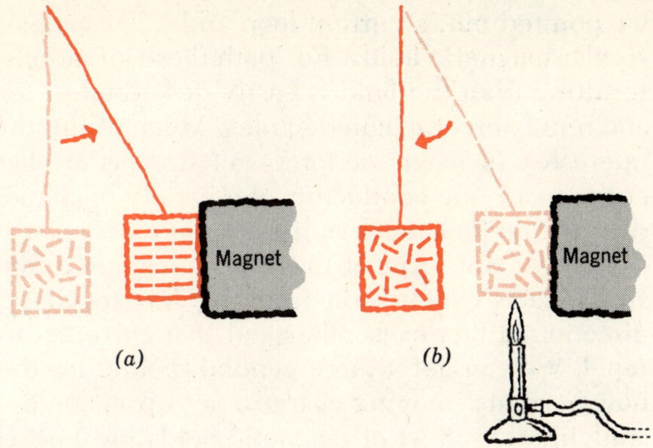

FIGURE 12-9 A magnetized piece of iron can be demagnetized by heating it to a sufficient temperature.

molecules and electrons inside the block. This vigorous random motion destroys the alignment of the small magnets in the block that has been set up by the magnetic field of the permanent magent. Without this order the block cannot act as a magnet and be attracted to the permanent magnet.

An electromagnet can be made from current-carrying loops of wire and an iron core. When current flows through the wire, the current loops act as a magnet, setting up a magnetic field. This field aligns the small magnets in the iron core, which then greatly strengthen the total field. When the current is turned off, these small magnets tend to become disordered again. In electromagnets, the core is made of a material that is easily magnetized by a magnetic field but retains little of this magnetization when the field is turned off.

We have illustrated in this section the principle that *all* magnetic fields arise from moving charges—whether they are moving in the ordinary electric currents or the hidden electric currents inside matter. The magnetic force actually arises from an interaction between moving charges. Wherever there are moving charges there are magnetic effects.

The Magnetic Force on Moving Charges

All magnetic effects are electrical in origin. A magnetic field is not the same as the electric field produced by a stationary charge but is an *additional* field produced whenever the charge is in motion. As we pointed out, a current loop and a bar magnet are surrounded by similar magnetic fields. For both these situations, the magnetic force resulting from the field is easily described in terms of the attraction and repulsion of magnetic poles. Much of the time, however, we are interested in magnetic forces on streams of charged particles or on current-carrying conductors of a variety of shapes. The pole concept does not help us much in these situations.

The general form of the law of magnetic force is complex. Two simple facts which come from this law are that currents in the same direction attract each other and that currents in opposite directions repel. We can get a more general feeling for these forces by seeing how currents (moving charges) set up magnetic fields, and then by studying the effect of magnetic fields on the behavior of other currents (moving charges). Accordingly, we describe the basic character of the magnetic force as follows.

With one exception, whenever a charge is moving in an external magnetic field it will experience a force at right angles to both the direction of its motion and the direction of the external field. That is, a moving charge (or a current in which many charges are moving through a conductor) has a magnetic field around it. When it passes through the magnetic field of some other system, it experiences a deflecting force at right angles to its velocity. The exception occurs when the charge is moving parallel to the direction of the magnetic field. In this one case, there is no deflecting force.

We can illustrate this basic description of the magnetic force in several ways. For example, suppose that we have a beam of electrons moving in a straight line (see Fig. 12-10). The electrons pass through a tube and strike a fluorescent screen, such as is used in an ordinary television set. The electrons produce a luminous spot on this screen

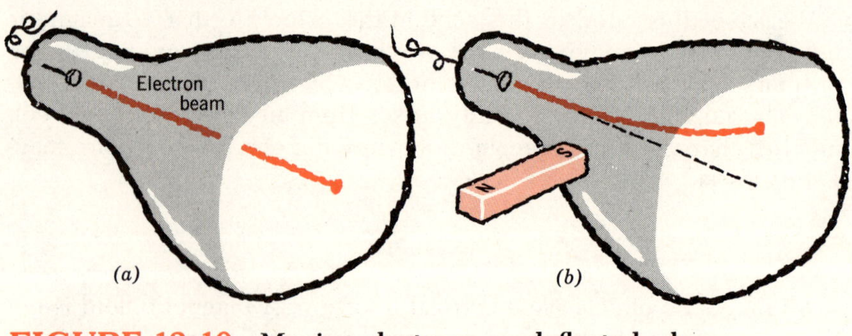

FIGURE 12-10 **Moving electrons are deflected when moving in a magnetic field.**

when they strike it. When a magnet is brought near, the electrons are deflected, and they strike a different spot on the screen.

The motion of an electron, or other charged particle, in an unchanging magnetic field is actually rather curious. (We assume that there are no electric or gravitational fields present.) The particle curves around the field lines, changing its direction but not its speed (see Fig. 12-11*a*). The magnetic field alone does not increase the speed of the moving charged particle because it acts at right angles to its velocity. A simple kind of motion occurs when a particle is moving at right angles to a constant uniform magnetic field. Such a field is not changing with time, and it is the same in magnitude and direction over the region of space in which we are interested. In this case the particle moves in a circle (see Fig. 12-11*b*). The magnetic force supplies the centripetal force necessary to maintain this circular motion.

Applications of the Magnetic Forces

The Electric Motor. The magnetic force can be used to make a current-carrying wire move or accelerate. Figure 12-12 shows a simple conducting wire wrapped around a frame and connected to a battery. (As far as we are concerned here, the battery is simply a device which keeps a current flowing through the wire.) The wire runs between the poles of a magnet and is therefore in a magnetic field. When the switch is closed so that a current flows through the wire, the wire moves, either up or down depending on the direction of the current. The electrons moving through the wire are deflected at right angles to their motion. This makes the whole wire move perpendicular to the magnetic field lines and to itself.

(a) (b)

FIGURE 12-11 **Two typical paths of an electric charge moving in a magnetic field.**

FIGURE 12-12 An electric current in a magnetic field experiences a force at right angles to both the direction of the current and the direction of the magnetic field.

The force illustrated in the above experiment can be used to operate an electric motor. Figure 12-13 shows one kind of electric motor, in which electrons flow through a loop or coil of wire in a magnetic field. The force on the electrons flowing in the coil causes the coil to turn, and thus turns the motor. A drive shaft attached to the coil can be used to operate an egg beater, a grinding wheel, a vacuum cleaner, and other devices.

If the current were to flow in the opposite direction through the loop, the motor would turn the other way, since the force would be in the other direction. To prevent this we set up the two curved plates on the ends of the wire of the loop as shown. This arrangement insures that the incoming current in the wire always enters

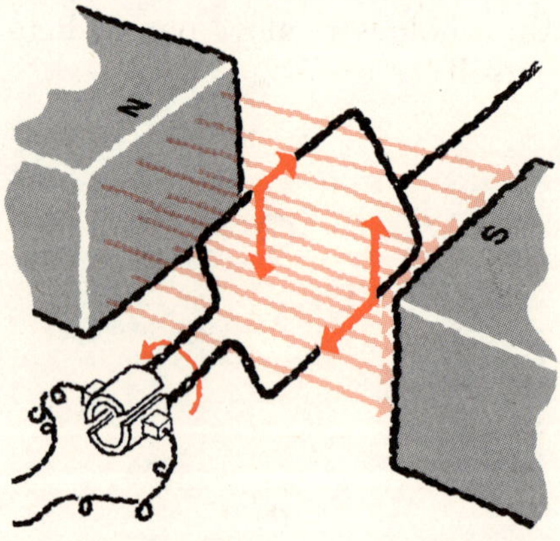

FIGURE 12-13 Diagram illustrating the operation of an electric motor.

whichever side of the loop happens to be close to the north pole of the magnet as the loop rotates. Thus the motor will keep running in the same direction. Of course, practical motors are more elaborately designed so that they run more smoothly and efficiently.

The Cyclotron. The American physicist Ernest O. Lawrence (1902–1958) invented a device which uses a combination of electric and magnetic fields to accelerate charged particles, such as protons. This device, called a cyclotron, is illustrated in Figure 12-14. Two hollow D-shaped conductors are oppositely charged, producing an electric field between them. There is also a magnetic field going through the two D's, perpendicular to them and to the electric field. The proton starts out near the center of the space between the two conductors and is accelerated by the electric field, from one D to the other. Inside the conducting D's the electric field is weak. Because of the magnetic field the proton turns in a semi-circle and comes back out into the space between the two D's. When it emerges into this space it finds that the electric field has been reversed while it was inside the D. Therefore it is accelerated again, enters the other D, and turns around again—but this time it has more speed and turns in a larger circle. This procedure continues, with the electric field increasing the speed of the particle again and again while the magnetic field turns the particle around and keeps it from getting out of the cyclotron. When the particle reaches the outer edge of the D's it is deflected out with its acquired high energy (Fig. 12-14*b*). A wide

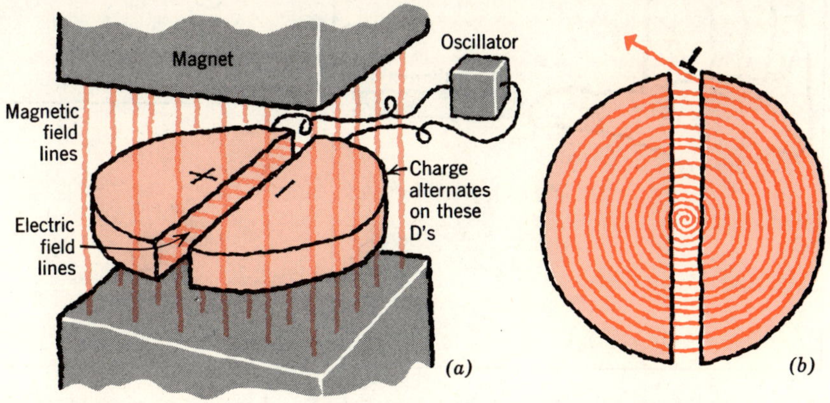

FIGURE 12-14 **Diagram illustrating the operation of a cyclotron.**

 variety of similar accelerators has been designed to obtain high-energy particles for use in industry and research. These devices use magnetic fields to keep the particles moving in a circle or close to it so that electric fields may be used repeatedly to increase the particles' energy.

Directions and Magnitudes of Magnetic Forces and Fields

Figure 12-15*a* illustrates a "right-hand rule" for determining the direction of the magnetic field around an electric current or a moving charge. If the thumb of the right hand points in the direction of the current, then the bent fingers of the right hand point in the direction of the magnetic field. The direction of the current is the direction in which (positive) charge is moved. If the current is caused by the motion of negative charges, such as electrons, then the direction of the current is opposite to the direction of motion of the negative charges (see Fig. 12-15*b* and *c*).

The direction of the field in a current loop can also be found by this rule. However, we can use a different, more convenient, rule for this special case (see Fig. 12-16.) The direction of the current in the

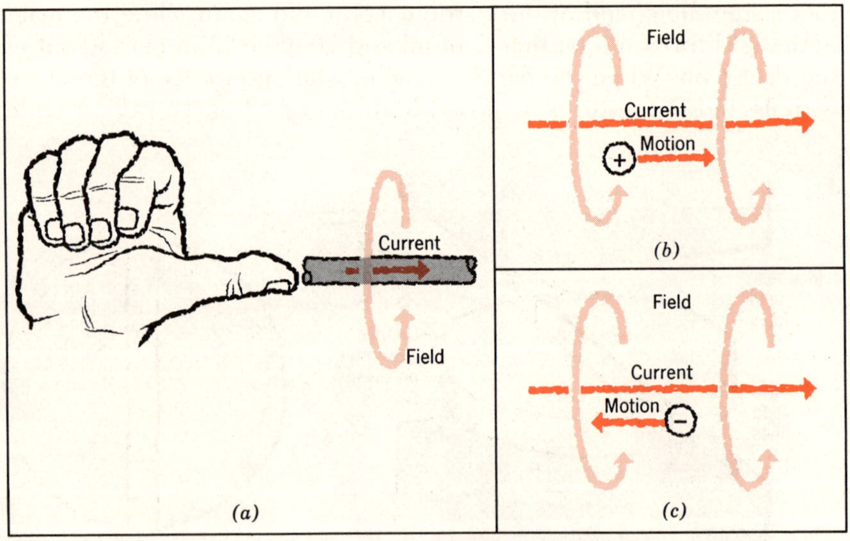

FIGURE 12-15 **The right hand rule for determining the direction of the magnetic field lines around an electric current or a moving charge.**

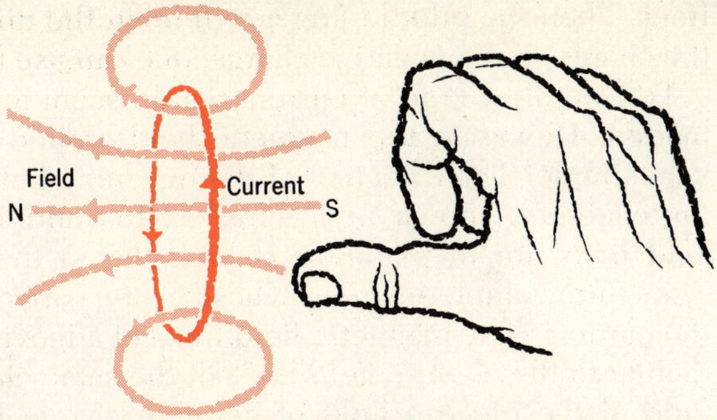

FIGURE 12-16 **The right hand rule for determining the direction of the magnetic field through a current carrying loop.**

loop is indicated by the fingers of the right hand, and the direction of the magnetic field lines through the middle of the loop is in the direction of the thumb. The north magnetic pole of this current loop is in the direction of the thumb.

The next rule tells us the direction of the force on a moving charge in a magnetic field. We shall consider a moving positive charge; the force on a moving negative charge will be in the opposite direction. Figure 12-17 shows how the directions are related. We point the thumb of the right hand in the direction of motion of the positive charge. We point the first finger in the direction of the magnetic field

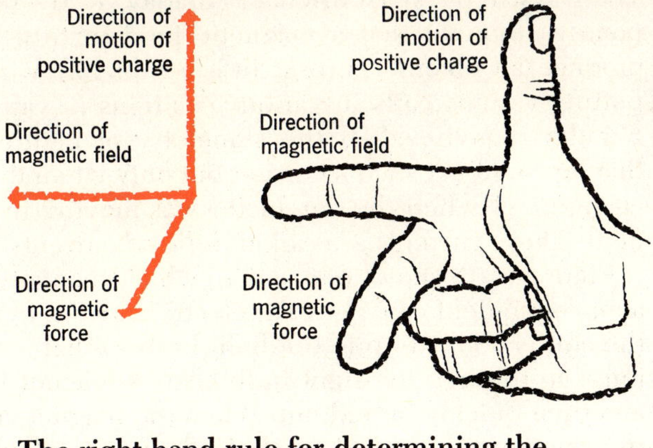

FIGURE 12-17 **The right hand rule for determining the direction of the force on a moving charge in a magnetic field.**

lines. Then the middle finger will be in the direction of the force on the charge. For a negative charge, one can use the left hand.

The simplest kind of current would seem to be a single charge in motion. However, the magnetic field around such a charge is not very simple. The direction of the magnetic field lines around a moving charge is shown in Figures 12-15*a* and 12-15*b*. The magnetic field lines are in planes perpendicular to the direction of motion. They form complete circles whose centers are on the line of motion of the particle. The magnetic field lines of a moving charge are perpendicular to the electric field lines of the same charge. Like the magnitude of the electric field of a single charge the magnitude of the magnetic field of a moving charge falls off with distance in such a way that the field is inversely proportional to the square of the distance from the charge. Unlike the electric field of a stationary charge, however, the magnetic field at a given point of space caused by a moving charge depends on both the distance and direction of the point from the charge. The magnetic field is essentially zero in the line of motion directly behind or in front of the moving charge.

Induced Electric Currents

The Law of Induced Currents

We have seen that an electric current produces a magnetic field. The English physicist Michael Faraday (1791–1867) and his contemporaries thought that a magnetic field in turn ought to be able to produce an electric current in a conductor. Various scientists tried putting various coils in various positions in various magnetic fields. Faraday, convinced of the symmetry of nature, finally discovered that he could get such a current but only when the magnetic field was changing or when the conductor was moving in certain ways in the field. These currents are called *induced* currents.

Figure 12-18 shows one way in which the motion of a conductor in a magnetic field can cause an electric current to flow. The wire from the coil is hooked up to the light bulb so that any current in the coil must go through the light bulb also. A magnet is placed in the coil and then quickly pulled out. When the magnet and the coil are moving relative to each other, the light glows, showing that current is flowing in the coil. When the relative motion stops, the light goes out. The faster the relative motion, the brighter the light, showing that the quicker the motion, the larger the current induced in the

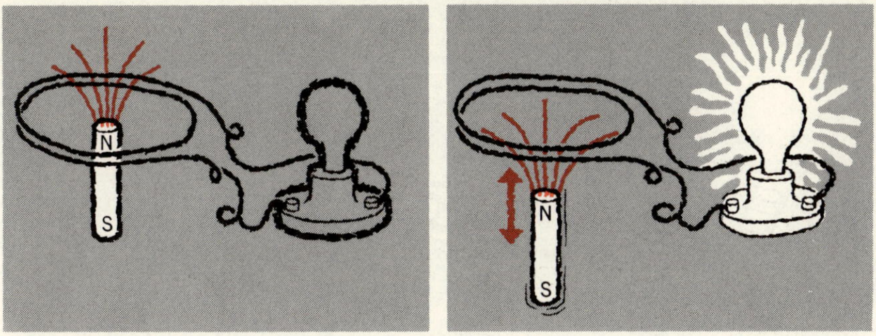

(a) (b)

FIGURE 12-18 Changing the magnetic field through the coil induces an electric current.

coil. Whenever the total number of magnetic field lines through a coil is changing, a current is induced in the coil. The rate at which this change takes place is proportional to the current induced.

The Electric Generator

Now that we have discussed induced currents, we can explain how an electric generator works. Figure 12-19 is a diagram of such a device. As the water wheel turns, the coil moves in the magnetic field. This changes the total number of magnetic field lines through the coil. As long as the coil turns, the field through the coil is changing. This change generates an induced current in the coil, which flows out through the terminals and causes the bulb to light. This is a simple device for obtaining an electric current from mechanical motion. It is very similar to the electric motor, in which electric current generates mechanical motion. These are opposite processes, but the devices which accomplish them are very similar.

We know that electrical machinery is capable of doing work. Therefore there is energy associated with electric currents. The motor and the generator are devices that can be used for transforming electric energy into some other forms or obtaining it from other forms. The energy is transformed by means of work, either done on the generator or done by the motor.

The Direction of Induced Currents

An induced current has a magnetic field of its own. This field opposes whatever change induced the current in the first place. If it

FIGURE 12-19 **Diagram illustrating the operation of an electric generator.**

was the motion of a conductor that caused the change, the induced current will be in such a direction that the force on it opposes the motion. This enables us to state a general law that determines the direction of induced currents.

The direction of the induced current is such that this current opposes the change that caused it.

The Electromagnetic Interaction

Although we have discussed electric and magnetic forces separately in this chapter, we have seen that these two kinds of force are very intimately associated with each other. In fact, together they account for that interaction between two charged objects which is due to their charge. Therefore, instead of considering the forces separately, we should properly speak of the *electromagnetic interaction*, which includes them both. If the two charged objects are at rest, then the magnetic forces are zero, and only the electric forces need be considered. If the charged objects are in motion, however, then both electric and magnetic forces should be considered.

Since the electric and magnetic forces are intimately related, their fields are also intimately related, and we should properly speak of the *electromagnetic field.* A charge at rest gives rise to an electromagnetic field comprised of an electric field alone. A charge in motion gives rise to an electromagnetic field which includes both an electric and a magnetic field. According to the special principle of relativity, a charge can be considered either at rest or in motion, depending on the reference frame, despite the fact that the fields are different in the two cases. The two parts of the electromagnetic field are very intimately associated indeed.

Thus far we have introduced two interactions, the gravitational interaction and the electromagnetic interaction, and their fields. These two interactions account for nearly all the forces we are ordinarily aware of in daily life. The gravitational interaction keeps us on the earth. Nearly all other forces we are aware of arise (directly or indirectly) from the electromagnetic interaction. We know, of course, that much of our machinery and instrumentation is electrical. However, even the ordinary pulls and pushes of one object against another involve electrical forces. The atoms are held together by the electric attraction between the positive protons and the negative electrons. An understanding of the behavior of these forces on the atomic level, however, is impossible with only the laws discussed in this chapter. The principles of quantum physics, which are discussed in Chapter 16, will help us to understand better how the electric forces hold the atom together.

FUNDAMENTAL IDEAS

I. FUNDAMENTAL LAWS AND PRINCIPLES

A. MAGNETIC FORCES EXIST BETWEEN CHARGES IN MOTION. We do not give the exact law but rather discuss it in terms of the origin and action of magnetic fields. The electric force described in the previous chapter and the magnetic force comprise the electromagnetic interaction, the second fundamental interaction we have discussed.

1. The magnetic field produced by moving charges.
   a. Straight current.
   b. Current loop.
2. The magnetic force on moving charges.

B. THERE ARE NO FREE MAGNETIC POLES

C. THE LAW OF INDUCED ELECTRIC CURRENTS

II. APPLICATIONS

A. APPLICATIONS OF MAGNETIC FORCES AND FIELDS

1. Model. Use of the magnetic forces between currents to explain the behavior of permanent magnets.
2. The Electric Motor.
3. The Cyclotron.

B. APPLICATIONS OF INDUCED CURRENTS

1. The Electric Generator.

QUESTIONS

Magnetic Fields

1. What is the difference in origin between electric and magnetic fields?
2. Is there any detectable difference between the magnetic fields produced by a permanent magnet and by moving charges?
3. Is there any net charge on a permanent magnet?
4. Can you explain the existence of a permanent magnet in terms of moving charges?
5. Why might magnetism in an iron bar be reduced by heating the bar?
6. Is the north magnetic pole of the earth near the North Pole?
7. Why is a core of magnetic metal used in electromagnets?
8. We have three identical iron rods. Two of them are magnetized, with poles at the ends. How would you find out which one was not magnetized without using any other objects?
9. Answer Question 8 for the case in which only one of the rods is magnetized.

The Magnetic Force on Moving Charges

10. A magnet can be used to distort a television picture. What can you deduce from this fact?
11. What is the difference between the electric force and the magnetic force?
12. Design an electric motor.
13. How does a cyclotron work?
14. If the aurora were caused by charged particles from the sun, why would they be found near the North and South Poles?

15. What is the difference between the effects of an electric field and a magnetic field?

16. Using the information about the forces on moving charges in magnetic fields, show that currents in the same direction attract each other and that currents in opposite directions repel each other.

17. Show that the forces between currents mentioned in Question 16 account for the attractions and repulsions of magnetic poles.

Induced Electric Currents

18. Design a device that obtains electricity from the turning of a crank.

19. What is the difference between an electric motor and an electric generator?

20. Figure 12-20 shows a curious device known as the "flying doughnut machine." The "doughnut" is made of a good electrical conductor. When we start the current in the coil of the electromagnet, the doughnut flies up and off. Why? If the doughnut is cut as shown it does not fly off. Explain.

General

21. State what happens in each of the following demonstrations and then explain it in terms of fundamental principles.
    (a) A coil hooked to a light globe is pulled through a magnetic field.
    (b) An electric egg beater is plugged in.
    (c) A magnet is brought near a beam of electrons.
    (d) A proton is released in a cyclotron.

22. A negative charge is propelled toward a current-carrying wire on the opposite side of which is a positive charge. What kinds of field are present, and what sort of motion is expected of the negative charge?

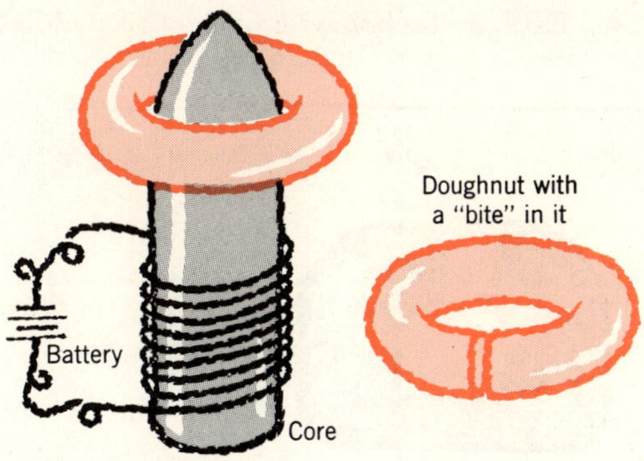

FIGURE 12-20 See question 20.

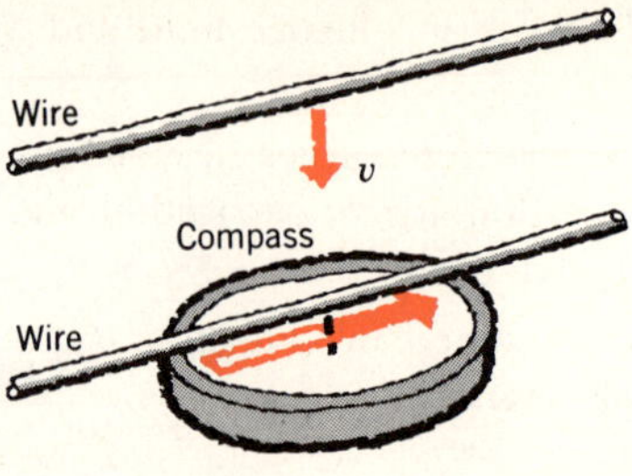

FIGURE 12-21 See question 23.

23. A current is passed through the lower wire in Figure 12-21. What effects would you expect to observe in terms of electric and magnetic fields and currents in wires and space, and what would happen to the compass needle?
24. How can objects be set in motion by using electrical and magnetic forces?
25. How may electricity be obtained?
26. You are stranded on a deserted island that has forests, streams, waterfalls, outcroppings of magnetite (a magnetic iron ore), native copper, etc. How would you go about building a modern technical industrial civilization?
27. List and explain all the interconnections you can think of to justify the grouping together of electric and magnetic forces into the electromagnetic interaction and of electric and magnetic fields into the electromagnetic field.
28. In Figure 12-22 I allow the paddle to swing down between the poles of the magnet. If I use a solid paddle, it stops between the poles. If I use a paddle with slots in it, it swings right through. Both paddles are made of a nonmagnetic material that is, however, a good electrical conductor. Explain the behavior of the two paddles.

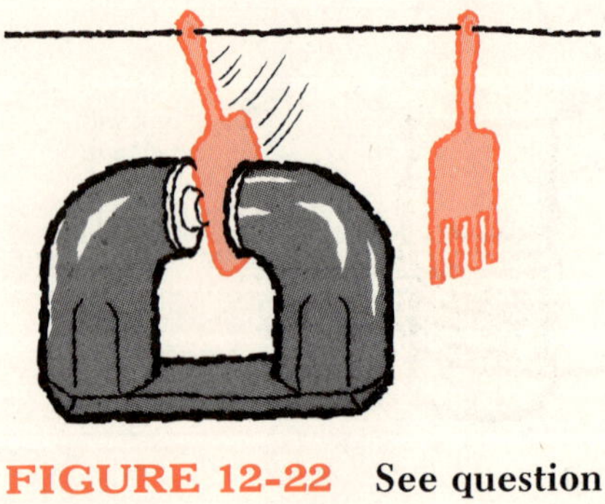

FIGURE 12-22 See question 28.

FIGURE 12-23 See question 30.

29. Explain why field lines can never cross.

30. The conducting rod shown in Figure 12-23 is pushed between the poles of a magnet. What forces act on the electrons in the rod?

31. Two positive charges are moving in straight lines parallel to each other. What kinds of force are acting on them? Show that no matter how fast they are moving, the forces between them which attract them to each other can never be strong enough to overcome the electric repulsion between them.

32. Electric and magnetic fields obey a principle called the principle of superposition. According to this principle, the field at a given point due to a number of charges is equal to the sum of the fields at that point due to each charge separately. Since the field has a magnitude and a direction at each point, the fields are to be added as vectors. Show that a field of the form shown in Figure 11-5 around two opposite charges (equal in magnitude) separated by a short distance can be obtained by adding together the fields for each charge separately. What would the field look like if the charges were equal?

13 Waves

Nearly everyone is familiar with waves in water and in other materials. If a rock is tossed into still water, the rock creates a disturbance at the point where it hits the water. However, this disturbance does not remain at that point but spreads out in waves. In this chapter we shall discuss the general behavior of waves.

It is convenient to begin by describing the waves in a long spring (see Fig. 13-2). One end of the spring is fastened to a support, and the other is held in the man's hand. When the man moves his hand up and down, he disturbs the spring. The disturbance does not remain in place but travels away from his hand to the other end of the spring and is reflected back. Of course the metal in the various parts of the spring does not move from one end of the spring to the other; it merely participates in the motion caused by the wave as the wave passes by, and then returns to rest. The wave is a disturbance, with some kind of form, which passes along the spring. In this example, the motion of each small part of the spring is mainly in a different direction from the direction of the wave. The wave moves along the spring horizontally, but each small region of the spring moves mainly up and down as the wave goes by. This

FIGURE 13-1 Waves in water. Courtesy of John S. Shelton.

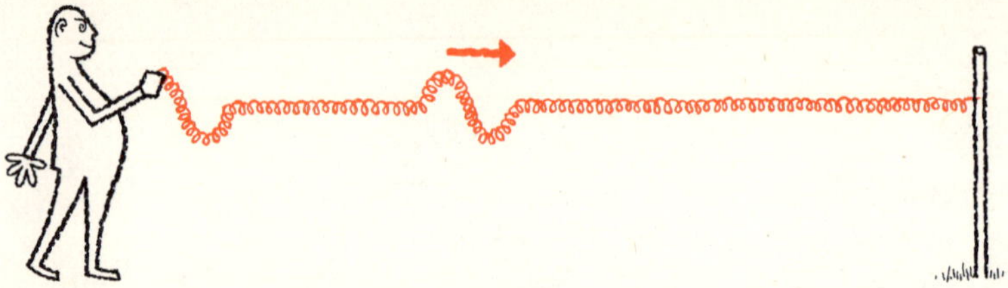

FIGURE 13-2 Waves on a spring.

behavior is similar to the behavior of water (or floating objects) when water waves go by. As a smooth wave goes by in water, floating objects bob around but remain more or less in place.

These examples illustrate the behavior of waves. First we have a medium, such as water or a spring. Second, we have some device that creates a disturbance in that medium, such as a rock striking the water, or a hand moving the spring. This disturbance may act once, sending out a pulse or two, or it may act continually, in a more or less regular manner. The source of the disturbance is usually called an *oscillator*, particularly if it is capable of acting continually. For example, we can refer to the man's hand in Figure 13-3 as an oscillator. If he moves his hand up and down continually, he will send out, not just one or two waves, but a whole series of waves along the spring. We call such a series of waves a *wave train*. The disturbance created by the oscillator does not remain near the oscillator but moves out through the medium in wave form. The particles of matter that make up the medium move in their own neighborhood when the wave passes by, helping to support and propagate the disturbance.

The wave disturbance must carry energy with it, since it causes the matter of the medium to move wherever it goes. The wave receives energy from the oscillator and transmits this energy through the medium. This energy may be given up by the wave elsewhere or slowly dissipated by friction in the medium.

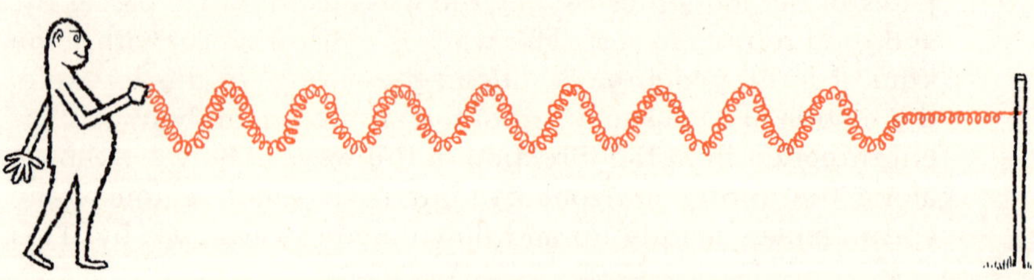

FIGURE 13-3 A wave train on a spring.

Wave Description

Properties of Waves

The form and motion of simple waves can be described in terms of their speed, frequency, wavelength, and amplitude.

Each vibration, or back and forth movement, of the oscillator sends a wave down the spring. The speed at which the waves travel down the spring is the *speed* of the wave.

The series of waves in Figure 13-4 is made up of repeated waves, each of the same form. The length of each of these repeated wave forms is called the *wavelength.* The wavelength is the distance between similar points on successive waves. Within each wavelength of these simple waves there is an upward pulse (or crest) and a downward pulse (or trough). Each complete vibration of the oscillator sends a wave, one wavelength long, down the spring.

The *frequency* of the series of waves is the number of complete waves that pass a fixed point in each second. The oscillator also has a frequency, which is the number of complete vibrations the oscillator makes in each second. If the location of the oscillator does not change, this frequency is the same as the frequency of the wave.

There is a useful relationship that holds among the frequency, speed, and wavelength of a series of waves:

Speed = frequency multiplied by wavelength

For example, suppose that the frequency of the wave in Figure 13-5 is three waves per second and the wavelength is two meters. If the wave train starts at *A*, as in Figure 13-5*a*, then one second later three waves will have gone by, as in Figure 13-5*b*. During this second, the beginning of the wave train has moved three wavelengths, or six me-

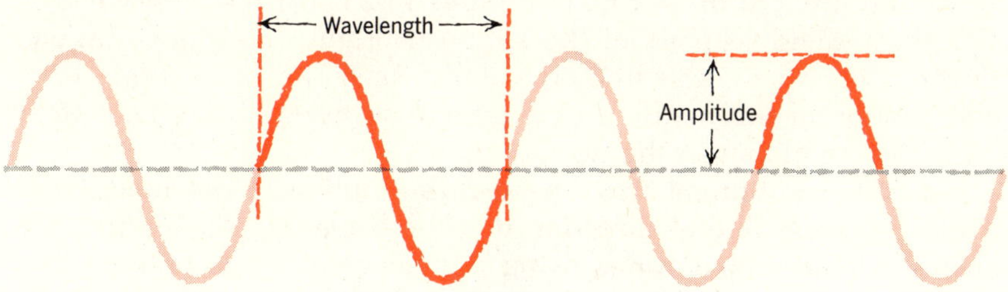

FIGURE 13-4 Diagram illustrating wavelength and amplitude of a wave.

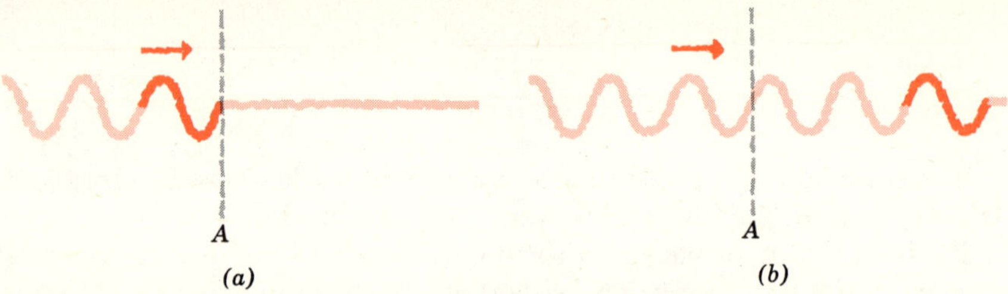

FIGURE 13-5 **In the time interval between (*a*) and (*b*) the wave has moved three wavelengths to the right.**

ters, to the right. The speed of the wave is, therefore, six meters per second.

The simple waves we have discussed in this section are periodic—that is, the wave form repeats itself. It is, of course, possible for the wave form to be very irregular. In such cases the idea of a single frequency and a single wavelength does not apply.

The *amplitude* of a wave is the maximum distance through which the material of the medium is displaced from its rest position when the wave passes by. Figure 13-4 shows the amplitude of a wave on the spring. The maximum distance the spring rises as each part of the wave goes by, the high point of the wave, gives the amplitude. Of course we could just as well obtain the amplitude by taking the maximum distance that each part of the spring drops. The energy in a wave is generally proportional to the square of the amplitude.

Types of Waves

There are two types of waves that we wish to describe: *transverse* and *longitudinal*. We have already described one kind of transverse wave, the wave in the spring in Figures 13-2 and 13-3. We have seen that the individual parts of the spring in Figure 13-2 move up and down as the wave goes by horizontally; such waves are transverse. The motion of the medium as a transverse wave passes is at right angles to the motion of the wave.

When a longitudinal wave passes, the motion of the medium is along the same line of direction that the wave travels. Figure 13-6 shows a longitudinal wave, using the same spring as before. This time we start the wave by pushing and pulling on the spring along its own direction. The push puts a small region of compression into the spring, which continues as a wave down the spring. The pull stretch-

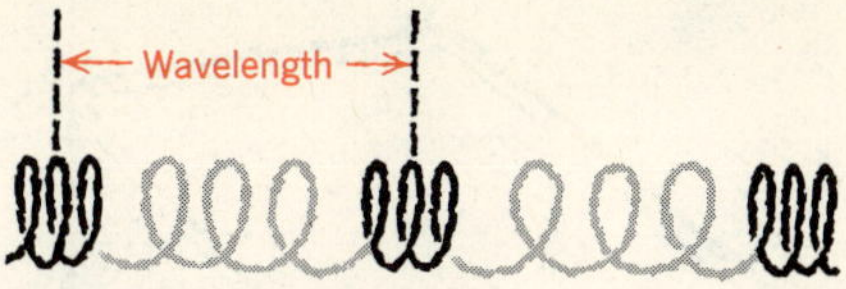

FIGURE 13-6 **A longitudinal wave on a spring.**

es a small region, and a region of "stretch" travels down the spring. In this type of wave, the motion of the metal of the spring is back and forth along the direction of the spring. Thus the wave and the material of the spring move along the same line, and we have a longitudinal wave. For the longitudinal wave the amplitude is still given by the maximum distance each part of the spring moves away from the rest position as the wave goes by. In this case, however, this distance is along the length of the spring.

Some waves, such as waves in water, are partly transverse and partly longitudinal. When a large water wave goes by, the molecules of water move not only up and down but also back and forth along the direction of the wave.

Wave Phenomena

There are certain characteristic kinds of behavior that are common to nearly all waves. Sound and light exhibit this behavior, and scientists have concluded that sound is transmitted as waves in the air and other material, and that light is transmitted as waves in the electromagnetic field. Light and sound will be discussed in more detail later; here we give some examples of the behavior of waves. Interference and diffraction are especially characteristic of waves.

Reflection. When a wave reaches the edge of the transmitting medium, or indeed, any abrupt change in the medium, part or all of the wave is reflected. We have already seen that a wave on a spring is reflected back from the end of the spring. Reflected sound waves are heard as echoes. Reflected light waves are responsible for one's reflection in a mirror.

Refraction. When waves travel from one region into another in which their speed is different, refraction occurs. This involves the slowing down (or speeding up) of the wave and generally a change in its direction of travel. If there is a definite boundary between the two regions, the direction of travel will change at the boundary unless the direction of the wave is perpendicular to the boundary. Very

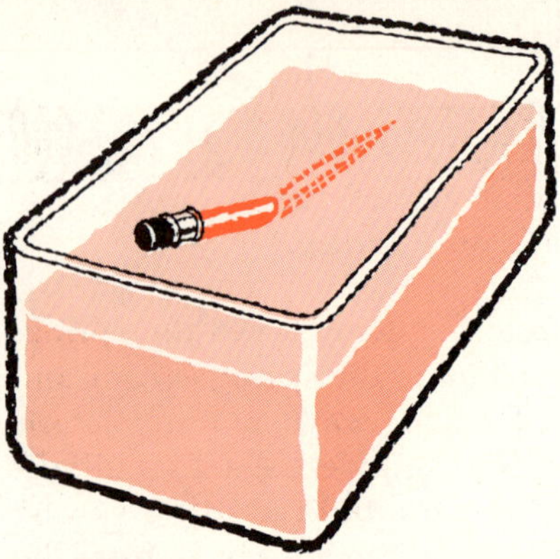

FIGURE 13-7 The pencil appears bent because of the refraction of the light as it passes through the surface of the water.

often, in such a case, there will be a wave reflected from the boundary and also a wave refracted at the boundary entering the new region. One example of the refraction of light waves is shown in Figure 13-7. Light waves from the partly submerged pencil are bent, or refracted, as they leave the water. This causes the pencil itself to appear bent.

Diffraction. A very important property of waves is that they bend around corners. This is obvious in the case of sound, since we can

FIGURE 13-8 The diffraction of sound waves around a corner.

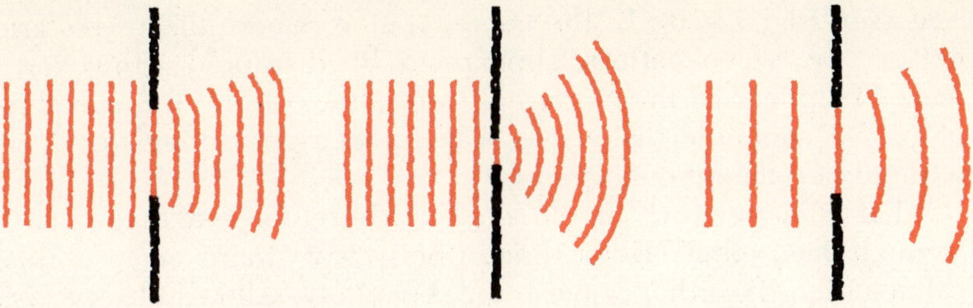

FIGURE 13-9 **Diffraction depends on the size of the opening and the wavelength.**

hear around corners (see Fig. 13-8). This property is illustrated for water waves in Figure 13-9. The waves bend to some extent around the edges of the barrier. Notice that this diffraction (or bending around corners) is less obvious in Figure 13-9 *a*, where the hole in the barrier is much larger than the wavelength of the wave. Where the hole is not much bigger than the wavelength, as in Figures 13-9*b* and 13-9*c*, diffraction effects become very important. We can make diffraction effects more important either by increasing the wavelength, as in Figure 13-9*c*, or by decreasing the size of the aperture through which the waves move, as in Figure 13-9*b*.

Interference. What happens when more than one wave train is moving through the same medium? To investigate this question, we set up in water two wave trains that move through intersecting chan-

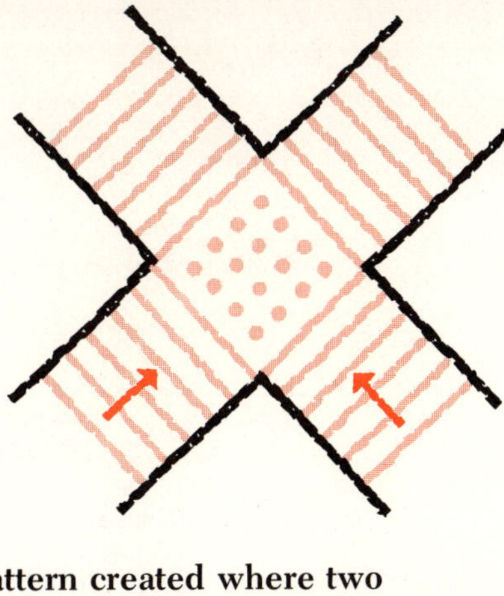

FIGURE 13-10 **Interference pattern created where two wave trains cross each other.**

nels (see Fig. 13-10). In the center region, where the waves are together, the wave pattern appears to be disrupted. However, the waves reappear in their original form after passing through this region. We conclude that waves can travel right through each other without permanent distortion.

What is happening in the region where the waves are passing through each other? Even though both wave trains are present, we cannot identify either wave train separately. All we observe is the net cooperative effect of the combined, or superimposed, waves. We call the wave pattern in this region the *interference pattern* set up by the two wave trains. (It would be just as appropriate to call this a *cooperative* pattern.)

To see what happens when two waves come together to form an interference pattern, let us look at Figure 13-11*a*. This figure shows two pulses moving toward each other on a spring. When the pulses come together, they momentarily add up to create a large pulse; then each pulse reappears in its original form on the other side of this region. We call this pattern *constructive interference*, because the two small pulses build up to (construct) one large one. Similarly, if two downward pulses come together they result in one large downward pulse; this is also called constructive interference.

Figure 13-11*b* shows what happens when an upward pulse and a

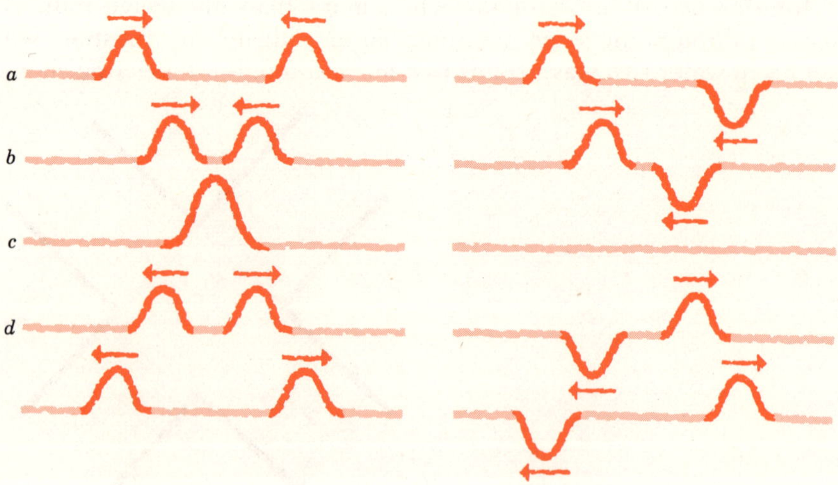

(*a*) Constructive interference

(*b*) Destructive interference

FIGURE 13-11 **Constructive and destructive interference between wave pulses on a spring.**

FIGURE 13-12*a* **Diagram illustrating the two source interference pattern—the interference pattern caused by waves from two oscillators vibrating at the same frequency. (See Figure 13-12*b* also.)**

downward pulse come together. Where the pulses meet, they momentarily cancel each other out. If the pulses were not of the same size, they would be only partly cancelled out. This type of interference is called *destructive interference.*

We can summarize all this by saying that at each point where two waves come together, the resulting disturbance is found simply by adding the disturbance caused by one wave to that caused by the other. In the region where the two wave trains of Figure 13-10 come together, therefore, the interference pattern is caused by a combination of constructive and destructive interferences occurring at different points.

A very important type of interference pattern is produced by two oscillators that are sending out simple waves of the same frequency and the same amplitude (see Fig. 13-12*a* and *b*). In the vicinity of certain lines, as indicated in the figure, there is destructive interference

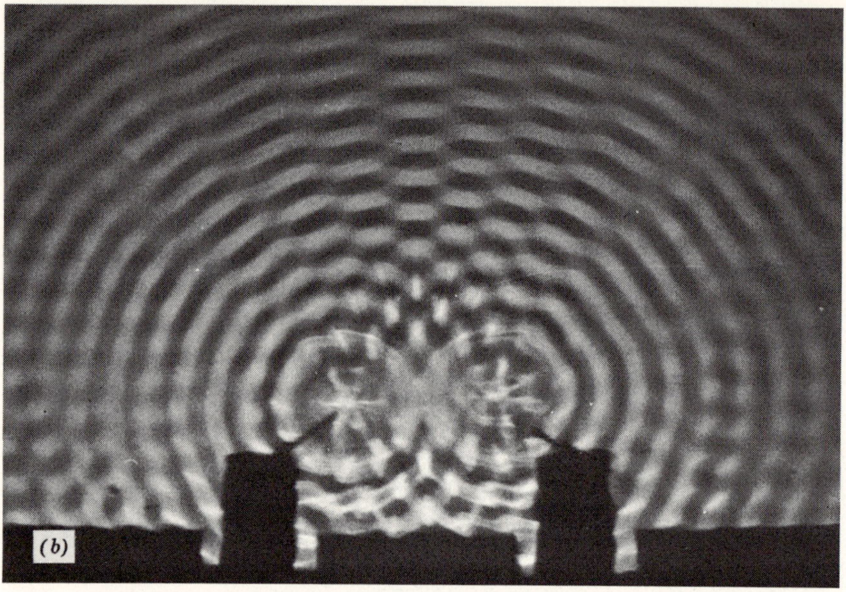

FIGURE 13-12*b* **Two source interference pattern of water waves. Photograph from *PSSC Physics*, D.C. Heath and Company, Boston, 1965. Used with permission.**

at all times. In these positions the waves from the two oscillators always cancel each other out. Between these lines the waves coming from the two oscillators interfere to produce waves moving outward.

Sound Waves

The Nature of Sound

Sound is transmitted by waves through solids, liquids, and gases. In this section we are interested in sound waves moving through the air. These are longitudinal waves that are started in much the same way as the longitudinal waves in the spring illustrated in Figure 13-6. Figure 13-13 shows how a vibrating tuning fork can cause sound waves in air. As the prong of the fork moves, it momentarily compresses the air in front of it, causing a small region of high pressure. This compression causes a small pulse of high pressure to move out-

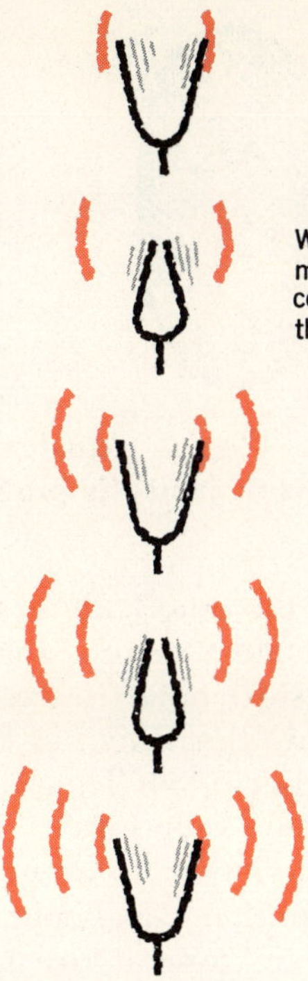

FIGURE 13-13 **Sound waves being generated by a tuning fork.**

ward in a wave. As the prong of the tuning fork moves back, it leaves a region of rarefied air behind it, causing a pulse of low pressure to move outward in a wave. In this way we get a series of pulses of compressed and rarefied air.

Sound waves move through the air at a speed of 331 meters per second when the air temperature is at the freezing point of water. This speed increases with temperature but does not change very much with air density. It is nearly the same for all audible frequencies. (If this were not true, you can imagine how an orchestra would sound at the back of an auditorium: the notes of different frequency would arrive at our ears after different time lags.)

That sound is indeed composed of pressure waves is strongly suggested by a simple experiment using the wheel illustrated in Figure 13-14. As the wheel is rotated, the jet of air blows against it and through the holes as they come by. As each hole passes in front of the jet, the air jet passing through the hole causes a pressure pulse in

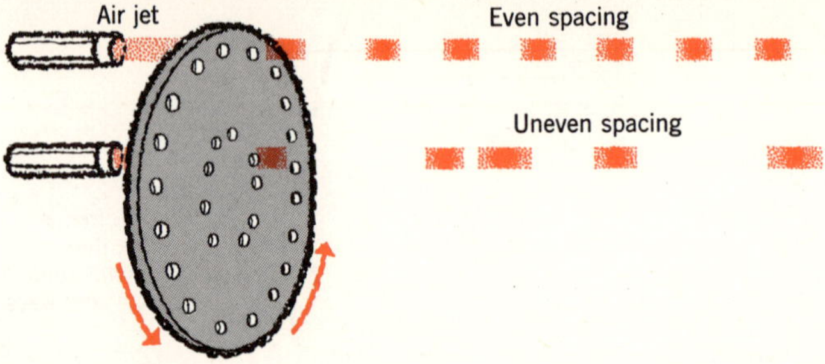

FIGURE 13-14 **Musical notes are produced by evenly spaced pulses. Noise is produced by unevenly spaced pulses.**

the air on the other side of the hole. If the holes are equally spaced we will obtain a wave with a definite frequency, which sounds to us like a musical note. If we increase the frequency of the wave by speeding up the wheel or by spacing the holes closer together, we hear a higher pitch. The pitch of a sound wave depends on the frequency of the sound wave. The higher the frequency, the higher the pitch. If the holes are irregularly spaced on the wheel, we no longer hear a definite musical note, but what we would call noise instead.

Musical instruments are oscillators that are designed to vibrate in a regular way to produce musical notes and not noise. The tuning fork, in particular, is designed to produce a simple wave of definite frequency. Such devices as the reeds of woodwind instruments, the air columns of flutes and organ pipes, and the strings of pianos and violins are all oscillators that vibrate in a somewhat more complex manner.

FIGURE 13-15 **An oscillator.**

Interference of Sound Waves

Beats. When two musical tones of nearly the same pitch are sounded together, *beats* are heard. These beats may be heard as a rhythmic rise and fall in the loudness of the sound. If the two tones are far enough apart in frequency, the beats may occur in such rapid succession that they are not heard individually but as an additional musical tone. Figure 13-16 shows how these beats arise. Actually the waves are longitudinal, but we have drawn them as transverse in order to show the effect more clearly. These waves might arise from two tuning forks of slightly different frequencies that are sounded at the same time. Figure 13-16*a* shows two waves of slightly different frequencies. Their wavelengths are also slightly different, because the speed is the same for both waves. The actual wave that we hear is the sum of these two waves, as shown in Figure 13-16*b*. When the two waves interfere constructively, the tone is louder. When they interfere destructively, the tone is diminished. When the wave of Figure 13-16*b* comes by the ear, the ear hears these alternating loud and soft tones, or beats.

Standing Waves. The principle of wave interference helps us to understand why musical instruments (as oscillators) produce musical tones instead of noise when they are plucked, struck, blown, or scraped with a bow. We shall confine our discussion here to the musical tones produced by a stretched string and a metal plate, but the

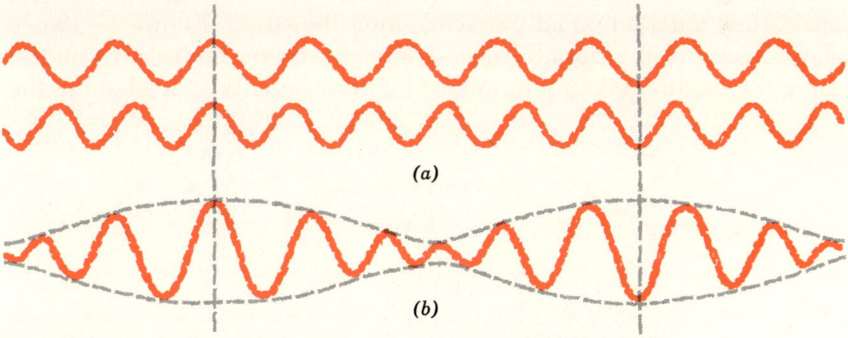

FIGURE 13-16 **Audible beats are produced by the interference of two wave trains of somewhat different frequencies.**

principles are applicable to a wide range of physical phenomena, and not merely to musical instruments or sound waves.

Figure 13-17*a* shows part of a stretched string of infinite length. On this string two trains of transverse waves of equal amplitude are traveling at the same speed in opposite directions with the same frequency and, of course, the same wavelength. The pattern formed when these two trains come together and interfere is illustrated in Figure 13-17*b*. The waves in this pattern appear to vibrate in place. For this reason they are called *standing waves*. At certain points, called *nodes*, the string does not move up and down at all. Between these nodes, the string vibrates up and down with the same frequency as the original waves. The nodes are seen to be one half wavelength apart.

The strings used in musical instruments are not infinite, so we shall discuss what happens when the stretched string is of some definite length, fixed at the ends, as in Figure 13-18. Since any waves traveling on the string are reflected at the ends, a disturbance will eventually produce waves traveling in both directions. It is possible, therefore, to produce standing waves on such a string. However, as we shall see, standing waves will not be produced for all frequencies on a given stretched string. Rather, a particular stretched string will have standing waves of certain characteristic frequencies.

In order for standing waves to occur on a stretched string, there must be a node at each end of the string, since each end is fixed and cannot move. There may or may not be nodes elsewhere on the string. Since the nodes, as we have seen, must be a half wavelength apart, they must be equally spaced along the string. Figure 13-18*a*–*c* shows some of the possibilities. There can be no nodes, two nodes, three nodes, indeed, any number of nodes, equally spaced along the

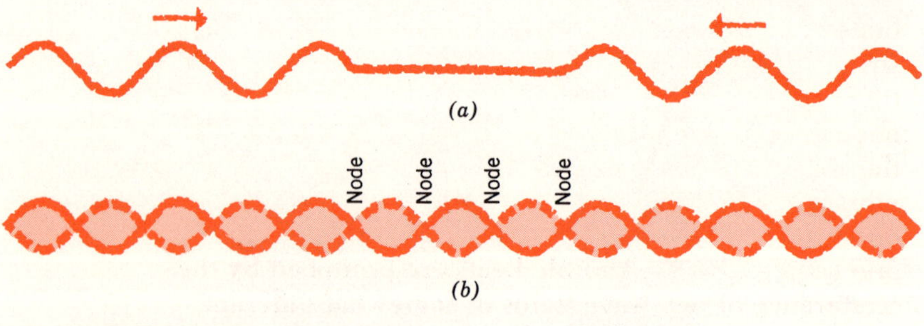

FIGURE 13-17 **Two waves on a stretched string traveling in opposite directions interfere to produce standing waves.**

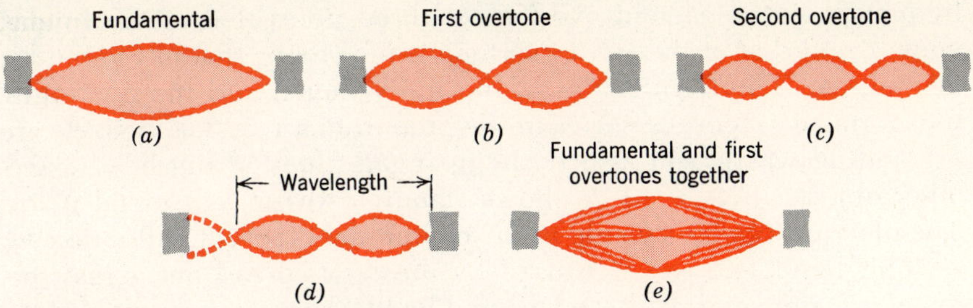

FIGURE 13-18 Standing waves (fundamental and overtones) on a stretched string.

string in addition to the two nodes at the ends of the string. Figure 13-18*d*, on the other hand, shows a wave that *cannot* occur. The wavelength of this wave is too long to form two nodes (in addition to the endpoints of the string) and too short to form just one. With such a wavelength, it is impossible to form nodes *equally* spaced along the string. Therefore, standing waves on a particular stretched string can have only certain wavelengths. Since the wavelength is related to the frequency of the wave, there will be certain permitted frequencies also, corresponding to the permitted wavelengths. (Waves of different wavelength have very nearly the same speed along the string. This speed may be changed by tightening or loosening the string, as in tuning a guitar or a violin.)

These permitted frequencies are the frequencies heard when guitar strings, for example, are plucked. They are called the *natural*, *characteristic*, or resonant frequencies of the stretched string, and they include the *fundamental* frequency and various *overtones*. The *fundamental* frequency is heard when the string vibrates as a whole, as in Figure 13-18*a*. The first *overtone* is heard when the string vibrates as in Figure 13-18*b*. The second overtone is heard when the string vibrates as in Figure 13-18*c*. In general, the number that identifies the overtone (first, second, etc.) corresponds to the number of nodes on the vibrating string (not counting the ends) that produces the overtone. For example, the seventh overtone is produced by a vibrating string with seven nodes, excluding the end nodes, and so on. The string may vibrate in such a way as to produce more than one overtone at once. For example, Figure 13-18*e* shows a string vibrating to produce both the fundamental and the first overtone at the same time.

For the vibrating string, fixed at the ends, it can be shown that the frequency of the first overtone is twice that of the fundamental. The frequency of the second overtone is three times that of the fundamental, and that of the third overtone is four times that of the fundamental, etc. The frequencies of the fundamental and the overtones thus form an ascending series in the proportions 1, 2, 3, 4, 5, 6, etc.

Standing waves can also occur in organ pipes, drum heads, and other objects. Figure 13-19 shows standing waves in a metal plate. The plate is held by a clamp in the center and is set into vibration by a violin bow. The waves caused by the bow spread out across the plate, are reflected at the edges and at the center, and combine to produce standing-wave patterns. Standing waves for two of the natural frequencies of the plate are shown in the figure. On a surface, such as this plate, the nodes are *lines* where no vibration occurs, rather than points as on a string. To help us locate the nodes, salt has been spread over the plate. It is shaken off by the vibration of the plate, except at the nodes.

The simple numerical relations among the natural frequencies of the string make it pleasing to the ear when more than one frequency sounds at once. The natural frequencies of the plate, however, are not so simply related, and it does not sound "harmonic" when more than one overtone is heard at the same time. This same harsh lack of harmony is heard in the clash of cymbals.

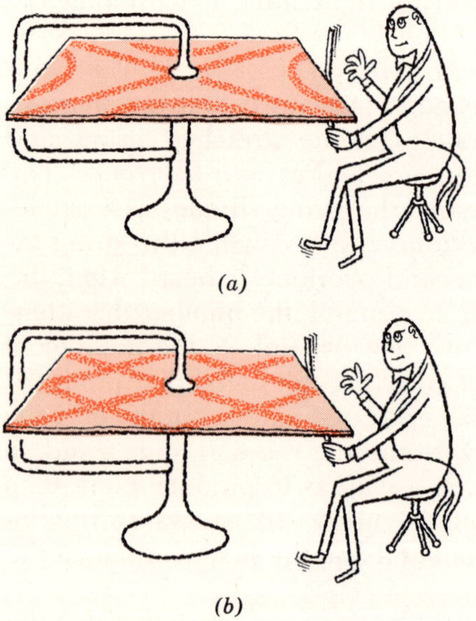

FIGURE 13-19 **Standing waves on a plate.**

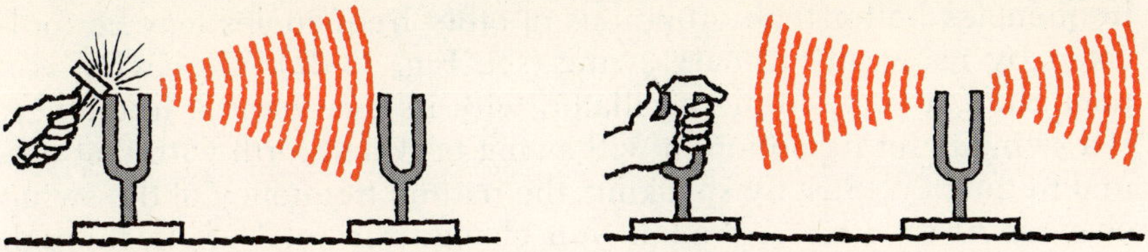

FIGURE 13-20 **A tuning fork most easily picks up energy from a wave whose frequency is equal to its own natural frequency.**

Circumstances often occur in which waves or other disturbances strike an oscillator, setting it in motion. Some of the energy in the waves is transferred to the oscillator, and it begins to vibrate. Energy is more readily absorbed from the wave by the oscillator if the frequency of the wave is the same as, or very close to, one of the natural frequencies of the oscillator. For example, a violin string will begin to vibrate if sound waves reach it that have the same frequency as one of its natural vibrations. Each oscillator, in this way, tends to "pick up" vibrations at certain frequencies. This effect can be observed by using two tuning forks, as shown in Figure 13-20. If one of the tuning forks is set into vibration, it will produce sound waves that will travel to the other tuning fork and start it vibrating. The effect will be hardly noticeable for forks of different pitch, but for those of the same pitch (frequency) it will be marked. In some cases an oscillator picks up so much energy from a wave at its resonant frequency that it is damaged (see Fig. 13-21).

The reason an oscillator can pick up vibrations of its own natural

FIGURE 13-21 **A sound wave of the resonant frequency can shatter the glass.**
Courtesy of Memorex Corporation.

 frequencies better than vibrations of other frequencies may be indicated by using an ordinary swing (see Fig. 13-22). The swing is a pendulum, a very simple oscillator, with its own natural frequency. If a swing is set in motion, it will swing back and forth with that natural frequency. (Strictly speaking, the natural frequency of the swing changes to a certain extent if you change the total angle through which the swing is swinging.)

The man in Figure 13-22*a* gives the swing a push each time it comes back to his position. He pushes the swing just after it starts back on its downward path, pushing it in the direction in which it is moving. Thus, as he pushes it, he puts energy into the swing; this causes the swing to go higher and higher. The important point is that his pushes come at the same frequency as the natural frequency of the swing. Each time the swing makes one trip back and forth it receives one push. Furthermore, the pushes come at just the right time to increase the motion of the swing rather than oppose it.

Suppose, however, that the pushes were at a higher frequency than the frequency of the swing (see Fig. 13-22*b*). Then a push would sometimes increase the speed of the swing and sometimes oppose it. Sometimes the push would put energy into the swing and at other times it would take energy out. Therefore, the pushing would be about as likely to retard the motion as to increase it. The most efficient way to get energy into the swing is to time the pushes

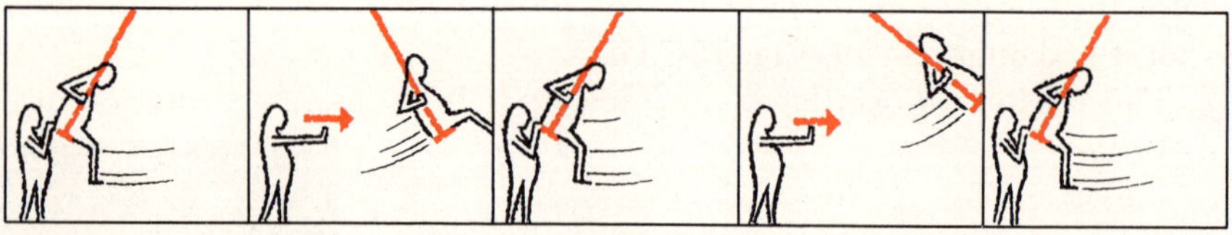

(*a*)

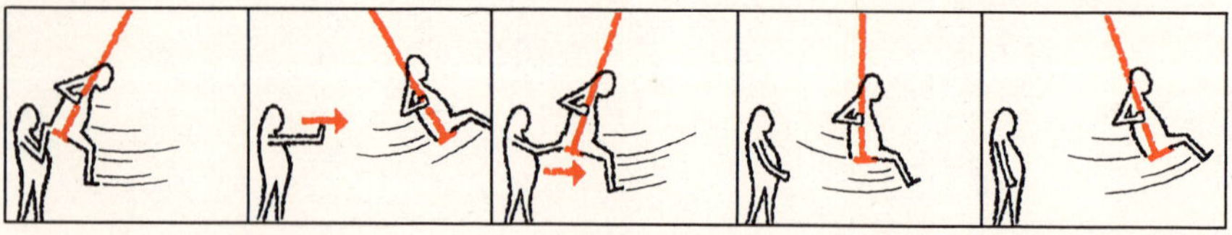

(*b*)

FIGURE 13-22 **An efficient way to get energy into a swing is to give it pushes at the same frequency as its natural frequency.**

so that they occur at the same frequency as the oscillations of the swing.

Energy can also be taken *out* of the swing very effectively by a series of pushes at its natural frequency. If, instead of pushing on the swing just as it is going away, the man pushes on it just as it is approaching him, energy will be removed from the swing each time.

The example of the swing illustrates the fact that an oscillator can most effectively acquire energy from an outside disturbance if the disturbance has the same frequency as one of the natural frequencies of the oscillator. It will not acquire energy so efficiently from disturbances of other frequencies. Therefore, if the disturbance contains several frequencies, the oscillator will respond most vigorously to its own natural frequency, thus appearing to select that frequency from the others. It is by this same principle that the receiving electric circuits of a radio or television set which has been tuned can respond to incoming radio waves of a certain frequency, and thus select a definite channel or station.

The Doppler Effect

The sound of a low-flying jet airplane changes in pitch as it passes overhead. This change cannot be accounted for by the change in the plane's distance from us, nor does it indicate any change in the performance of the engine. Other sources of sound, such as racing cars and train whistles, give the same effect. The pitch of the sound is higher when the source is approaching us than when it is not moving. When it is going away from us, the pitch is lower than when it is not moving. The greater the speed at which the source is moving toward or away from us, the greater this change in pitch. Similar changes of pitch occur when it is the listener and not the source that is in motion. When the source and the listener are approaching each other, the pitch is higher. When they are moving apart, the pitch is lower. This phenomenon is called the Doppler effect, after the Austrian physicist who explained it.

We can explain the changes in pitch (frequency) in the Doppler effect by analogy with the situation shown in Figure 13-23. Instead of a source emitting waves which move through the air, we have a source of pies that move along a belt.[1] The man at the left is putting

[1]We have noticed that Albert Baez has previously used a similar example in *The New College Physics—A Spiral Approach,* San Francisco, W. H. Freeman, 1967, p. 229.

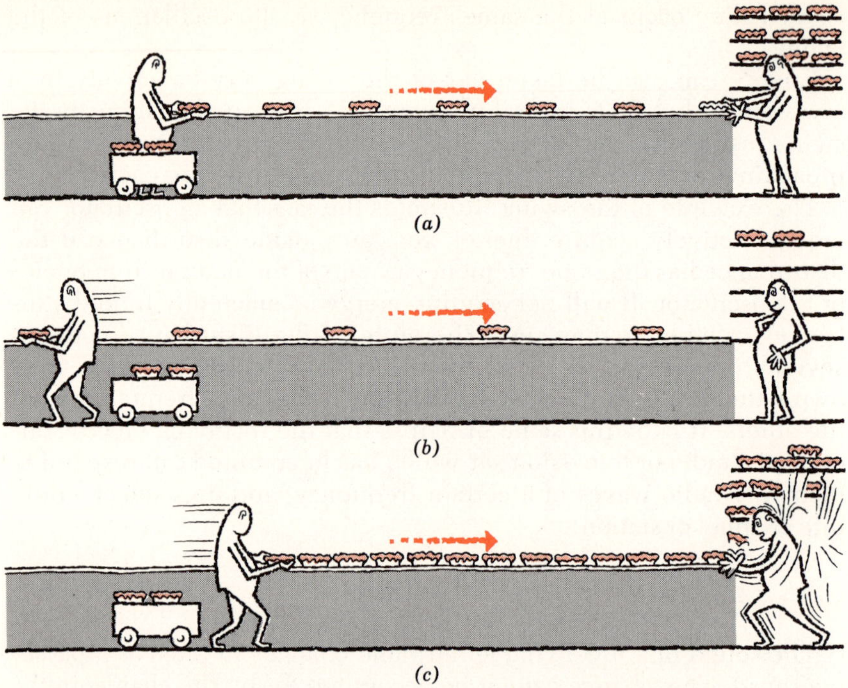

FIGURE 13-23 **Illustration of the Doppler effect. The frequency at which the man at the shelf receives the pies is greater when the other man is approaching him and less when he is walking away.**

pies on the belt at a certain frequency—say, one every five seconds. The man at the right is trying to remove them at the same frequency. If the first man moves to the left along the belt as he puts down the pies (Fig. 13-23*b*), the pies will be more widely spaced along the belt, and the man at the other end will receive them at a lower frequency. (The speed of the pies has not changed, of course.) If the man at the left moves in the opposite direction (Fig. 13-23*c*), still putting down pies at the same frequency, the pies will be closer together on the belt, and the receiver will receive them at a higher frequency. In the Doppler effect, it is waves rather than pies that are being sent out and received, but the principle is the same. The Doppler effect occurs for light waves also, although the theory is somewhat different.

Earlier in this chapter we emphasized the fact that an oscillator and the waves it emits have the same frequency if the oscillator is stationary. Because of the Doppler effect, the frequency of the waves may not be the same as that of the oscillator if the oscillator is in motion. Of course, if the person who observes the waves is moving at the same speed and in the same direction as the oscillator he will receive the waves at the same frequency as the oscillator. If there is relative motion between oscillator and the observer, however, the frequency at which the waves will be received in the reference frame fixed to the observer will be different from the rate at which they are emitted in a reference frame in which the oscillator merely oscillates and has no net or average velocity.

FUNDAMENTAL IDEAS

I. THE NATURE OF WAVES

- **A.** PROPERTIES OF WAVES
  1. Wavelength.
  2. Frequency.
  3. Speed
  4. Amplitude.
- **B.** TYPES OF WAVES
  1. Transverse.
  2. Longitudinal.
- **C.** BEHAVIOR OF WAVES
  1. Reflection.
  2. Refraction.
  3. Diffraction.
  4. Interference
     - a Standing waves.
     - b. Beats.
  5. Doppler effect.

II. APPLICATIONS: The wave model explains behavior of:

- **A.** SOUND
- **B.** WAVES in springs, water, and stretched strings.

QUESTIONS

Oscillators, Properties, and Types of Waves

1. What is an oscillator?
2. What is a wave?
3. Why does the frequency of an oscillator equal the frequency of the waves it emits? Are there any apparent exceptions? Explain.
4. Why, in each wavelength of the simple waves in Figure 13-3, is there one upward pulse and one downward pulse? Why not one upward pulse in one wavelength and one downward pulse in the next?
5. What information must we have in order to specify a wave completely?
6. How do we know that waves carry energy?
7. What is the difference between a transverse wave and a longitudinal wave? Which kind would you expect to find in solids, liquids, and gases? Explain.

Diffraction and Interference Including Standing Waves

8. Two wave forms are moving toward each other on a spring as shown in Figure 13-24. What happens when they meet?
9. When the two wave pulses in Figure 13-11 cancel each other out, what happens to their energy?
10. Since waves are coming from both oscillators, how does it happen that there are lines in Figure 13-12 where there appear to be no waves?
11. Why do we believe that the two pulses in Figure 13-11 actually pass through each other rather than just bounce back off each other?
12. In some auditoriums there are spots in which it is hard to hear the music, although there are spots all around it where it is easy to hear. Explain.
13. Why does a violin string have certain natural tones and not others? How do the natural frequencies relate to each other?
14. Why are the nodes on a vibrating string one half wavelength apart?
15. How many overtones does a stretched string have?

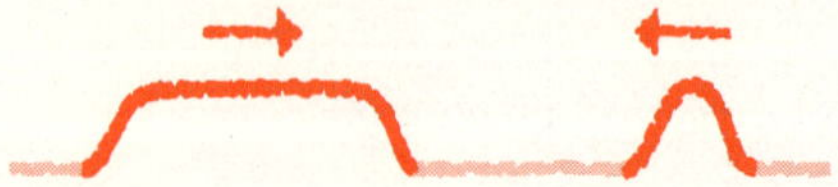

FIGURE 13-24 See question 8.

FIGURE 13-25 See question 19.

Doppler Effect

16. Does the Doppler effect hold for water waves? Explain.

17. Explain why the Doppler effect occurs when the source of waves is stationary and the listener (receiver) is moving.

Sound

18. Why do we think that sound waves exist?

19. Air is evacuated from the bell jar in Figure 13-25 while the bell is ringing. Why can we no longer hear the bell after the air is pumped out?

20. Why can we call a musical instrument an oscillator?

21. Why does the speed of sound in air increase with temperature?

22. Why does the speed of sound in air *not* vary much with pressure, for normal pressure changes?

23. Why does a tuning fork stop vibrating?

24. If I hold the handle of a tuning fork solidly against a table, the sound from the tuning fork becomes louder. Why? How will this affect the length of time the fork keeps vibrating? Explain.

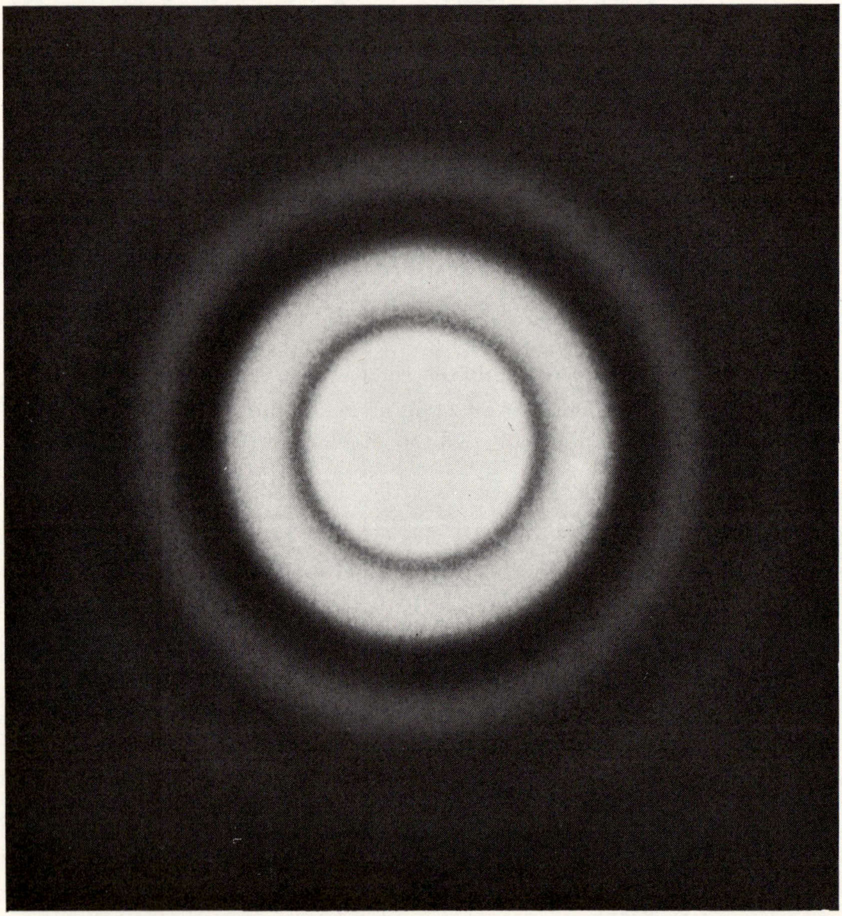

14 Light and Its Wave Behavior

The nature of light has long been a perplexing puzzle. Many scientists, influenced by Newton, pictured light as a beam of tiny particles. The Dutch physicist Christian Huygens, a contemporary of Newton, strongly supported the idea that light was essentially wavelike.

An Italian Jesuit, Francesco Maria Grimaldi (1618–1663), had already performed an important experiment (see Fig. 14-2). He made a tiny hole in a window shutter facing the sun and allowed the light coming through to fall on a second screen that also had a small hole in it. Beyond this second hole he placed a white paper. He found that a region larger than the second hole was illuminated on the white paper. In fact, the smaller the second hole, the larger the illuminated region on the paper. In a second experiment, Grimaldi removed the second screen and placed a small object in the beam of light coming through the hole in the first screen. He found that the shadow this object cast was smaller than the object itself. This showed that light does not travel in straight lines and cast sharp shadows but diffracts, or bends around corners.

☞ **FIGURE 14-1** **Light waves passing through a very small hole spread out into a diffraction pattern like this one. Courtesy of Dr. Brian J. Thompson.**

FIGURE 14-2 **Grimaldi found that light passing through a small hole diffracted to form a spot of light larger than the hole it passed through.**

Although Newton knew of this experiment, he did not believe that it necessarily disproved his own particle theory of light. Huygens, although he believed in the wave nature of light, could not account for diffraction because of errors in his theory, and he did not accept Grimaldi's results. Later, after interference as well as diffraction was observed by Thomas Young (1773–1829), the wave theory of light propagation became well established.

The Speed of Light

A Danish astronomer, Olaf Roemer, obtained a value of the speed of light from astronomical data in about 1676. More direct methods of measurement, similar to that illustrated in Figure 14-3, were used by later experimenters. As indicated in the figure, the light beam passes through one notch of the wheel and returns through another when the wheel is at rest. When the wheel is turning, the light that has been sent through the first notch may be blocked when it returns because the second notch has moved. By knowing how fast the wheel must rotate in order to block or admit the returning light, we can find the total time it takes for the light pulse to travel back and forth. From this and the total distance we can find the speed of light.

The speed of light in a vacuum is one of the fundamental constants of nature. It is usually represented by the letter c and has a value of

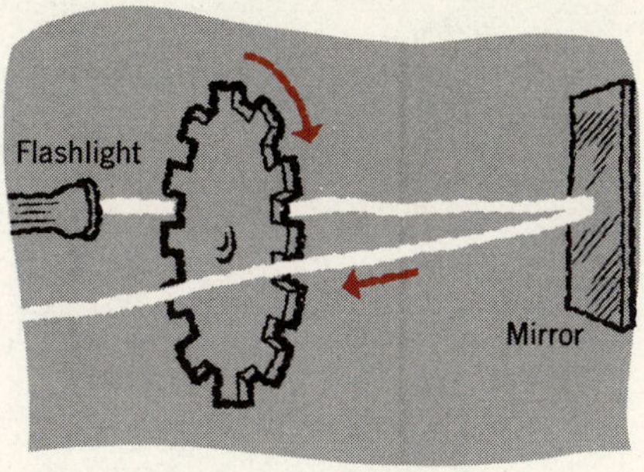

FIGURE 14-3 **Diagram illustrating a method by which the speed of light can be measured.**

$$c = 2.99793 \times 10^8 \text{ meter/second}$$

As far as we know, no material object, no energy, and no information can travel faster than this speed.

Interference and Diffraction of Light

Rarely in our everyday experience do we notice the wave nature of light. The colored bands observed in thin films of oil spread on the surface of water are due to light interference effects. For the most part, however, we think of light rays and beams for everyday effects. However, there are several experiments which can be performed which clearly demonstrate wave behavior for light.

The Two-Slit Experiment. About 1798 an English physician, Thomas Young, performed the experiment diagramed in Figure 14-4. Young used two pinholes in a card, but we shall use two slits that are very close together. The light passing through the slits falls on a screen. If the light beam directed at the two slits were composed of very small particles, we would expect to see just two illuminated regions where the particles of light coming through the slits struck the screen. Actually, however, the pattern observed is very similar to the interference pattern seen in Figure 13-12. As the reader will recall, this pattern was caused when water waves were emitted at the same time by two oscillators of the same frequency.

Of course, in this two-slit experiment we do not see the whole wave pattern in space, but merely the illumination caused by these waves when they strike the screen. We see only a series of light and

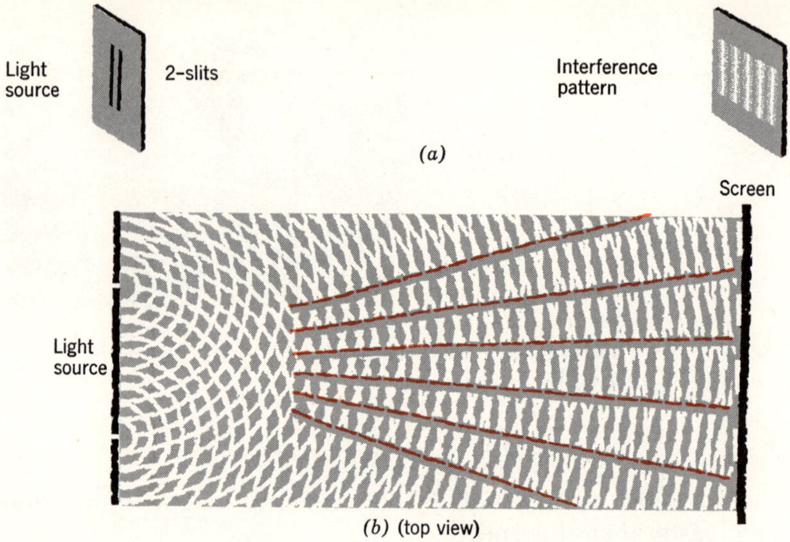

FIGURE 14-4 **Interference of light in the two-slit experiment.**

dark strips on the screen. The brightest parts of the pattern on the screen are about in the middle of the light strips. These are the regions of maximum intensity. The darkest parts of the pattern are about in the middle of the dark strips. These are the regions of minimum intensity. The pattern is made up of a series of these maximums and minimums, one after the other. The closer the slits are to each other, the greater the distance between successive maximums and minimums. This is a general rule for this type of pattern.

Figure 14-4*b* allows us to compare the water-wave interference pattern from Figure 13-12 with the light-wave interference pattern caused by light coming through two slits. In the earlier experiment we noticed that there were lines in the pattern along which there was always destructive interference. In Figure 14-4*b* we see that the regions of darkness occur where these lines reach the screen. Between them, the waves are flowing outward. These waves cause the bright maximums where they reach the screen. This similarity between the behavior of light and the behavior of water waves is strong evidence that light indeed travels as a wave. At least, it certainly exhibits wave like behavior. As we shall see, light of different frequencies (or wavelengths) appears to the eye as light of different colors. The wavelengths of visible light range from about 7.4×10^{-7} meter for very red light to about 4.1×10^{-7} meter for violet light.

If we know the distance between the maximums and the minimums on the screen, the distance between the slits, and the distance to the screen, we can actually find the wavelength of the light. Figure 14-5 illustrates how this can be done. The light from the two slits must interfere constructively at each maximum on the screen. We shall discuss one of the maximums nearest the center maximum on the screen.

In order to have constructive interference at this maximum, the light from one of the slits must travel a path that is a full wavelength longer than the path of the light from the other slit. This ensures that the two light waves will combine at that point on the screen to form a maximum, and not a minimum or something in between. Therefore, the wavelength of the light can be found from the difference between these two path lengths. This difference is hard to measure since light has a very small wavelength, but it can be calculated from the other quantities mentioned above.

The fact that the wavelength of the light can be calculated from the position of the first maximum in a two-slit interference pattern means that the position of this maximum depends on the wavelength. Different wavelengths (colors) of light will have their interference maximums at different positions on the screen. This fact enables us to separate the wavelengths (colors) in a beam of light. Actually the two-slit arrangement is not very effective for this purpose; we usually

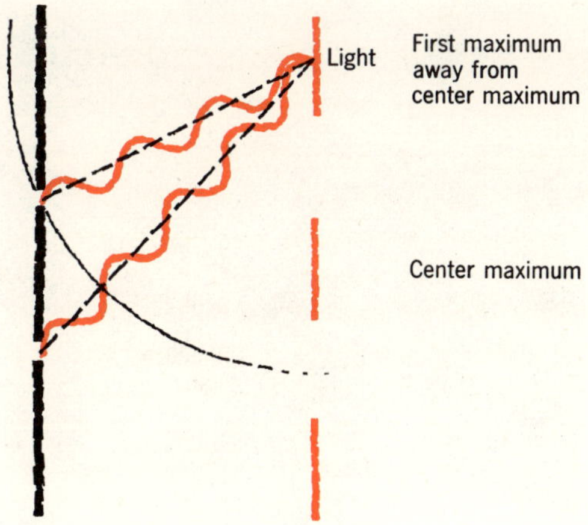

FIGURE 14-5 **In the two slit experiment, the first maximum away from the center occurs at a place which is one wavelength closer to one slit than it is to the other.**

use a *diffraction grating*, a device with about 15,000 slits per inch. The interference pattern of this device is much the same as that of the two-slit arrangement, except that the additional slits allow more light through and also, it turns out, produce some very narrow maximums. Since these maximums are so narrow, we can more easily separate the maximums caused by the different wavelengths. These separated maximums appear as different colors on the screen.

The diffraction grating is a very useful instrument because it enables us to determine the wavelengths (colors) present in a beam of light. Figure 14-6 illustrates the use of such a grating to analyze the light emitted by an ordinary light bulb. The central maximum of the interference pattern does not separate the colors because it is in the same position regardless of wavelength. The first maximum on each side of the center, however, is spread out to form what looks like a rainbow. Since the position of this maximum is different for different colors, light of each color strikes the screen at a different place. This rainbow is called a *spectrum*. The fact that the spectrum of the light bulb resembles a rainbow proves that the light from the bulb contains all visible colors. In later chapters we shall have occasion to discuss light that does not contain all colors and that therefore has a spectrum in which certain colors are missing.

Our discussion here has focused on the first maximum away from

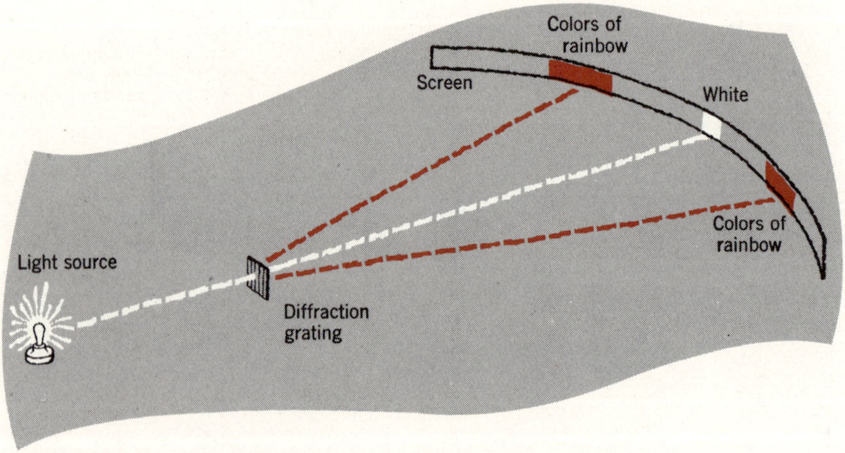

FIGURE 14-6 **A diffraction grating may be used to obtain a spectrum (rainbow).**

the center of the interference pattern. Spectra are also obtained from the maximums farther from the center.

The One-Slit Experiment. In Figure 13-9 we saw that diffraction occurs when a water wave passes through an aperture between two barriers. We can perform a similar experiment for light. We simply use one slit instead of the two we used before. Figure 14-7 shows the diffraction pattern we obtain when we pass light through a single slit. The presence of maximums and minimums in this pattern shows that we are getting interference effects from the light that comes through different parts of the same slit. We shall not discuss these interference effects further, except to note that they do exist. As the slit is made more narrow, the light pattern on the screen retains the same form, but becomes broader. This is a general property of patterns caused by diffraction. The smaller the object or aperture causing the diffraction, the larger the actual diffraction pattern. Of course, if the aperture is so big that diffraction is unimportant, this rule does not apply. A slit six meters wide will permit a wider beam of light to pass through than a slit a half meter wide; but for these wide slits, diffraction is not the important consideration.

We have omitted one effect from our discussion of the two-slit interference pattern. Diffraction, as well as interference, occurs in the two-slit experiment as well as the one-slit experiment. In fact, waves from both slits must diffract—that is, bend around corners—in order to be present in regions away from the center of the pattern. The light from each slit separately would form its own one-slit diffraction pattern. Because the two slits are usually very close together, these two patterns are nearly on top of each other and may be represented together as the one-slit pattern in Figure 14-8. The two-slit interfer-

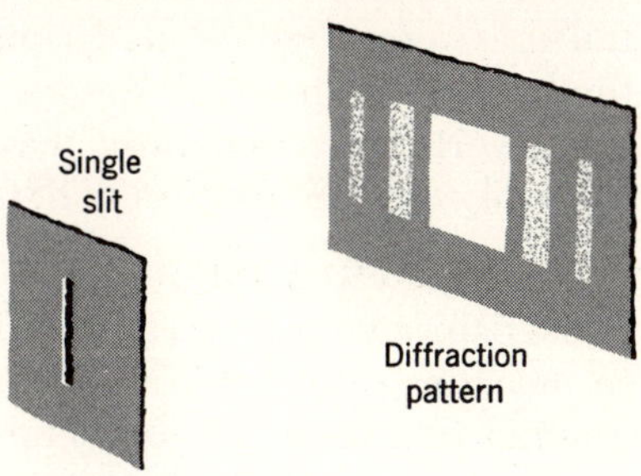

FIGURE 14-7 **Diffraction pattern formed by light passing through a single slit.**

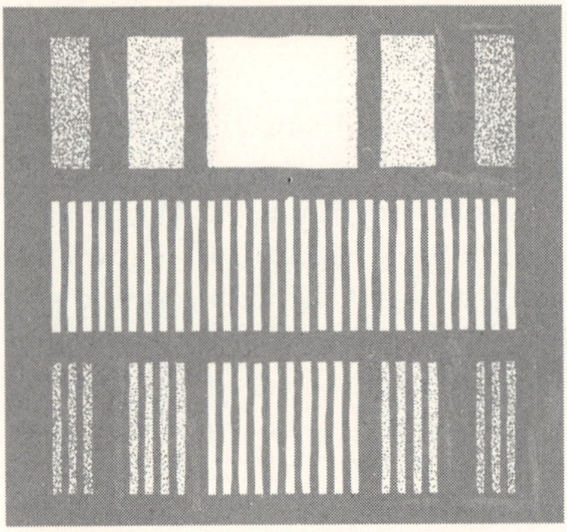

FIGURE 14-8 Only where light diffracted from each of the two slits strikes the screen can we see the two slit pattern. Therefore the two-slit pattern can only be seen in the maximum regions of the one-slit patterns.

ence pattern cannot occur except where light from both slits reaches the screen, and this will only be in the maximums of the one-slit pattern. Therefore, the actual two-slit pattern is as shown at the bottom of Figure 14-8. The two-slit interference maximums appear only within the broader regions where the one-slit pattern has its maximums.

The one-slit diffraction pattern, with its alternating light and dark regions, is similar to the diffraction patterns obtained when light is allowed to pass through small holes or around objects. For example, Figure 14-9 shows the type of pattern that Grimaldi (see Fig. 14-1 and 14-2) should have obtained when he let light pass in a straight beam through a small round hole. Once again we have the bright central maximum, which, as Grimaldi noticed, was actually larger than the hole through which the light passed. In addition, however, there are a series of concentric maximums and minimums, which were probably too dim for him to detect.

The Michelson Interferometer. The principle of light interference is used in a very accurate instrument, the Michelson interferometer, to measure either distance or wavelengths. The instrument is diagramed in Figure 14-10*a*. The light beam is split by a partially reflecting mirror. Half the beam continues through the mirror, and half is reflected. Both of these beams are reflected back to the partially reflecting mirror, and part of each of these reflected beams arrives at the point *A*. Thus the instrument takes a beam of light from a single

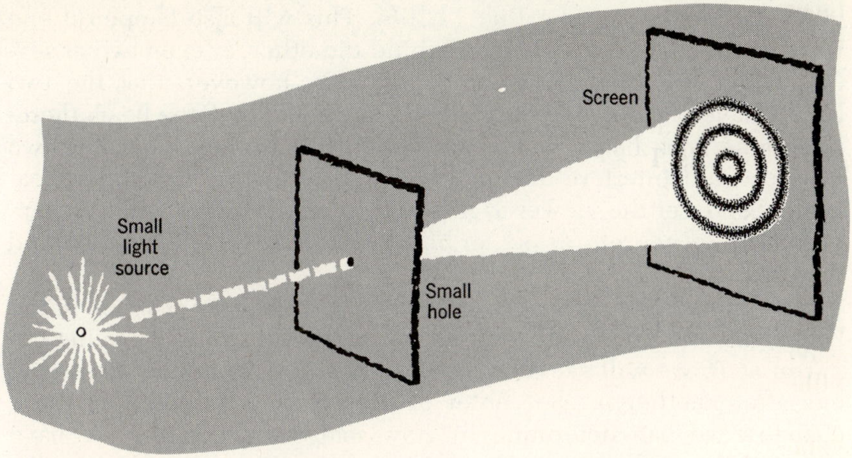

FIGURE 14-9 **Diffraction pattern caused by light passing through a small hole.**

source, divides it into two parts, sends each of these parts along a different path, and then reunites a portion of each of these two parts. If these separated beams have traveled the same distance at the same speed, then there will be constructive interference when they come back together again, and a bright spot will appear in the center of the

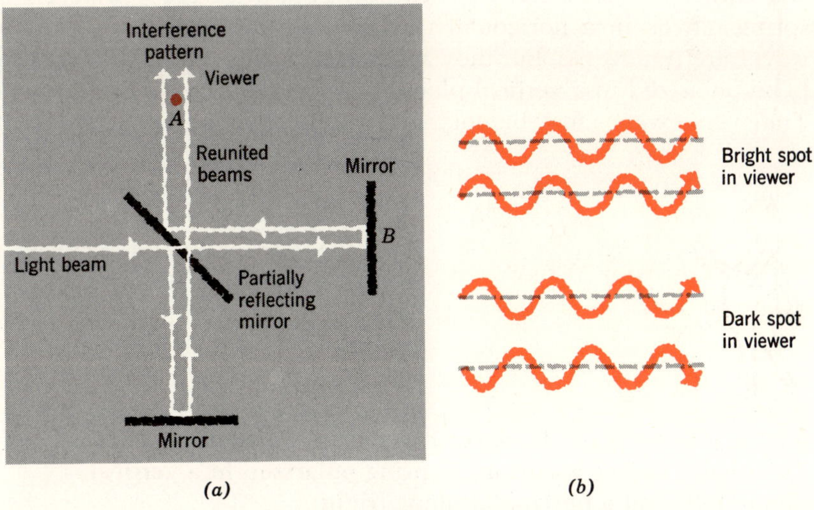

FIGURE 14-10 **Diagram of a Michelson interferometer.**

viewer located at *A* (see Fig. 14-10*b*). This will also happen if one wave returns a full wavelength behind the other, or even two or several whole wavelengths behind. Suppose, however, that the two beams have not traveled the same distance and that one beam therefore came back half a wavelength behind the other. Then the two waves would interfere destructively, and a dark spot would appear in the center of the viewer at *A* (see Fig. 14-10*b*). This will also happen if one wave returns 3/2, or 5/2, or 7/2, etc., wavelengths behind the other.

If we move the mirror at *B*, we change the distance that one of the beams travels. If we watch the viewer at *A* while slowly moving the mirror at *B*, we will see alternating dark and bright spots appearing one after another in the center of the viewer. By counting these changes, we can determine by how many wavelengths we have changed the path length. From this information we can relate the wavelength to the distance the mirror has been moved. If we know either of these lengths, we can then find the other.

Polarization

One reason for believing that light waves are transverse is that they can be polarized. To explain what is meant by polarization, let us examine transverse waves more closely. Figure 14-11 shows transverse waves in two springs. In Figure 14-11*a* the spring moves up and down as the wave goes by. In Figure 14-11*b*, however, the spring moves in a horizontal direction as the wave goes by. Both waves are transverse, but they are not the same. The first one is said to be polarized in a vertical plane; the second, in a horizontal plane. Transverse waves may be polarized in other planes, since the oscilla-

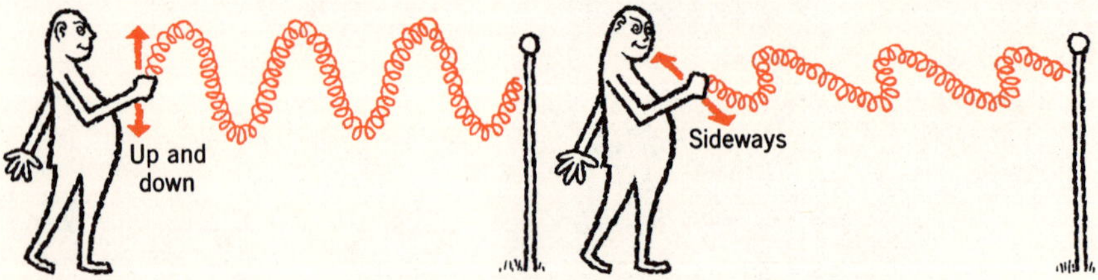

FIGURE 14-11 **Transverse waves may be polarized. This figure shows waves on a spring polarized in a vertical plane (left) and a horizontal plane (right).**

tion that starts the wave can be in any number of different directions. "Ordinary" light (from sunlight and ordinary lamps) may be thought of as containing many waves, polarized in different planes.

Light polarized in only one plane can be obtained from ordinary light by using a polaroid sheet (see Fig. 14-12). The polaroid material absorbs the light that is polarized in one plane while allowing the light to pass through that is polarized in a perpendicular plane. For example, the polaroid sheet at the left in Figure 14-12*a* allows the light that is polarized in the vertical plane to pass through but stops light polarized in a horizontal plane. Thus, the light that emerges from this polaroid sheet is dimmed, since some of the light has been absorbed. This light is not dimmed to any great extent by the second polaroid sheet, because the light is already polarized in the proper plane to pass through this sheet. If this second sheet of polaroid is turned through a right angle, as shown in Figure 14-12*b*, then the light coming from the first polaroid will not pass through it. Thus, by assuming that light consists of transverse waves, we are able to explain the effect of the polaroid sheets. Purely longitudinal waves could not account for this effect.

Light can also be polarized by other methods. Light reflected from

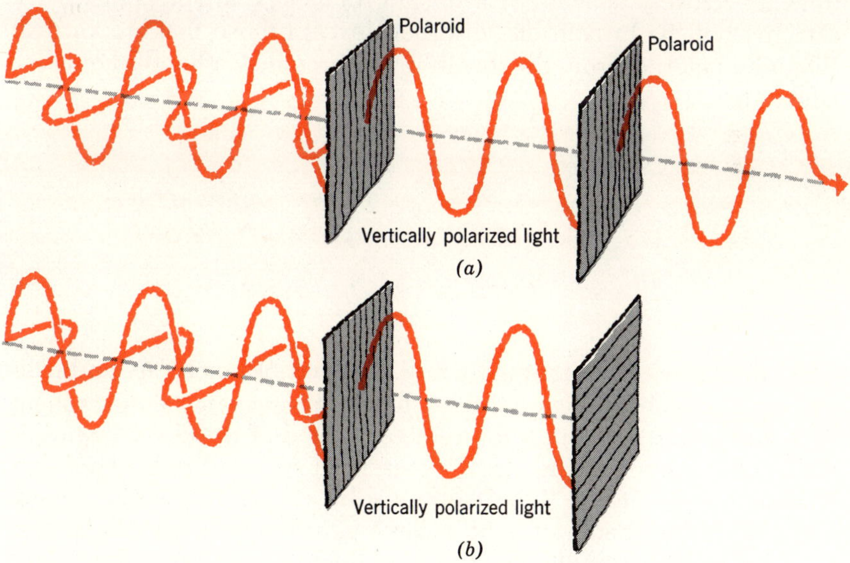

FIGURE 14-12 **Diagram illustrating the polarization of light.**

FIGURE 14-13 **Very rough analogy of polarization.**

a surface is often strongly polarized. Polaroid sheets are often used in sunglasses and oriented to absorb much of this polarized reflected glare. Of course, they absorb some of the direct light also.

A rough analogy of the action of polaroid sheets in polarizing a beam of light is shown in Figure 14-13. This figure shows a wood gatherer striding through a forest of tall straight trees. The long brittle poles he holds in his hand are oriented in all directions. If the trees are closely spaced with no low branches, all the poles he is carrying will be broken off (absorbed) except those that are vertical. When he emerges from the forest he has, in effect, a handful of polarized poles.

Light as an Electromagnetic Wave

Electromagnetic Waves

On the basis of his formulation of the laws of electricity and magnetism, Maxwell predicted the existence of waves in the electromagnetic field. Furthermore, the predicted speed of these waves turned out to be the same as the already measured speed of light. Therefore it seemed evident that light itself must be a wave in the electromagnetic field. The theory of light became part of the theory of electricity and magnetism. Light waves, ultra-violet waves, infrared waves, X rays, radio-broadcasting waves, radar waves, and others are all waves of different frequencies in the electromagnetic field (see Fig. 14-14).

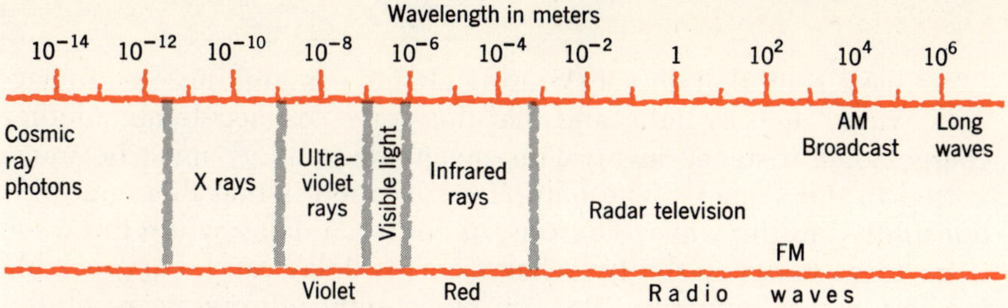

FIGURE 14-14 **Electromagnetic waves and their wavelengths.**

It was indeed an impressive intellectual achievement to include so much of nature in one organized theory.

Electromagnetic waves such as light involve both the electric and the magnetic field in combination. We can see this by asking how an electromagnetic wave can get started. Suppose that we have an electron or some other charge sitting in space. We are going to use this electron as an oscillator. This means that we will make the electron move back and forth. When it moves, its electric field must move also. Thus we have created a disturbance in the electric field. Moving the electron also causes a current, which sets up a magnetic field. As the electron is accelerated, its velocity, and therefore the current changes, changing the magnetic field. Therefore, by means of this oscillating charge, we have created a disturbance in both the electric and the magnetic fields. This disturbance produces electromagnetic waves moving outward at the speed of light. All of this is in accordance with the laws of electricity and magnetism which state that an *accelerating* electric charge should radiate electromagnetic waves, but a stationary or uniformly moving charge should not. This radiation occurs in many situations. The oscillator can consist of the electrons moving back and forth in a radio-broadcasting antenna, emitting radio waves. Or the oscillator can be an electron in an atom, emitting light and other waves. Or it can be an electron accelerated by crashing into the metal target of an X-ray tube, to emit X rays.

What happens when these electromagnetic waves run into another charged particle, as, for example, another electron? The electron could be in a radio antenna, an atom, or elsewhere. The electron will feel the accelerating effects of the oscillating electric and magnetic fields in the wave. Therefore it will be set into an oscillating motion by the wave it receives.

Energy in the Electromagnetic Wave

We have seen that an accelerating charge can emit an electromagnetic wave, such as light, and that this wave can accelerate another charge some distance away. This means that energy must be transmitted in the process. The natural assumption is that this energy is transmitted in the wave. There is, of course, a delay while the wave is traveling between the two charges. This delay can be appreciable in some cases; light from the sun takes eight minutes to reach the earth. Where is the energy during those eight minutes? It would seem obvious that the energy is in transit, carried in the light wave. Momentum is also carried with the wave. Light can exert pressure on an object and cause it to move.

These conclusions may seem easily acceptable. Yet we must recall that light is a wave in the field—the field that was invented to explain magnetic and electric effects. The field did not itself appear to be material. Now we find that it can support waves, and that it contains energy and momentum. This makes it seem a little more real. Both particles and fields seem to be fundamental realities.

The Ether

Many people felt that some material medium was required to support the field and its waves. After all, who had ever heard of a wave without a material medium to wave in? What is waving, if there is no medium? Therefore, the "ether" was devised to support the field and its waves. This ether, of course, has nothing to do with the ether of medical practice; rather, it was supposed to be a material medium which filled all space. The planets had to move through it without appreciable resistance. Furthermore, it had to transport transverse waves of various frequencies. The ether could hardly be a normal liquid or gas and allow transverse waves to pass through it. In fact, even steel is not rigid enough to support a transverse wave moving as fast as light. Therefore, the ether, although apparently necessary to support electromagnetic waves, presented several difficulties. We shall consider some of them in Chapter 15.

FUNDAMENTAL IDEAS

I. BEHAVIOR OF LIGHT WAVES

A. DIFFRACTION

B. INTERFERENCE
  1. Diffraction grating
  2. Interferometer
  3. Single slit
  4. Double slit

C. POLARIZATION

II. FUNDAMENTAL CONSTANT

A. The speed, c, of light or of any electromagnetic wave in a vacuum. $c = 2.99793 \times 10^8$ meter/second.

III. FIELDS

A. WAVES, including light, in the electromagnetic field.

B. ENERGY in the electromagnetic field.

C. ETHER to support the electromagnetic field and the waves traveling in it.

QUESTIONS

General

1. Why do we think that light waves exist? How do these reasons differ from our reasons for believing that other kinds of wave exist?
2. Would you try to measure the speed of light with a stopwatch? How would you measure it?
3. What is a shadow?
4. Why, if light is a wave, do we speak of light *rays*?
5. How can the result of Grimaldi's experiment be explained by the wave theory of light?

Interference

6. Describe the interference and diffraction patterns characteristic of one-slit and two-slit experiments.
7. What is the reason for the dark places in the two-slit interference pattern?
8. By what combination of experimental results and reasoning can we show that the color of light depends on its frequency?
9. What method or device can be used to separate a white beam of light into its colors?
10. In terms of the wavelength, what is the difference in the distances traveled to the second maximum from each slit in the two-slit experiment? To the third maximum?

11. In the two-slit experiment, what is the path difference in terms of the wavelength for waves from the two slits to the first minimum? To the second minimum?
12. We mentioned in the text that the first maximum away from the center maximum will be at different positions for different colors of light. Will this be true of the next maximum out also? The next one after that? The center maximum? Explain.

Polarization

13. What is meant by polarization?
14. How would you try to convince someone that light is a transverse wave?
15. Would you be surprised if you found a polarized beam of sound waves? Why, or why not?

Electromagnetic Field and the Ether

16. Give a reason for believing that the electromagnetic field can contain energy.
17. Why does a radio or television set use an antenna?
18. Give a reason for believing that electromagnetic waves involve both the electric and the magnetic fields.
19. Why was the ether invented?
20. What difficulties does the ether present?

15 Modern Relativity

The luminiferous ether, that is the only substance we are confident of in dynamics. . . . One thing we are sure of, and that is the reality and substantiality of the luminiferous ether. . . . You may regard the existence of the luminiferous ether as a reality of science.

Lord Kelvin[1]

Einstein's Special Theory of Relativity

Ether and the Field

The existence of the electromagnetic field raised interesting problems in physics. The field, as well as particles of matter in motion, was necessary to describe the physical world. It was natural to think that waves in the field, such as light, would need some sort of material medium in which to travel, and that indeed the field itself must have some material basis. Thus the idea of the ether, an unobserved substance filling all space, was introduced. The ether was thought to be a form of matter, prob-

[1]Lord Kelvin, *Popular Lectures and Addresses* (1891), Vol. I, pp. 317, 334, 354.

FIGURE 15-1 Albert Einstein in his student days. Courtesy of Lotti Jacobi.

ably composed of particles. Light waves, coming through the void from the remote stars, were believed to travel through the ether. The ether filled the void.

The concept of the ether raised at least two problems. First, a transverse wave such as light seemed to require a solid substance through which to travel. However, the earth and the planets appeared to travel through the ether without the slightest resistance.

The second problem involved the special principle of relativity. Ordinary objects can appear to be moving at different speeds if they are observed from reference frames in relative motion. As we pointed out in Chapter 3, for example, the stewardess of an airplane walking down the aisle appears to be moving at a considerably different speed if she is observed from the ground than from a seat in the airplane. Naturally it was assumed that the speed of a beam of light in a vacuum could also appear to be different in different frames. However, Maxwell's laws of electricity and magnetism predicted a single definite value for the speed of light in a vacuum; if this speed were actually different in different reference frames, then Maxwell's laws themselves would have to be different in different frames. This conclusion, of course, would appear to violate the special principle of relativity. Strictly speaking, the problem could be resolved, if somewhat uneasily, by assuming the existence of a material ether. Then the speed of light predicted by the laws of electricity and magnetism could be the speed observed in a reference frame at rest with respect to the ether. These and other considerations convinced many scientists of the existence of the ether and may help to account for the now obsolete claim expressed in the quotation from Lord Kelvin at the beginning of this chapter—a claim which is totally contrary to our present understanding of physics.

If the ether really exists, then the earth should be moving through it at a definite speed. The American scientists A. A. Michelson and E. W. Morley tried to measure that speed in 1887, using the Michelson interferometer (see p. 276). If the earth is moving through the ether, then from the earth the ether should appear to be moving past the earth. By the same principle, if we are riding in an open car on a calm day, a wind appears to be blowing by the car. Actually it is the car, not the air, that is "moving," but we feel the relative motion as a wind. Similarly, there should be an ether "wind" blowing by the earth. Michelson and Morley attempted to detect this "wind."

The Michelson-Morley experiment is explained in more detail in the next section. The result was that no ether wind, no velocity relative to the ether, could be detected. In fact, surprising as it appeared, the results of this and other careful experiments failed to uncover any convincing evidence for the existence of the ether. The speed of light in a vacuum appeared to be always the same, regardless of the motion of the observer.

The Michelson-Morley Experiment[2]

To explain the principle on which the Michelson-Morley experiment depends, we refer to Figure 15-2, which shows a boat race. The boats are going to travel the same total distance relative to the ground, but under different conditions. One boat goes a certain distance upstream and then back downstream to the starting point. The other boat goes across stream the same distance and then back. Both boats move with the same speed with respect to the water in which they are traveling. They will, however, have different speeds with respect to the river bank. The first boat will be slowed by the motion of the water when it is going upstream, and it will be speeded up by the water as it goes downstream. The second boat, in going across stream and back, will be slowed a little in both directions because it must be aimed a little upstream in order to keep the current from pulling it off course. It turns out that the boat going across stream wins every time. The total result of the changes in speed caused by

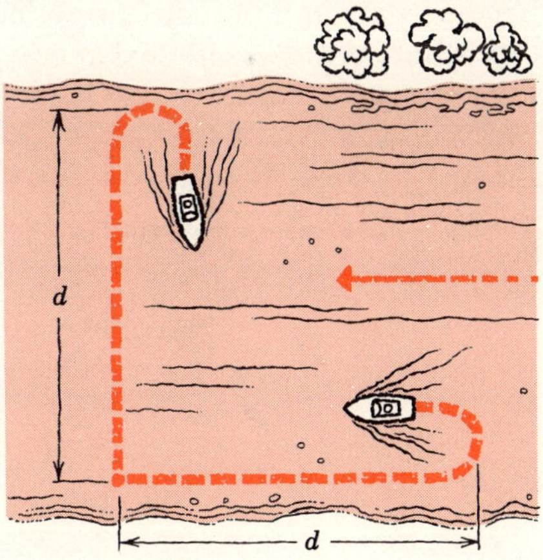

FIGURE 15-2 The boat going cross stream and back will win the race.

[2]This section may be omitted if it is not desired to treat the Michelson-Morley experiment in detail.

the moving water is that the time for the round trip is less for the cross stream journey.

The Michelson-Morley experiment is diagrammed in Figure 15-3. Here we have a situation similar to the boat race. Instead of the flowing water, we assume that we have the ether flowing by the earth. Instead of the boats, we have the waves in two light beams that have been split from the same beam by a partially reflecting mirror. One of these goes across the ether stream and back, and the other goes upstream and down. If the ether is really flowing by, and the speed of light is constant with respect to the ether, then the waves going across and back should return before the waves going upstream and back. That is, the waves in beam B should return before the waves in beam A in Figure 15-3*a*.

How can we tell that one set of waves gets back before the other? We know that we get interference effects when the two beams of light come back together. Whether there is an ether there or not, some pattern will be set up; for example, there might be a bright spot in the viewer. Let us now rotate the instrument until it is in the position shown in Figure 15-3*b*. In this situation, it is now beam *A* that goes across the stream and beam *B* that goes downstream and back. Therefore the waves in beam *A* will now return before the waves in beam *B*. Because of this shift, the interference pattern should change as we rotate the interferometer. The bright spot in the viewer could become dark and perhaps bright again. This result would signify that this ether "wind" really existed.

In fact, no significant result was observed; this, of course, raised a

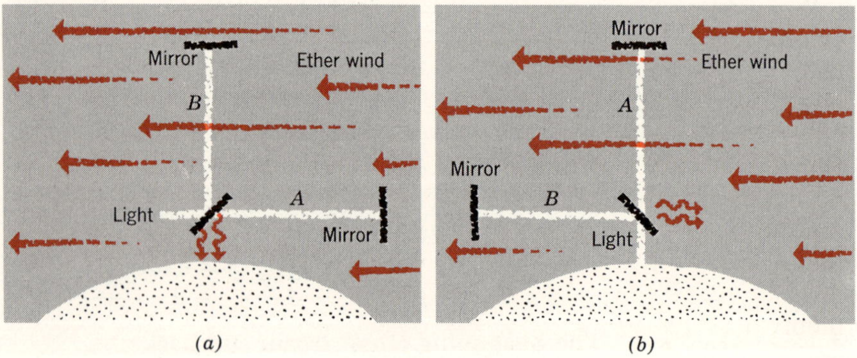

FIGURE 15-3 The Michelson-Morley experiment was unable to detect the expected effects of the ether on the speed of light.

problem. Unless the earth were sitting motionless in the ether, there should be some effect. Perhaps the earth just happened to be moving with the ether on that particular day, so that there was temporarily no relative motion between the earth and the ether. However, the earth, in traveling around the sun, is always changing the direction of its motion. Six months later the earth would be traveling in the opposite direction and should then have a speed with respect to the ether. Experiments were performed to check this possibility, but again there was no effect.

Einstein's Two Postulates

A true scientific revolutionary, Albert Einstein (1879–1955) was also independent and courageous in political and educational matters. As a boy he was informed by one of his teachers that he would never amount to anything. Of studying for examinations later in his career he himself said:

> It had such a deterring effect upon me that, after I had passed the final examination, I found the consideration of any scientific problems distasteful to me for an entire year.[3]

About five years later, however, after managing to find a steady job as a patent clerk, he published a series of scientific papers. One, on relativity, created the subject from a modern point of view. Another, on Brownian motion, we mentioned in Chapter 9. A third, which we shall discuss in the next chapter, helped lay the foundation for quantum physics.

We have seen how the special principle of relativity seemed to be challenged by the laws of the electromagnetic field. Yet experiments had been unable to measure the earth's speed with respect to the ether. Nor could experiments find any speed for light in a vacuum other than the value predicted by Maxwell's electromagnetic theory. Einstein decided that the facts should be accepted at face value; why not assume them to be true instead of trying to explain them away by some theory? The basic reality was the observed phenomenon; the theory was made up to make things simple or calculable. Einstein therefore chose the following postulates as the basis for his special theory of relativity.

[3]Quoted in Martin J. Klein, "Einstein and Some Civilized Discontents," *Physics Today*, **18** (January 1965), pp. 38–44. Mr. Klein also comments on some of Einstein's political opinions.

1. The velocity of light in a vacuum will be observed to be the same in all reference frames moving at a constant speed in a straight line.

2. All the laws of physics will be observed to be the same in all reference frames moving at a constant speed in a straight line.[4]

(Notice that the special *theory* of relativity and the special *principle* of relativity are not the same. The special *theory* of relativity includes two postulates, *one* of which is the special *principle* of relativity (see Chapter 3, pp. 38); the other postulate concerns the constant speed of light in a vacuum.)

These two postulates simply recognize the continued failure of certain experiments and state those failures as principles. No one had been able to measure a speed relative to any "preferred" reference frame, in the ether or elsewhere, so the special principle of relativity must hold. The second postulate, therefore, is just the special principle of relativity. Only one speed had been found for light in a vacuum, regardless of the reference frame, so the first postulate states that the speed is always the same. The reference frames mentioned in Einstein's two postulates do not include reference frames that are being accelerated; for example, a reference frame fixed to a merry-go-round would be excluded. In fact, the theory is called the *special* theory rather than the *general* theory because it excludes such reference frames. The postulates in the special theory refer only to reference frames that are either at rest or in uniform straight-line motion. According to the special principle of relativity, the laws of physics are the same in all such reference frames. In accelerated reference frames, however, the laws of physics may appear to be different.

The first law of motion, which is not altered by Einstein's special theory of relativity, affords a test as to whether a given reference frame is or is not accelerating. The first law of motion is the same in all nonaccelerating reference frames, as indeed it must be according to the special principle of relativity. In an accelerating reference frame, however, the first law of motion will appear to be violated. For example, a man standing still, watching a merry-go-round, is obeying the first law of motion. However, if we watch him from a reference frame fixed to the merry-go-round, he will appear to be going around and around, without benefit of any centripetal force. Thus he will appear to be violating the first law of motion. Actually,

[4]See page 38 Chapter 3. Compare with Albert Einstein and Leopold Infeld, *The Evolution of Physics*, New York: Simon & Schuster, 1938, p. 186.

of course, this apparent violation is because of our choice of an accelerating reference frame. It is our reference frame that is going around, not the man (in the special theory of relativity).

Some Consequences of Einstein's Two Postulates

The Speed of Light. Einstein's two postulates may seem innocent enough but they have some very surprising consequences. For example, imagine a space station out in space, far away from any other object (see Fig. 15-4*a*). We may consider it as stationary if we please. Suppose that there is a rocket moving past the station at half the speed of light—that is, 1.5×10^8 meters per second—and sending out a forward beam of light waves. These light waves will have the speed of light, about 3×10^8 meters per second, with respect to the rocket. What speed will they have as observed from the space station? One might think that the answer should be the sum of the two speeds, or 4.5×10^8 meters per second, but it is not. It is 3×10^8 meters per second, just the same speed as if the waves had been sent out by the station and not by the rocket. This is one of the curious results of modern relativity theory.

The example we have just discussed is an extreme case of the more general situation illustrated in Figure 15-4*b*. In this figure the rocket fires a projectile straight ahead, at some speed less than that of light. Ordinarily, to find the speed of the projectile with respect to the space station, we would simply add the speed of the rocket with respect to the station to the speed of the projectile with respect to the rocket. As an example of this rule, suppose that the rocket is travel-

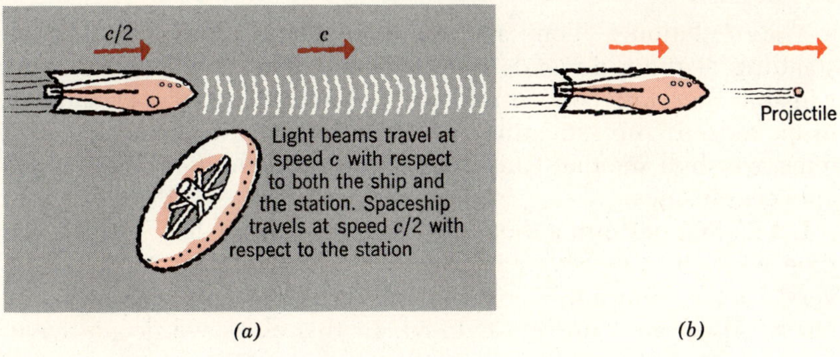

FIGURE 15-4 **The speed of light in a vacuum does not depend on the reference frame in which it is measured. The speed of a projectile does.**

 ing four meters per second with respect to the station and that the projectile is moving 30 meters per second with respect to the rocket. Then, according to our rule, the speed of the projectile with respect to the station should be the sum of these two speeds, or 34 meters per second. According to the special theory of relativity, however, our rule does *not* give the correct answer. Actually, the projectile will be going less than 34 meters per second, by an extremely small amount. For such low speeds, our rule is very nearly true, but it becomes more and more inaccurate as the speeds approach the speed of light. If it were possible for the projectile to reach the speed of light, we would have the same situation as we did for the light beam in Figure 15-4*a*. The projectile would then have the speed of light in any reference frame. The speed of light acts as a natural speed limit for material objects. In fact, as far as we know, the speed of light in a vacuum is the maximum speed at which matter, energy, or information can be transferred from one place to another.

Recently it has been suggested that particles may exist, called tachyons, which travel faster than the speed of light in a vacuum. Such particles could never travel slower than the speed of light in a vacuum, and thus the speed of light is a lower limit for tachyon speeds rather than an upper limit. Such particles would have exceedingly strange properties, and so far none have been discovered.

As we have illustrated, we cannot obtain a relative speed for an object greater than the speed of light by adding up relative speeds that are less than or equal to the speed of light. If this is true, and we believe that it is, then something must happen to ordinary time and space measurements when we compare two systems in relative motion. In fact something *does* happen, as we shall now see.

Time Dilation.[5] Time and space are abstract notions. Our understanding of them depends on our measurement of time and space. Einstein therefore concentrated on the behavior of clocks and measuring rods. In investigating the behavior of clocks in moving systems, we shall see that Einstein's postulates lead us to some remarkable conclusions.

Let us first perform a thought experiment to see how a clock keeps time when it is moving past us at a high speed. We shall choose a very simple, but impractical, kind of clock—the so-called "light clock," illustrated in Figure 15-5*a*. In this clock we do not have to

[5]The discussion which follows is taken largely from our previous text, *Conceptual Physics, Matter in Motion*, John Wiley & Sons, Inc., New York, 1969, pp. 400f. Similar remarks, of course, apply elsewhere in this book.

worry about the springs, gears, etc., that we find in ordinary timepieces. The light clock consists of two parallel mirrors with a pulse of light bouncing back and forth between them. Each time the light pulse is reflected from one of the two mirrors, we shall imagine that one tick of the light clock has occurred.

Now we hand the light clock to someone else, and have him race by us at a high speed. (Since this is only a thought experiment, we might as well have him moving at about half the speed of light.) As he races by, he holds the clock as shown in Figure 15-5*b*. As far as he is concerned, in his reference frame the clock is behaving quite normally. The light pulse is simply bouncing straight back and forth between the two mirrors. It looks the same to him as it did to us before we gave it to him. He thinks it is still keeping perfectly good time, and in his reference frame it is.

From our reference frame, however, things are different. According to our observation, the light bouncing between the mirrors does not move vertically as the runner moves but travels the diagonal path shown in Figure 15-5*b*. Because the mirrors are moving to the right in our reference frame the light pulse must move to the right also as it bounces back and forth so as to keep up with the mirrors. The pulse actually travels a longer distance as it moves between the mir-

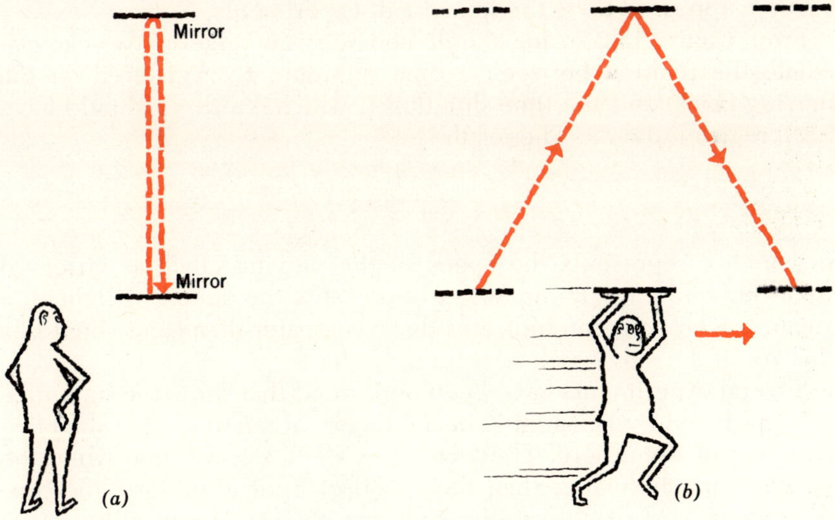

FIGURE 15-5 A moving light clock runs slow as determined by a stationary observer. In fact, all moving clocks run slow.

 rors in our reference frame than in that of the runner. Therefore, the pulse also travels for a longer time between the mirrors in our reference frame than in his, since the speed of the pulse of light is the same in both reference frames. As we see it from our reference frame, there is a longer time lag between ticks when the clock is in motion than when it is stationary. Therefore, the moving clock appears to run slow. This is what is meant by the term *time dilation.* Clocks in motion run slow. The faster they move, the slower they run. If it were possible to accelerate a clock so that it reached the speed of light, it would appear to stop completely.

The person moving with the clock, however, does not observe any slowing down of the clock. On the contrary, he thinks his clock is operating normally. In his reference frame the light clock is stationary, with the beam of light bouncing back and forth between the mirrors at its characteristic speed, *the same for all reference frames.* However, if he checked a clock in our reference frame, he would find our clock to be running slow.

Although this has only been a thought experiment, it correctly describes the behavior of any kind of clock, as far as is known—all moving clocks run slow. This curious fact is, of course, the result of the assumption that the speed of light is the same in all reference frames. We were led to this assumption by the fact that the speed of light in a vacuum appeared to be the same in all experiments.

From Figure 15-5, using simple geometry and algebra, we can calculate the relation between a time duration, t_0, registered on the moving clock, and the time duration t, which we think should have been registered on it. The result is:

$$t = \frac{t_0}{\sqrt{1 - s^2/c^2}}$$

in which s represents the speed of the moving clock as observed from our reference frame, and c represents the speed of light in a vacuum. The equation indicates that t is greater than t_0, as should be the case if the moving clock is running slow.

Several experiments have been performed that show the time dilation predicted by Einstein's special theory of relativity; we shall discuss one of them here. The factor $1 - s^2/c^2$ is nearly one when the speed is much smaller than the speed of light; therefore, the time dilation is hard to observe for ordinary objects. We usually do not have clocks whipping through the laboratory at half the speed of light or thereabouts. However, there are many small particles which do indeed move at speeds comparable to that of light. We have discussed two such particles, the electron and the proton. There is another fundamental particle, called the muon, which is almost exactly like the electron but has much larger mass. Muons are formed high above the earth when cosmic rays, made up of high-energy protons and other particles, crash into the atmosphere. These muons are continually being formed at about the same rate in the upper atmosphere. They are thus continually descending through the atmosphere to the surface of the earth, at speeds nearly that of light. While they are descending, some of them change into other particles. There is no way of telling how long an individual muon will last before changing. However, for a large group of muons, there is a definite average lifetime, which is characteristic of muons. By studying this average lifetime, Bruno Rossi and his collaborators obtained the first experimental evidence for the time dilation.

Rossi's muon experiment may be performed using the equipment illustrated in Figure 15-6 located on the top of a mountain.[4] The shield illustrated in the figure is so constructed that only muons traveling at about 95 percent the speed of light or faster can pass through it. All others are absorbed. Detector 1 will count all muons entering the can. Muons with speeds greater than 95 percent the speed of light will not stop in the can, so detector 2 is needed to count them. The difference between the numbers of muons counted by the two detectors will tell us exactly how many muons with speeds very close to 95 percent the speed of light are caught in the can. A third detector, which is not shown, is able to measure the time between the entry of a muon into the can and its change into other particles. In this way we can determine the lifetime of these muons. From the lifetimes we can predict how many of them would have lived long enough to reach the bottom of the mountain had they been allowed to do so. These lifetimes are so short that we would expect very few muons to reach sea level.

Next we take our equipment down to sea level and adjust the shield to account for the fact that the additional air between sea level and the mountain top would absorb some of the muons. Then we repeat our experiment and count the muons that would have had 95 percent of the speed of light at the mountain top. We find that we get almost as many as before. In other words, the muons lasted longer in flight than they did at rest in the laboratory. When we measure how much longer, we get a value which agrees with the time-dilation

[4]*Time Dilation: An Experiment with Mu-Mesons*, a motion picture distributed by Educational Services, Inc., 39 Chappell St., Newton, Mass., 1963.

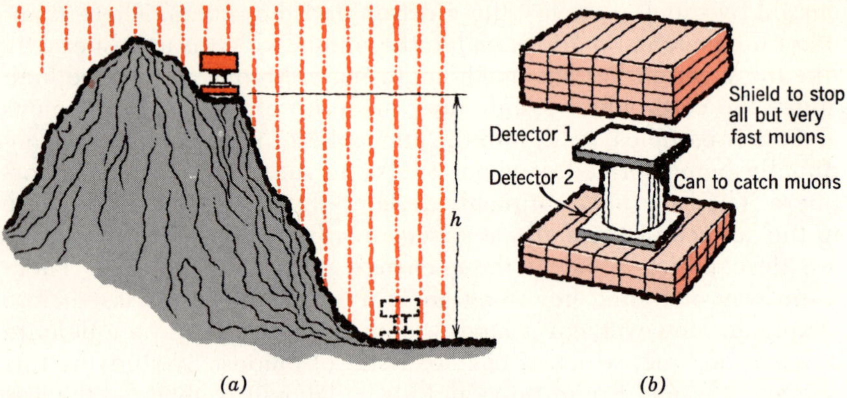

FIGURE 15-6 **Moving muons have a longer average lifetime than muons at rest, thus exhibiting the time dilation.**

formula we have just obtained from Einstein's postulates. Thus it appears that Einstein's theory is correct, or at least on the right track.

The Apparent Paradox of the Twins. The time dilation predicted by the special theory of relativity has given rise to a paradox that is still discussed among physicists. It seems paradoxical that each of two observers in relative motion will claim that the other's clocks are running slow. However, no contradiction arises as long as each maintains his velocity, since they will meet at most once, and then recede from each other forevermore. What would happen, however, if some means were found to bring them back together again? Whose clocks would really have been running slow?

Putting the paradox in its usually stated form, suppose that there are two twins, one of whom remains home on earth while the other takes a high-speed journey to a star (see Fig 15-7). After arriving near this star, the traveling twin reverses his direction and returns to earth. During his outward journey, and during his return journey as well, his clocks will appear to be running slow as observed from the reference frame of his stay-at-home brother. Since this slow down should apply to all processes, including biochemical processes, the traveler will have aged less than his brother, and therefore should appear to be the younger of the two when he returns, even though they were both the same age when he left. This would be a remarkable result, but in itself it contains no contradiction.

However, we might also look at the matter from the reference frame of the traveler. Since he is at rest in his own reference frame,

FIGURE 15-7 **When he returns to earth the traveling twin is younger than his brother.**

he notices no slowdown of his clocks. It appears to him, rather that the earth and his brother are moving; therefore he observes that his brother's clocks are running slow. When he returns to earth, then, he will conclude that he is the elder, that it is his brother who has aged less than might be expected. This result is directly contradictory to the result we obtained when we looked at the matter from the reference frame of the other twin. Thus we have an apparent paradox—which twin really would be the elder?

The traveler, of course, cannot have been traveling at a constant speed in a straight line during his entire journey. Therefore his reference frame is not one of those referred to in the postulates of the special theory of relativity. The stay-at-home's reference frame has not been accelerated and therefore is acceptable, but the traveler's frame is not acceptable. The traveler must have been accelerated to start on his journey, he must have been accelerated to turn around and start home, and he must have been accelerated (negatively) in order to stop when he finally returned home. According to the special theory of relativity, the laws of physics need not be the same in an accelerated reference frame; therefore we need not give the same status to the traveler's view that we do to that of his brother, and the contradiction expressed in the last paragraph was not validly obtained.

Even if we restrict ourselves to the reference frame of the stay-at-home twin, however, we still need to consider the possible effect of acceleration on the clocks of the traveler. The theory needed for such an analysis would take us beyond the level of our present discussion; however, we can say that according to currently accepted theory the acceleration apparently does not eliminate the effect of time dilation.

 The traveling twin would indeed return to earth younger than his brother. Moreover, recent experiments seem to justify this conclusion:

Atoms in crystals are in continual vibration, because of the kinetic energy associated with the temperature of the crystal. Experiments that can be performed on small particles within certain atoms indicate that time dilation occurs for them despite their acceleration as they vibrate back and forth. Similar experiments can be performed on these atoms as they are moved in a circle on a turntable.[6] Again the time dilation occurs. Therefore, the traveler, despite his acceleration, should indeed find when he returns home that he has aged less than his brother. It is indeed curious that the stay-at-home spends more time watching the journey than the traveler spends taking it. The traveler and his twin start out on earth at the same place at the same time. They also end up again on earth at the same place and the same time, but at a time later, of course, than the departure. In between these two meetings, they have, not surprisingly, traveled different distances. Surprisingly, they have traveled for different periods of time, also.

Lorentz Contraction. Let us return to the problem of the muons. discussed on page 297. According to the special principle of relativity, we should be able to consider the whole process from the muon's reference frame as well as from our own. In our reference frame, the muon is moving down to earth at a certain speed. The muon, in its reference frame, sees the earth moving up at the same speed. We have already found that the "normal" lifetime of the muon is not long enough to allow very many muons to travel the vertical distance from the top of the mountain to sea level, even though they are moving at nearly the speed of light. We have also seen that in fact the muons *do* successfully cover this distance. We solved this problem in our own reference frame by showing that the lifetimes of the moving muons are stretched out because of time dilation. This solution will not work, however, if we look at the problem from the reference frame of the muon.

The muon is at rest in its own frame and would observe no time dilation. Therefore, the muon will not survive the trip to seal level if it has to travel the whole distance. The only solution is that the muon must see the distance from the mountain top to sea level as shorter

[6]For discussion of this experiment, see J. Bronowski, "The Clock Paradox," *Scientific American* (February 1963), p. 134–144.

than we see it. It must see the distance shortened in exactly the same way that we see the time stretched:

$$L_0 = \frac{L}{\sqrt{1 - s^2/c^2}}$$

In this formula, L is the height of the mountain that the muons see, and L_0 is the height of the mountain that we see. The muons see a contracted length, a length that is shorter along the direction of motion than the length we see. This shortening, called the *Lorentz contraction*, is another result of Einstein's two postulates. The length of any moving object should be shortened along its direction of motion. The faster the object moves, the shorter it becomes. If it were possible for the object to be accelerated to the speed of light, its length along the direction of its motion would collapse to zero.

In order to clarify the Lorentz contraction, we refer to Figure 15-8. Here we have two measuring rods moving at a high speed. The one that is parallel to the direction of motion is shortened by the Lorentz contraction. The other, which is moving sideways, is not shortened by this contraction. However, it *is* made thinner. The Lorentz contraction alters the length *along* the direction of motion, and not perpendicular to it. Of course, someone moving with the rod would not notice any shortening of the rod, since the rod is at rest with respect to him. Instead, he would think the rod in *our* system was shortened by this contraction.

The Lorentz contraction and the time dilation are very small at ordinary speeds. Therefore, they may be ignored for most of our ordinary activities in everyday life. They become very important, however, for objects moving at nearly the speed of light.

Our discussion has centered on the behavior of clocks and measur-

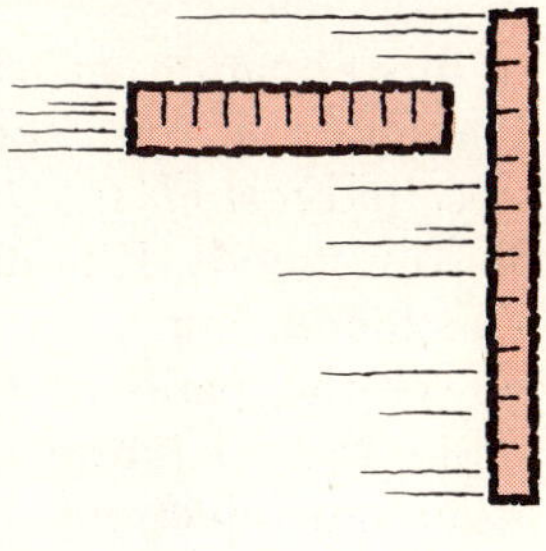

FIGURE 15-8 The Lorentz Contraction: A moving object is shortened along its direction of motion.

 ing rods. Yet we have spoken of time and distance. Because the time dilation and the Lorentz contraction appear to affect all clocks and all measuring rods according to the same principles, we consider these principles as properties of time and space, and not just as properties of objects. After all, in physics, our precise knowledge of time and space is dependent on the devices we use for measurement. For this reason some physicists would claim that the properties of space and time are the properties of our measuring instruments.

The Increase of Mass with Speed. When we discussed the example of the rocket and the space station, we noticed that even a light wave emitted from the front of a rocket cannot go faster than light in the reference frame of the space station. This is a general principle. We cannot accelerate any object enough for it to reach the speed of light. We can, of course, apply a force to an object for a long time. As a result of this force, the object will continue to accelerate. That is, its speed will continue to increase as long as the force is applied. However, this increase, or even a series of such increases, cannot cause the object to reach the speed of light.

What can prevent a moving object from reaching the speed of light when a constant force is exerted on it? If the speed were to continue increasing at a constant rate, it would eventually equal and even surpass the speed of light. The acceleration of an object, however, depends not only on the force but on the mass as well. If the mass is increasing with speed, then the force will have less and less effect as the speed increases. In fact this is what happens, according to the theory of relativity. The mass actually depends on the speed, according to the equation

$$m = \frac{m_0}{\sqrt{1 - s^2/c^2}}$$

In this equation, m is the mass of a particular object at any speed s. The letter m_0 represents the mass the object would have at rest, the so-called *rest mass*. The increase of mass with speed is not important for low speeds. It is important, however, for electrons in a cyclotron. The increasing mass due to the increasing speed of the electrons causes the paths of the electrons to change in the cyclotron and spoils the operation of the device. Other devices must be used to accelerate electrons to very high energies. This circumstance, although a practical nuisance in accelerators, is another experimental fact which supports Einstein's theory.

The momentum of a moving particle, according to the special theory of relativity, is still equal to the mass of the particle m multiplied by the velocity v. The mass, however, is the mass m given in the previous equation. Therefore, the momentum p is given by the equation

$$p = mv = \frac{m_0 v}{\sqrt{1 - s^2/c^2}}$$

E = MC². The results of the last section also have another consequence. In trying to accelerate a very fast particle to the speed of light, we are pushing on it with a force while it is traveling through space. Thus, we are applying force through a distance, or doing work on the particle; therefore its energy should increase. But the speed cannot increase very much since the particle cannot go faster than light. The energy must show up in some other way. At the same time, we know that the mass of the particle is also increasing. This leads us to suspect that the mass itself must represent energy, and it turns out that this is correct. In fact, the total energy of the particle is

$$E = mc^2$$

The particle still has some mass when it is at rest—its rest mass. If the equation is correct, then the particle should have some energy when it is at rest, because of this rest mass. In fact, this energy should be

$$E_0 = m_0 c^2$$

This form of energy, rest energy, is at present mostly inaccessible in most objects. However, in atomic-bomb explosions and similar processes, rest energy can be transformed into other forms which are "useful." Such processes, therefore, provide experimental evidence that rest energy really exists.

In fact, according to all relevant experiments that have been performed, the equation $E = mc^2$ appears to hold for all forms of energy and all mass. Energy, in whatever form, has mass. Mass, wherever it may be, represents energy. It appears that mass and energy are really the same thing but are measured as different quantities in different units. It is incorrect, then, to say that mass is a form of energy. "Mass" and "energy" relate to the same real entity, sometimes called mass-energy. These two great central concepts of nineteenth-century physics turn out to refer to the same basic thing. This fact is often called "the equivalence of mass and energy." The mass-energy equivalence combines two of the great conservation laws. Both mass and energy are still conserved. Since they basically represent the same thing, however, we often speak of one conservation law—the conservation of mass-energy.

What we are accustomed to regard as a very large amount of energy does not have very much mass. This explains why the equivalence between mass and energy was not discovered earlier. The total energy of a particle is equal to the mass multiplied by c^2, and c^2 is very large. Even a small mass multiplied by the very large c^2 can yield an enormous amount of energy. For example, in the atomic bomb and other processes, a rather small change in rest mass releases tremendous amounts of energy.

It should be clear that the theoretical and experimental results of the theory of relativity have greatly modified our ideas. Since the ether had no observable properties, the idea of the existence of the ether has largely been abandoned. Waves in the field travel through space without its help. Because of the equivalence of mass and energy, the electromagnetic field and its waves, which we know contain energy, must also contain mass. Thus we have two types of "matter" in motion—the fundamental particles and also the field with its waves. Mass-energy and momentum are conserved. The mass of an object changes with speed. Our ideas of space, time, and matter are greatly changed, and thus nearly all of physics is also changed. When these changes are made, it turns out that Maxwell's laws in their original form are still correct. It is Newton's law of gravitation that must be modified, and the second law of motion requires some attention also.

Further Implications of Einstein's Special Relativity

Light Velocity, Information, and Cause and Effect. We have pointed out that the fundamental constant c, the speed of light in a vacuum, acts as nature's speed limit for the transfer of energy, information, and causes. This means that material objects as we know them cannot travel faster than this speed. Suppose that there is an explosion on the sun, and then a disturbance of compass needles all over the earth. Could the explosion on the sun be the cause of the compass disturbance? Only if the time between these events is as long as or longer than the time required for light to travel from the sun to the earth. If the time is not this long, then the explosion could not be the cause of the disturbance. A "cause" cannot be propagated through space faster than light. Even the "news" of the explosion on the sun could not get to the earth any faster than light. The maximum transmission speed of signals, causes, information, energy, and material objects is the speed of light in a vacuum, according to Einstein's special theory of relativity.

Simultaneous Events. Two events that happen at the same time when observed in one reference frame do not necessarily happen at the same time when observed in another reference frame moving with respect to the first. We can illustrate how this follows from Einstein's postulates by an example taken from one of Einstein's books.[7]

Suppose that there is an observer on the ground near a railroad track watching a train go by (see Fig. 15-9). At the instant that the train is halfway past the observer, it is struck by lightning at both ends. The light from each lightning flash travels to him, and he sees both flashes at the same time. Since he knows that his distances from the two flashes are the same, he concludes that the two flashes occurred at the same time.

Now suppose that a passenger on the train is located exactly halfway between the two ends of the train. He would not see the two flashes at the same time. He would see the light flash coming from the front of the train a short time before he sees the flash coming from the rear of the train, since he is moving toward the light wave coming from the front of the train and away from the light wave coming from the rear of the train. Therefore the light wave from the front of the train reaches him first and he sees it first.

What will the passenger conclude about the two flashes of light-

FIGURE 15-9 If the two lightning strikes happen at the same time for the "stationary" observer, then they will not happen at the same time for the "moving" observer.

[7]See Albert Einstein, *Relativity, the Special and General Theory*, trans. by Robert W. Lawson, Gloucester, Mass.: Peter Smith, 1931, and London: Methuen and Co., Ltd.

ning, as he observes them from his reference frame fixed to the train? He will decide that the flash he saw first happened first. Even if he takes into account the time the light takes to reach him, he will still decide that the one he saw first happened first. After all, both ends of the train are an equal distance away from him. Einstein's first postulate tells him that the speed of light is the same in his reference frame as it is in the other. Therefore he will conclude that the waves from the two flashes of light took the same time to reach him. Therefore, the flash he saw first occurred first.

This example shows that two events which happened at the same time for the observer on the ground did not happen at the same time for the observer on the train. The idea of simultaneity depends to a certain extent on the reference frame. This is rather disturbing; we would prefer to think of two events as simultaneous, or not simultaneous, in an absolute sense. The fact that we cannot do this without specifying a particular reference frame is a consequence of Einstein's special theory of relativity. Of course, if we refer all events to the same reference frame, we can avoid talking about differences in simultaneity. But we have no justification for giving one particular reference frame this special treatment. Which reference frame should we choose? Einstein's second postulate is the special principle of relativity. According to this principle, one reference frame is no better than any other that is moving uniformly. Until some reason is found to prefer one of these reference frames to another and abandon the special principle of relativity, simultaneity must be considered as relative to the reference frame.

The results of the preceding paragraphs may seem startling, but they really do not violate common sense as much as may first appear. For example, if two events occur at the same time and at the same place, they will be observed to be simultaneous in all reference frames. There will appear to be a time interval between two "simultaneous" events in some reference frames only when they are a distance apart. In this case, the observer can be present at only one of the events and must obtain his information about the other from a signal. Light is the fastest signal, but it still takes time to travel from one place to another. Observers in relative motion will disagree as to the distance between events and therefore as to the time it takes light to travel between them.

The Fourth Dimension. The theory of relativity is popularly associated with the so-called "fourth dimension," which is supposed to be time. Why is time the fourth dimension?

In physics the number of dimensions is the number of pieces of information necessary to specify a position, a condition, or an event, etc. In Chapter 3 we pointed out that three pieces of information are needed in order to locate a position in space with respect to a reference frame. For this reason we regard a point in space as a three-dimensional entity.

The pieces of information which specify a position in space are the distances of the point from the three axes of the reference frame. These distances are not unique, but can be changed together (or separately) by properly chosen changes of the reference frame (see Fig. 15-10).

An event may be specified by giving the position in space at which the event occurred, and also the time at which it occurred. Therefore an event is four dimensional regardless of the theory of relativity—four items of information are required, three to determine the position in space and one to determine the time. However, before Einstein it was felt that the three space dimensions were independent of

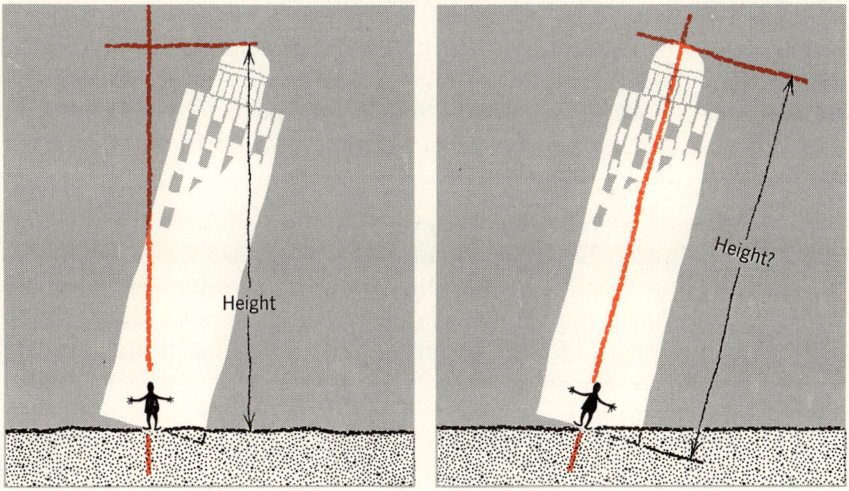

(a) (b)

FIGURE 15-10 **How tall is the leaning tower of Pisa really? (*a*) the leaning Tower as observed by an erect tourist, and (*b*) the leaning Tower as observed by a leaning tourist. This figure was suggested by a figure and caption in Roger Price *Droodles*, Simon & Schuster, New York, 1953, Copyright, 1953 by Roger Price, pp. 30, 31.**

the time dimension. There was no known way to change the reference frame in space that would force a change in the time dimension as well as the space dimensions. The discovery of the Einstein time dilation and the Lorentz contraction changed this situation. In Figure 15-11, not only is the rod in the moving reference frame shortened by the Lorentz contraction, but the clock experiences time dilation. Thus both space and time dimensions are altered together by observing a situation from a new reference frame moving with respect to the first. Time and space are thus intimately connected, and we speak of a four dimensional space time, instead of a separate space and time. As we have already seen, this alteration of the time dimension is such that events separated in space which are simultaneous in one reference frame will not be simultaneous in the other. Indeed, the time order of events separated in space may be reversed. However, in the special theory of relativity, there are some limitations on these changes—a cause cannot be changed into an effect by a change of the reference frame. Nevertheless, this mixing of time and space into a four dimensional *spacetime* was sufficiently impressive for H. Minkowski to point out:

> The views of time and space which I have set forth have their foundations in experimental physics. Therein is their strength. Their tendency is revolutionary. From henceforth space in itself and time in itself sink to mere shadows, and only a kind of union of the two preserves an independent existence.[8]

The theory of relativity causes a revolution in our ideas of time and space, a revolution based on theoretical insight combined with the success of careful experiments. The experimental effects of relativity are extremely small in ordinary motions familiar to us in daily life. In this respect they are like the experimental evidences for the rotation and revolution of the earth—small in many practical effects, but of deep philosophical import.

In the first chapter of this book we showed how the principle of relativity figured in the seventeenth-century revolution in science, even though it was not completely understood. The new physics of Newton, with the laws of motion and the forces arising from the grav-

[8]H. Minkowski, *Space and Time*, translated from an address delivered at the 80th Assembly of German Natural Scientists and Physicians, Cologne, September 21, 1908. From *The Principle of Relativity* by Albert Einstein, Hendrik A. Lorentz, H. Minkowski, and Hermann Weyl, Dover Publications, Inc., New York, 1923, p. 75. Reprinted through permission of the publisher.

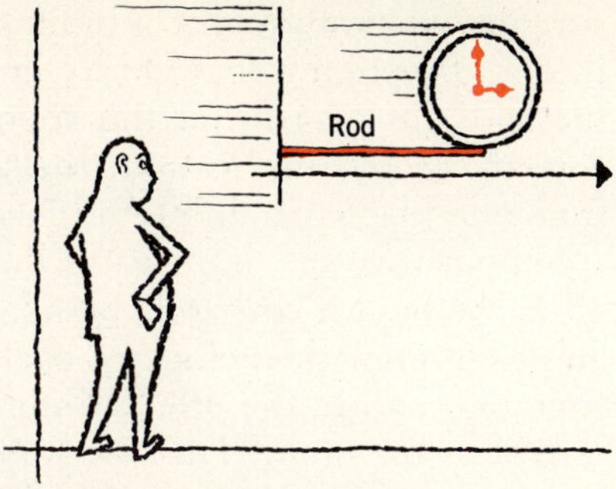

FIGURE 15-11 **Time and space measurements are different for reference frames in motion.**

itational interaction, actually obeyed this principle. Again, at the beginning of the twentieth century, in what must be considered another revolution in physics, we find the special principle of relativity once more deeply involved in a change in our basic ideas of the physical universe. This time it is the electromagnetic interaction, with its finite light speed, which has to be reconciled with the principle.

We may form some idea of the importance of Einstein's theory by reconsidering the revolution of ideas that was necessitated when it was discovered that the earth was not stationary but was actually rotating and revolving. As long as the earth was thought to be at rest in the center of the universe, the "up" and "down" directions had to be regarded as very different from any sideways direction. "Down" was the direction along which objects were pulled toward the center of the earth, which was also the center of the universe. With a moving earth, however, what was up or down at one time would not be in the same direction after the earth itself had turned and moved. Thus, there was no absolute "up" or "down"; the direction was relative to the position of the earth. Under the right circumstances, any direction could be up or down. Thus the abandonment of a stationary earth in favor of a moving earth destroyed the difference among the three dimensions of space. Similarly, the abandonment of the stationary ether destroyed part of the difference between the three dimensions of space and the one dimension of time. So revolutionary were this and later developments that when we speak of the new physics today, we are usually referring to relativity and the quantum physics to be discussed later rather than to Galileo, Kepler, and Newton.

The One-Way Speed of Light. There seems to be no way to measure the one-way speed of light, that is, the speed of light as it travels in one direction only.[9] Oddly enough, it appears that the suggested methods for measuring the speed of light actually measure, either directly or by implication, the average speed of light as it travels from one place to a different place and then back again.

Suppose we try to measure the one-way speed of light. In Figure 15-12 we have a very long plank, somewhere out in space. We station an observer with a clock at each end. We shall send a light signal from one end to the other, timing its departure by one clock and its arrival by the other. Thus we find the time it takes for the light signal to travel the length of the plank. We then measure the length of the plank and calculate the speed of the light signal.

Unfortunately for this scheme, there is no way to be sure that the clocks are synchronized once they are in place. We could, of course, send a signal from the first clock to the second, telling the observer with the second clock how the first clock was set. This observer, however, would have to know how long it took the signal to reach him, and he could not know this unless he knew the one-way speed of the signal. It would not do to try to use a light signal, since we do not yet know its one-way speed. However, we have a similar problem if we try to use any other signal—we cannot measure its one-way speed until the clocks are set, and we cannot set the clocks until we know the one-way speed of some signal.

How do we get out of this dilemma? We measure the two-way speed of light by finding how much time is required as registered on the first clock, say, while a light signal leaves that clock, travels to the other one, and then is reflected back to the first clock again. Such a measurement requires only one clock, and the problem of synchronization does not arise.

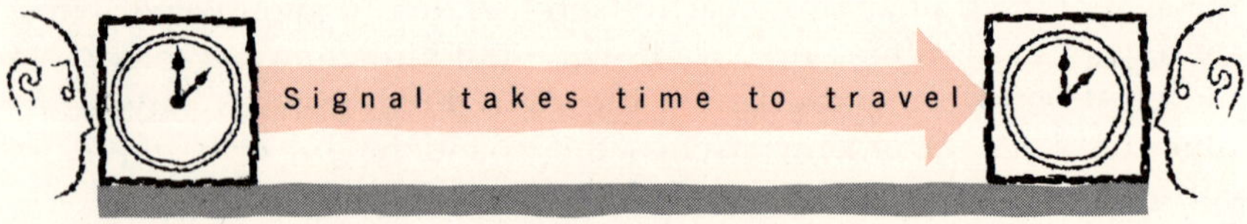

FIGURE 15-12 Two observers manning two clocks separated by a significant distance at each end of a plank somewhere in space. The plank is either at rest or in uniform motion.

[9]See Hans Reichenbach, *From Copernicus to Einstein*, trans. by Ralph B. Winn, New York: Philosophical Library, Inc., 1942, pp. 45-47.

Then, if we wish, we can decide to believe that light travels the same speed in both directions, thus defining the one-way speed of light. This is convenient, and it maintains the directional symmetry of space. This is the definition used by Einstein[10] but less convenient definitions would be just as valid. In this sense, the one-way speed of light does not exist—that is, until we arbitrarily define it.

The General Theory of Relativity

The General Principle of Relativity

The special principle of relativity states that the laws of physics should be the same in all reference frames that are not accelerating. We are naturally tempted to generalize this principle so that it will apply to all reference frames, however they move or accelerate. We would like to state a *general* principle of relativity:

The laws of physics are the same in any reference frame.

Einstein's general theory of relativity is an attempt to establish this principle. After all, why should nature care what reference frame we are using?

FIGURE 15-13 **To measure the one-way speed of the signal we use both clocks.**

[10]Albert Einstein, in *The Principle of Relativity*, by H.A. Lorentz, A. Einstein, H. Minkowski, and H. Weyl, Dover Publications, Inc. (originally published as a translation by Methuen & Co., Ltd., 1923), p. 39–40. Once we have the special theory of relativity to tell us how moving clocks keep time, we can set the clocks by moving another clock from one clock to the other. See Edwin F. Taylor and John Archibald Wheeler, *Spacetime Physics*, San Francisco: W. H. Freeman and Co., 1963, 1966, pp. 19, 72.

In our discussion of the general theory of relativity, we shall make use of a particular accelerating reference frame, that fixed to an "elevator" out somewhere in space illustrated in Figure 15-14*a*. This elevator, and the reference frame with it, is being accelerated through space at a constant rate by a constant outside force exerted "upward" on the bottom of the elevator.

If we were to release an object in this elevator, it would appear to drop to the floor of the elevator, just as if it were falling. (See Fig. 15-14*a*.) Actually, of course, the object would continue to move at a constant velocity, but the elevator would keep accelerating. Thus the floor of the elevator would catch up with the released object, making it appear as though the object had fallen to the floor under the influence of a gravitational force. If an elephant and a feather were released in the elevator at the same time, they would appear to fall together as the floor caught up with them, just as if they were falling together under the influence of a gravitational force. (Recall the discussion on p. 93.)

In fact, if we were inside one of the two elevators in Figure 15-14 we could not tell whether we were in the accelerated elevator out in space, or in the one resting on the earth. The effect of the acceleration of our reference frame would be indistinguishable from the ef-

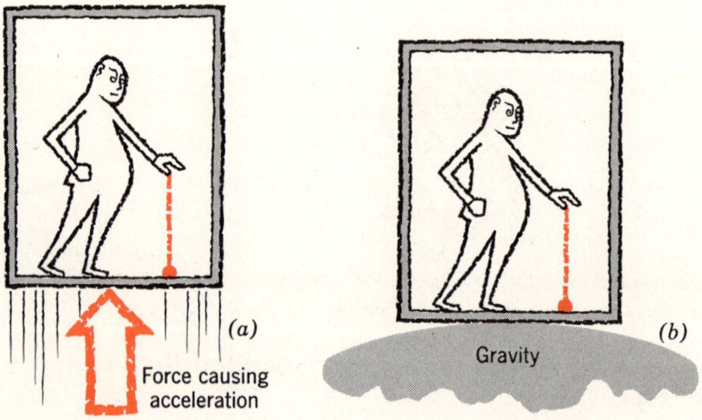

FIGURE 15-14 **The man inside the "elevator" cannot distinguish between situation (a) or situation (b). In (a) the elevator is somewhere in space being accelerated by a force pushing on the bottom. In (b) the elevator is at rest on the surface of a planet. (See Figure 15-15 for a slight modification of this principle.)**

fect of the gravitational field. This is a statement of what is called the *principle of equivalence*, which may be stated more precisely as follows:

There is no locally measurable difference between the effects of a gravitational field and the effects that arise from the acceleration of the reference frame.

That the term "locally" is necessary may be seen by examining Figure 15-15. The paths of the balls falling toward the earth are not strictly parallel, as are those in the accelerated elevator. We can make the paths as closely parallel as we please, however, by restricting ourselves to a small enough region in space.

The principle of equivalence enables us to think of the apparent forces in accelerated reference frames just as we think of gravitational forces in unaccelerated reference frames, at least locally. Thus Einstein was able to put accelerated reference frames on a par with unaccelerated reference frames, at least locally, and construct his general theory of relativity.

The theory has some very curious results which we shall mention but not discuss in detail. In this theory, spacetime, and space itself, turn out to be curved, the curvature becoming greater as one approaches massive objects. This sounds strange; what it means, for example, is that real space does not exactly obey the theorems of Euclidean geometry, which is studied in high school, but rather another kind of geometry similar to the geometry of curved surfaces.

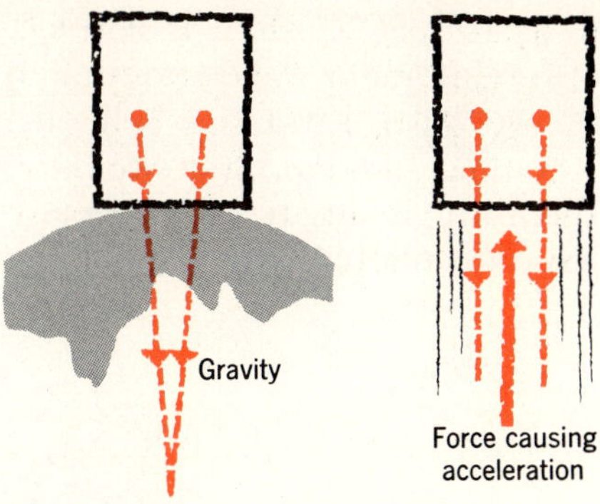

FIGURE 15-15 We can detect a difference in the paths traveled by two objects in the two different situations if the elevator is big enough.

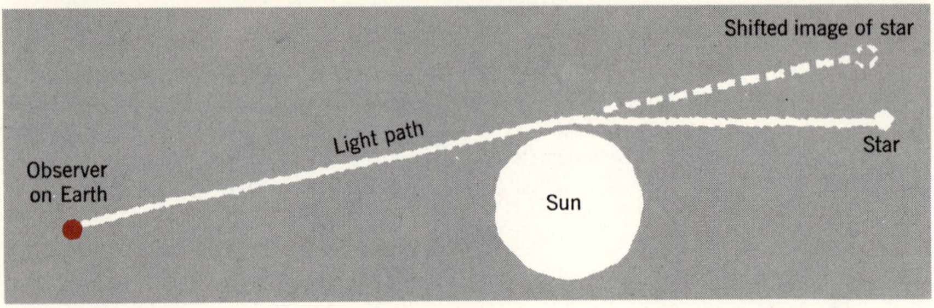

FIGURE 15-16 **Starlight is deflected in the gravitational field of the sun.**

Also in this theory the concept of curved spacetime replaces the concept of gravitational force. For example, a planet orbiting around the sun moves in a curved path, not because of any attraction to the sun, but because it travels in the curved spacetime surrounding the sun. It is like a path going through the mountains—the shortest path will not be straight because the ground is not flat.

Evidence for the General Theory of Relativity

Very careful experiments indicate that Einstein's general theory of relativity is superior to Newton's law of gravitation. We would expect light passing near the sun to be deflected, since the mass-energy in the light is attracted to the sun. (See Fig. 15-16.) Einstein's theory predicts twice as much deflection as Newton's. Observations of starlight passing very near the sun during total eclipses of the sun indicate that Einstein is correct. Similar, more recent experiments observing radio waves from radio sources in space favor Einstein's theory. Also, Einstein's general theory accounts for the very small variations in the orbit of Mercury, mentioned on p. 100. The general theory of relativity affords many unsolved problems, but it holds experimentally in every situation in which it has been tested.

FUNDAMENTAL IDEAS

I. FUNDAMENTAL LAWS AND PRINCIPLES

A. EINSTEIN'S TWO POSTULATES for the special theory of relativity.

1. The special principle of relativity.
2. The constancy of the speed of light, regardless of the reference frame.

B. PRINCIPLES IN GENERAL RELATIVITY
1. The general principle of relativity.
2. The principle of equivalence.

II. CONSEQUENCES AND EVIDENCES OF RELATIVITY

A. THE SPECIAL THEORY
1. Consequences
   a. Time Dilation.
   b. Lorentz contraction.
   c. Increase of mass with speed.
   d. $E = mc^2$
   e. Limiting nature of c.
   f. The fourth dimension.
   g. Problem of simultaneity.
2. Evidences.
   a. Michelson-Morley experiment.
   b. Lifetime of moving muons.
   c. Change of electron mass in a cyclotron.
   d. Atomic bomb and similar processes.

B. THE GENERAL THEORY
1. Consequences.
   a. Curved space.
   b. New theory of gravitation.
2. Evidences.
   a. Deflection of starlight near the sun.
   b. Path of Mercury

QUESTIONS

General

1. Suppose that the speed of light were 100 miles per hour.
   (a) Could you drive a distance of 500 miles in less than five hours? Explain.
   (b) Could you use the Lorentz contraction to help reduce? Explain.
   (c) Would the effects of car accidents at comparable speeds be worse or better than they are now?
   (d) Would it still be practical to send out time signals over the radio?

(e) Would radar be obsolete?

(f) Would it be necessary to have speed limits? If the speed limit were close to 100 mph, how would you suggest that the police measure speed?

2. Why, when we are interested in time and space, do we speak of clocks and measuring rods instead?
3. Explain what is meant by the statement, "Time is a property of clocks."
4. What is space? What is time?
5. Is the field material?
6. We have suggested that space is a property of our measuring rods. We also know that heat will cause rulers to expand. Why do we not consider, then, that heat contracts space?
7. How can we be sure that the speed of light is really the same in a moving system as in a stationary system, in view of the fact that the clocks and rulers we use to measure it may be affected by time dilation and the Lorentz contraction?

Ether

8. Why was the ether invented, and why was it abandoned?
9. Why did people think there was an ether "wind" blowing by the earth?
10. Why aren't people blown off the earth by the ether wind? (Give a reason other than the fact that there isn't any ether.)
11. Compare the results of the "boat race" and the Michelson-Morley "light race." Which contestant won each race and why?
12. A die-hard flat-earther claims that the earth is stationary, citing as evidence the results of the Michelson-Morley experiment. How might he argue, and how would you reply?
13. Why did many scientists believe in an ether?
14. Why did Einstein adopt his two postulates.

Speed of Light

15. If a hunter riding in an open jeep stands up and fires his gun in the direction in which the jeep is traveling, will the bullet travel faster than it would if he had fired when standing on the ground? Explain.
16. If a hunter riding in an open jeep at night stands up and directs a powerful searchlight in the direction in which the jeep is traveling, will the light travel faster across the ground than it would if he were standing on the ground with the searchlight? Explain.

17. How can we make a rocket move faster than light?
18. The speed of light c is often called a limiting speed. What does it limit?

Time Dilation

19. Explain the operation of a light clock when it is at rest and when it is moving.
20. What was measured in the muon experiment, and how does it support the predicted time dilation of special relativity?
21. A certain star is 1000 light years away. Could a human being live long enough to make a journey there, according to the laws of physics? (A light year is the distance that light travels in one year.)
22. A teenager claims to be the mother of a grayhaired old lady. Is this physically possible? Explain.
23. Why can we use a light clock in our thought experiment instead of an ordinary clock? Why should this special type of clock give the same results as any other clock? For that matter, why is this thought experiment valid—isn't an actual experiment required to prove something?
24. How can we be sure of the results of the muon experiment since the same muons are not used in both parts of the experiment?
25. A clock in either of two reference frames in relative motion appears to be running slow as observed from the other frame. Thus each of two clocks could appear to be running slower than the other. Can you resolve the apparent paradox.?

Lorentz Contraction

26. Could time dilation occur without Lorentz contraction? Explain.
27. A ruler in either of two reference frames in relative motion appears to be shortened when observed from the other frame. Thus each of two rulers could appear to be shorter than the other. Is this possible? Explain.

Mass Increase

28. If we keep applying a force on an object, why won't it eventually go faster than light?
29. What is the difference between rest mass and ordinary mass?

$E = mc^2$

30. Would the mass of the earth-moon system be greater or less if the moon and the earth were pulled closer together?

31. Do two marbles weigh more or less when they are apart than when they are together?

32. Is a hot biscuit heavier than a cold one? Would you expect to be able to detect the difference in weight?

33. Which of the following processes increase the mass of a body of matter and which ones decrease it: boiling, freezing, shining sunlight on the object, starting it in motion, cutting it in two, stretching it, bringing it to rest, cooling it, expanding it, contracting it, twisting it, spinning it?

34. Why is it difficult to measure the changes of mass resulting from the ordinary changes in an object's energy?

35. What has happened to the conservation laws as a result of the theory of relativity?

Simultaneity

36. Why was the notion of absolute simultaneity abandoned?

37. If clocks are set as Einstein's theory requires in two systems moving uniformly in a straight line with respect to each other, show why two events that are simultaneous in one system may not be simultaneous in the other.

38. What difficulty arises in trying to measure the one-way speed of light?

39. We (a) define or (b) assume the one-way speed of light in a vacuum to be equal to the two-way speed. Which word is correct and why? Why is the other choice incorrect?

The Fourth Dimension

40. Why is space three-dimensional?

41. Why is time called the fourth dimension?

42. What is space-time?

43. How many dimensions are required by the following quantities or combinations of quantities: (a) temperature, (b) the direction of an arrow, (c) the direction and position of an arrow in space, (d) the state of motion of a particle, (e) the state of motion of two particles, (f) the state of motion of two particles connected by a rigid rod, (g) the position and motion of a single particle.

The General Theory of Relativity

44. How does the general principle of relativity differ from the special principle of relativity?

45. Does the principle of equivalence state that a gravitational field is the same as the results of acceleration?

46. Does Newton's theory of gravitation account for the deflection of starlight passing near the sun? Why do we need Einstein's theory?

16 Wave Particle Duality

We now know how completely the quantum permeates all of existence. With the physicist it has become almost an obsession, haunting his every equation, dictating his every experiment, and leading him into long and not always fruitful argument with philosopher and priest on God and free will. Already its advent has revolutionized certain aspects of theoretical chemistry, and from chemistry it is but a short step to biology, the science of life itself.

Banesh Hoffmann[1]

Particles of Light

Planck's Constant

In 1900 Max Planck (1858–1947) used a revolutionary trick to solve a difficult problem concerning electromagnetic waves.

[1]From *The Strange Story of the Quantum* by Banesh Hoffman, second edition, Dover Publications, Inc., New York, 1947, 1959, p. 16. Reprinted through permission of the publisher.

FIGURE 16-1 A diffraction pattern due to the wave nature of electrons. The diffraction rings are not circles due to the effects of magnetic field on the moving electrons. Photograph courtesy of Dr. Darwin W. Smith.

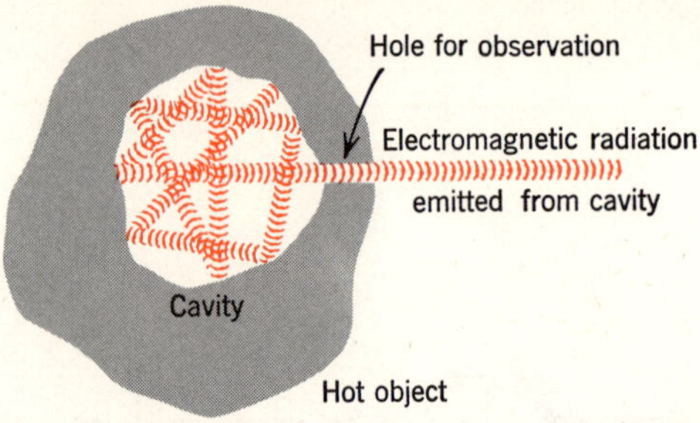

FIGURE 16-2 **Studying the radiation in a cavity, Planck found the first evidence for the existence of photons.**

The problem had to do with energy contained in the waves that should be found in an enclosed cavity in a material object at a given temperature (see Fig. 16-2). Since the object has internal energy, its molecules are vibrating and hence accelerating. Charged particles in the molecules also are accelerating and cause waves in the electromagnetic field. Therefore, in a cavity (with a small hole for observation) there should be electromagnetic waves bouncing around, being absorbed, reflected, and emitted by the walls of the cavity. The attempt to account theoretically for the energy contained in these waves according to the known laws of physics was unsuccessful—it kept turning out that there should be an infinite amount of energy in the waves, which of course was impossible. This was a theoretical catastrophe.

Planck avoided the catastrophe by assuming that the waves could be emitted only in discrete "bundles" of energy. The higher the frequency, the more energy there was in the bundle that had to be emitted. According to his calculations, this made it harder to emit at high frequency, so hard that one no longer got an infinite amount of energy in the waves. This assumption corrected the situation, and Planck was able to obtain a realistic law for cavity radiation.

For our purposes, it is not the problem of the cavity radiation itself that is important (we have discussed it only briefly and incompletely) but the fact that its solution led Planck to "discover" the discrete aspect of the electromagnetic field. Each of the bundles or packages of radiating waves has an energy given by the formula.

$$E = hf$$

E is the energy of the bundle and f is the frequency of the emitted wave. The constant h, called Planck's constant, has a value of

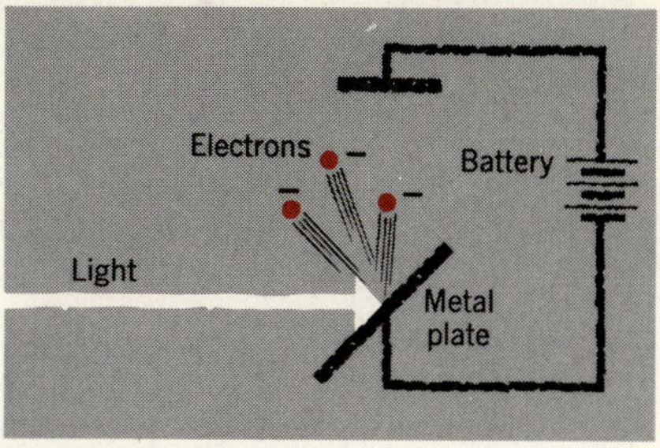

FIGURE 16-3 **The photoelectric effect: Light is absorbed as photons.**

6.63×10^{-34} joule second. This new fundamental constant was destined to be as important in physics as the constant c, the speed of light in a vacuum. Today we call such a bundle a *quantum* of light or a *photon*. The photon is a very important fundamental particle.

The Photoelectric Effect

Planck's constant was used by Einstein when he explained the photoelectric effect, which is the emission of electrons from a material surface, such as a metal, when light is allowed to shine on the surface. If a beam of light strikes the surface of a metal the energy of the light is transferred to some of the electrons in the metal, and they pop out (see Fig. 16-3). Now, light waves are waves in the electromagnetic field. As we have seen, these waves can disturb electrons or any charged particle. Therefore, it is entirely reasonable that these waves should disturb the electrons in a metal surface, and even disturb them so much that they are tossed out. However, P. Lenard (1867–1947) discovered two things about the way in which these electrons are tossed out that did not seem reasonable according to known physical laws. First, let us shine a light of a single color on the metal surface and measure the speed of the electrons that come out of the surface. Then let us make the light brighter, without changing its color. The frequency of the light is therefore still the same, although we would expect the amplitude of its wave to be greater. As a result of this brighter light, more electrons escape from the metal surface. However, the maximum speed with which the electrons leave the metal does not change. This result is rather surprising, since we would think that the more energetic wave (higher amplitude) would give the electrons a stronger toss. As an analogy, consider the bird on the clothesline shown in Figure 16-4. The bird could probably ride out a small-amplitude wave without being thrown off the rope. The big wave, however, should throw him off violently. Electromagnetic waves evidently do not have quite this effect on electrons. The amplitude of the wave apparently does not affect the speed with which the electrons leave the surface.

If we now change the color of the light, so that waves of higher frequency are striking the metal, we *do* increase the maximum speed with which the electrons leave the metal. The greater the frequency, the greater this speed. Below a certain frequency, the electrons do not have enough energy to leave the metal at all.

The known laws of the electromagnetic field were insufficient to give a convincing explanation of these two experimental facts. Einstein, in one of his three famous papers of 1905, suggested that the light waves were absorbed by the electrons in bundles of energy. The amount of energy in the bundle (or photon) depended on the frequency according to Planck's formula, $E = hf$. An electron got either the whole photon or nothing at all. If the photon had too little energy, the electron did not get enough energy to break free from the metal surface. If, however, the frequency were above some minimum,

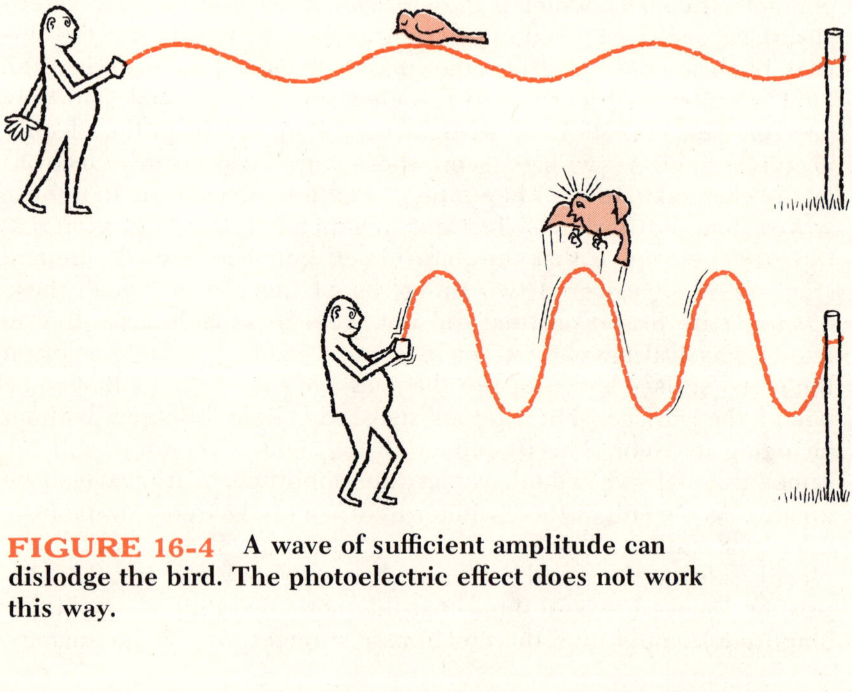

FIGURE 16-4 **A wave of sufficient amplitude can dislodge the bird. The photoelectric effect does not work this way.**

the electron would have enough energy to get out of the metal and still have some energy left as kinetic energy.

Increasing the brightness of the light did not increase the energy of the photons but merely supplied more photons. Therefore it did not increase the maximum energy with which the electrons were emitted. However, it *did* increase the *number* of electrons being emitted since there were more photons striking the metal. If, on the other hand, the frequency of the light was increased, the energy in each photon was greater. The more energetic photons supplied more energy to the electrons, and the electrons were emitted with a higher maximum energy.

We use the term *maximum energy* because some electrons lose more energy than others on their way out of the metal, by colliding with atoms or in other ways, and therefore come out with less kinetic energy than the maximum. They cannot come out with more than the maximum, however. When the electron is expelled with the maximum kinetic energy, the energy of the photon, hf, is accounted for as follows:

$$hf = \text{Maximum kinetic energy} + \text{work function}$$

The *work function* is the minimum energy the electron must have in order to escape from the surface of the metal. Later work showed that Einstein's equation described the photoelectric effect precisely.

The Compton Effect

Another experiment, performed with X rays by Arthur H. Compton (1892-1962), confirmed the existence of photons. X rays are high-energy electromagnetic waves, much more energetic than ordinary light waves. Their wave lengths range form 10^{-8} to 10^{-11} meter (the size of an atom is about 10^{-10} meter). One of the effects that occurs when X rays strike matter is the so-called Compton effect (see Fig. 16-5): A beam of X-ray waves of the same frequency (and wavelength) is allowed to fall upon a sample of matter. The X-ray waves interact with the individual electrons in the matter and are scattered. The frequency in the scattered waves is not the same as the frequency of the original wave but depends on the direction of the scattering

Compton found that he could predict the correct value for the frequency of the scattered waves if he assumed that the X-ray beam was a beam of particles (photons) instead of waves. He thought of the interaction of the wave with the electron as a collision between two

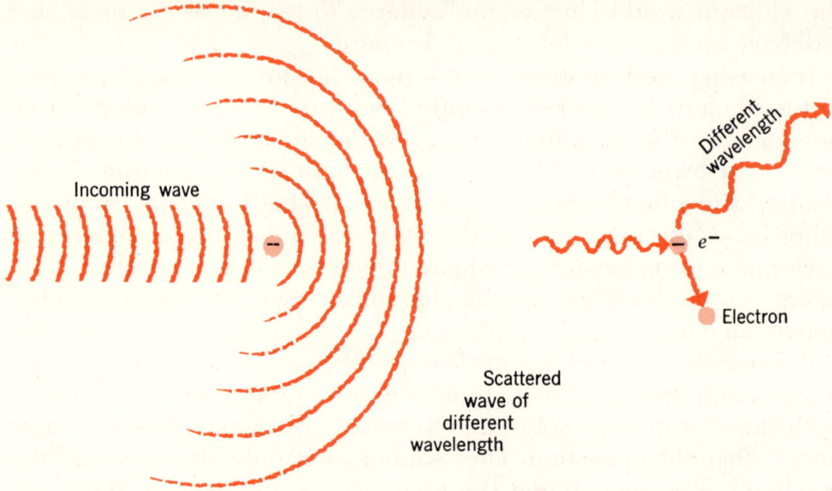

FIGURE 16-5 The change in wavelength of the scattered wave can be explained by the existence of photons.

particles, the photon and the electron. By applying the laws of conservation of energy and momentum to this collision, he could find the energy of a photon scattered in any direction. From this energy he could find the frequency of the wave by using Planck's equation, $E = hf$.

Thus Compton was able to treat the interaction between the wave and the electron as if it were an ordinary collision between two particles, the photon and the electron. This success supplied more evidence that electromagnetic waves often acted as if they were made up of the particles we call photons.

Waves in the Field as Particles—Photons

The experiments described above established the fact that electromagnetic waves, such as light, had a particle aspect. These waves act in many circumstances as if they were particles (photons) with definite energy and momentum. Yet light unquestionably behaved like a wave in such phenomena as diffraction and interference. Thus there was a problem—the wave-particle duality problem. The character of light (and the field) had these two proved (yet apparently contradictory) traits.

The Bohr Atom

An atom is made up of a very small, positively charged nucleus, with a cloud of electrons around it. (The discovery and composition of the nucleus will be discussed in Chapter 18.) The atom is held together by the forces of electrical attraction between the nucleus and the electrons. This fact raised two problems about the structure of the atom that needed to be solved.

First, the electrical forces between the nucleus and the electrons should cause the atom to collapse. The electrical forces would cause the electrons to be pulled straight to the nucleus if they were not moving. But there is good experimental evidence to show that they *are* moving. The magnetic properties of matter are explained by the moving charges within matter, and there are more direct experiments[2] as well. The atom, then, is pictured as a central nucleus with the electrons revolving around it (see Fig. 16-6). However, a revolving electron is an accelerating electron, and an accelerating electron should radiate according to the laws of electricity and magnetism, sending out electromagnetic waves. As it sends out these

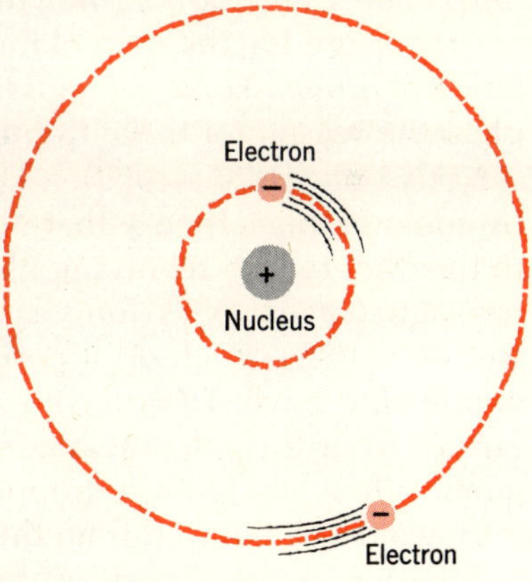

FIGURE 16-6 The Bohr model of an atom.

[2]The details of these experiments are beyond the level of this book. Briefly, an experiment of J. W. M. duMond and P. Kirkpatrick showed that an observed variation in the Compton frequency shift matched that predicted by the expected momentum of orbiting electrons in the Schrodinger theory.

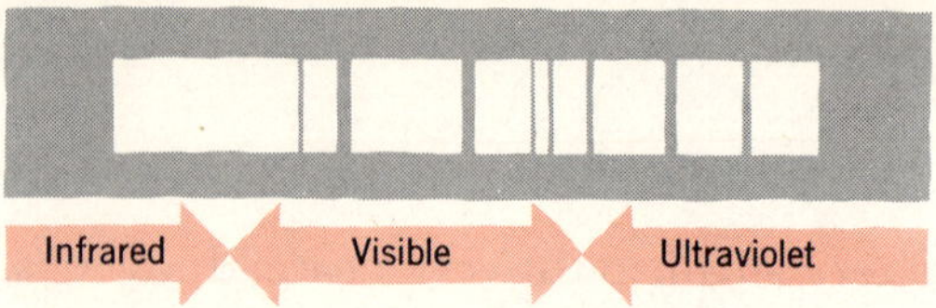

FIGURE 16-7 **The light spectrum of the element mercury. This spectrum can be obtained by allowing light from a sample of glowing gaseous mercury to pass through a diffraction grating and strike a photographic film. See Chapter 14, page 274.**

waves, it should lose energy and spiral into the nucleus, causing the atom to collapse. However, atoms do not collapse. They appear to be very stable structures.

When atoms do radiate, after various kinds of disturbances, the electromagnetic waves they send out are of certain definite frequencies. Each kind of atom has its own set of definite frequencies. Figure 16-7 shows an example of the frequencies emitted by one kind of atom. Each line represents a definite frequency of emitted light. One can identify the kind of atom, be it hydrogen, helium, oxygen, etc., by the characteristic frequencies of the waves it sends out. These frequencies must arise from the structure of the atom because they are the same for the same kind of atoms, and different for different kinds of atoms. How is it possible to discover this structure and explain the reason for these frequencies?

Neils Bohr (1885–1962), a Danish physicist, attacked this problem by constructing a theory that modified the laws of physics. According to this theory, the revolving electrons in the atom did not emit electromagnetic waves as long as they stayed in certain definite orbits (see Fig. 16-8). Each orbit corresponded to a certain energy that the atom could have. In order for an electron to move to a higher orbit, farther from the nucleus, energy would have to be supplied to the atom. When an electron dropped from one orbit to a lower orbit, energy would be emitted from the atom, perhaps as a photon.

Now we are in a position to understand why an atom emits only certain definite frequencies of light. When an electron drops from one level to a lower one, energy must be released. This may be in the form of a photon, containing just the amount of energy that has been removed from the atom. What energy can this emitted photon have? Only that lost by the atom when the electron dropped to the lower orbit. According to this theory, each photon that is emitted is

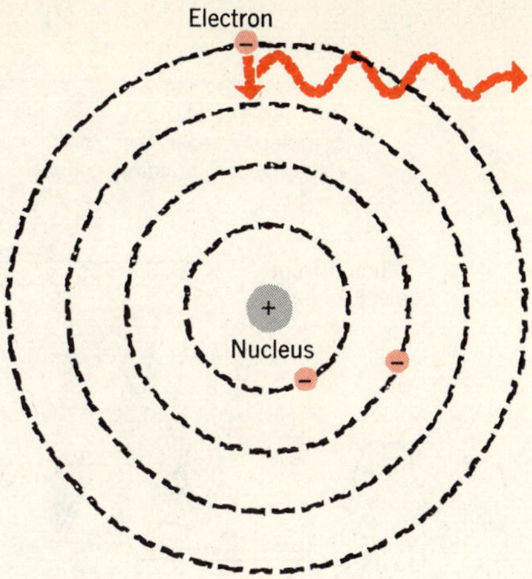

FIGURE 16-8 **A photon is emitted when an electron jumps to an orbit of lower energy.**

caused by an electron dropping from one orbit to another. Each atom, therefore, emits photons that represent differences in energy between two of its own characteristic orbits. Thus the emitted photons can have only certain definite energies. Since the energy of a photon determines its frequency, these photons can have only certain definite frequencies as well. This explains why certain frequencies are emitted from each kind of atom.

If the atom has an energy E_1 when the electron is in orbit and an energy E_2 after the electron has dropped to a lower orbit, then the energy of the emitted photon must be the difference in energy, $E_1 - E_2$. This means that we can find the frequency of the photon using $E_1 - E_2$ for the energy in the equation $E = hf$:

$$E_1 - E_2 = hf$$

Since there are only certain characteristic orbits for a given kind of atom, there are only certain energies that can appear on the left side of the equation. Therefore, for each kind of atom there will be only certain characteristic frequencies that can appear on the right.

The building shown in Figure 16-9 provides a rough analogy of these electron orbits and their energies. The various electrons live on the various floors, but they cannot live between floors. The floors represent the permitted orbits. Energy is required to hoist an electron from one floor to another. With sufficient energy, one might remove an electron from the building altogether and send it into space. The

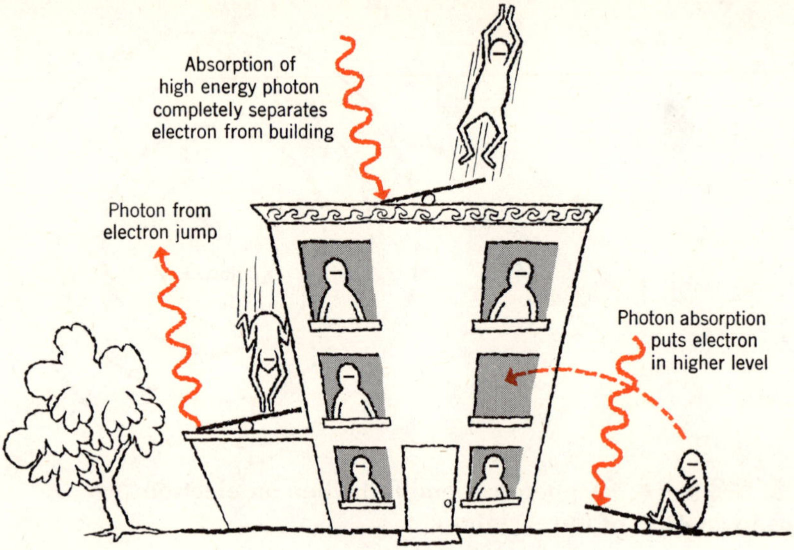

FIGURE 16-9 A rough analogy of energy changes of electrons in atoms.

first floor is right on the ground, representing the lowest-lying orbit in the atom.

Why don't all the electrons in the atom fall down to the lowest orbit? This would correspond to a situation in which all the people in a building lived on the ground floor. The explanation is given by the so-called *exclusion principle* discovered by Wolfgang Pauli (1900–1958). This principle tells us that no two electrons can ever be in exactly the same state or condition. The orbit does not determine the condition of an electron completely, but it does help to determine the condition. There can be only a certain number of electrons in each orbit. When an orbit has been filled, it will admit no new electrons unless it gets rid of some of the others.

We have seen that an electron dropped from one orbit to another can cause the atom to emit a photon with an energy equal to the difference in energy between the two orbits. The reverse may also occur if a photon strikes an atom. The photon may have enough energy to remove an electron from the atom completely. A particularly interesting process occurs if the photon energy corresponds exactly to the energy difference between two orbits. If there is an electron in the lower of these two orbits but none in the higher orbit, the atom can absorb the photon, causing the electron to jump to the higher orbit

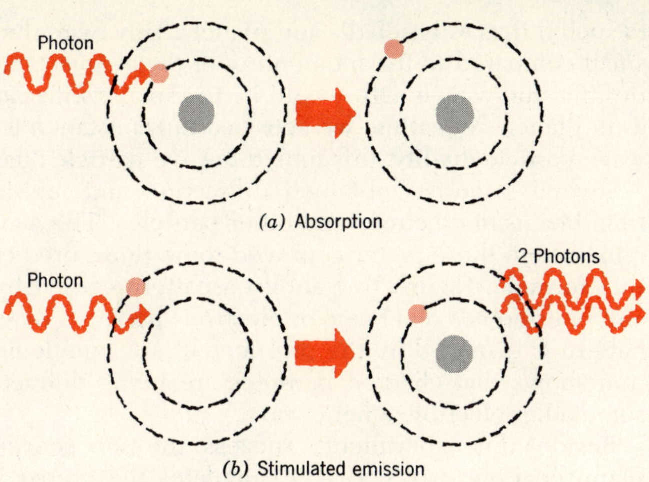

FIGURE 16-10 **Absorption and stimulated emission of photons.**

(see Fig. 16-10*a*). The photon, of course, disappears as it is absorbed, just as it appears when it is emitted. If there is already an electron in the higher orbit, however, and none in the lower orbit, then the presence of the photon can actually stimulate the emission of another photon of the same energy, as the electron drops to the lower orbit. This process is called the *stimulated emission* of radiation (see Fig. 16-10*b*). Thus, photons of the right energy can actually cause absorption or emission, depending on the state of the atoms they affect.

The Bohr theory of the atom was rather effective in explaining the frequencies emitted by the hydrogen atom, but it was unsuccessful for other atoms.

De Broglie Waves

In 1924 Louis de Broglie suggested that particles of matter, such as electrons and protons, should have a wave nature. The wavelength is given by Planck's constant h divided by the magnitude of the momentum of the particle:

$$\text{Wavelength} = \frac{h}{\text{momentum}}$$

(This same equation is true for photons.) The equation tells us that the greater the momentum of the particle, the smaller its wavelength. In principle the equation applies not only to small particles but also to such things as baseballs and planets. However, the constant h is so small compared to the momentum of these objects at normal velocities that the wavelength would be too small to measure. Once again, as in Planck's equation, we note that the constant h is involved in the wave-particle duality, this time the wave-particle duality of matter.

Several scientists obtained diffraction and interference patterns from beams of electrons and other particles. The wavelengths determined from these patterns proved to be those predicted by de Broglie's formula. Figure 16-1 shows a pattern caused by the diffraction and interference of a beam of electrons passed through a crystal. The pattern is distorted by the presence of a magnetic field. This distortion shows that charged particles are being diffracted, not light or some other electromagnetic wave.

Besides this experimental success, the new concept helped to explain the Bohr orbits. If one calculates the proper wavelengths for the electrons in Bohr orbits, these wavelengths fit into the circular path lengths of the orbits exactly (see Fig. 16-11). The Bohr orbits were just those orbits in which a wave electron would give standing waves, much like the waves on the strings in Figure 13-18 (p. 259).

However, a wave traveling around a central nucleus could not be expected to travel in a narrow circular path without spilling over into the rest of space. Therefore the Bohr orbits were approximations. What was needed was a mathematical treatment of the electron as a space-filling wave, not just as a wave along a curved line. This treatment was provided in 1926 by Erwin Schrodinger (1887–1961), who discovered the equation that describes these waves. Thus the Bohr

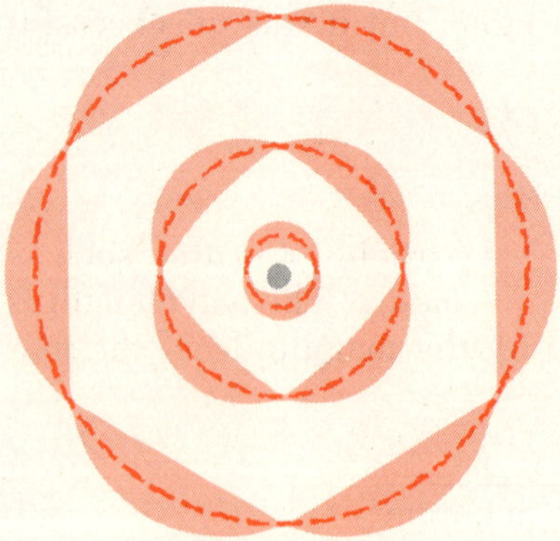

FIGURE 16-11 **The DeBroglie wavelengths fit into the Bohr orbits.**

orbits were replaced by "orbits" or states in which the electron wave was spread through the space around the nucleus in various ways, as illustrated in Figure 16-12. The Schrodinger theory also predicted other states, (or "orbits") which were missing in the Bohr theory. (Another theory of the atom, formulated in 1925 by Werner Heisenberg, was equivalent to Schrodinger's but far more abstract.) The original de Broglie waves around the Bohr orbit can be likened to the standing waves on a string that were discussed in Chapter 13. Space is not a string, however; it is more like the plate we discussed in Chapter 13 (see Fig. 13-19, p. 260). The more realistic waves of Schrodinger, therefore, are more like the standing waves on the plate. Even the waves on the plate, however, are confined to the plate, while the Schrodinger electron waves spread out into three-dimensional space.

There are important differences between these waves (introduced by de Broglie and Schrodinger) and the waves on the string or plate that we discussed in Chapter 13. Waves *in* matter are caused by the cooperative motion of many particles. Water waves, for example, arise from the collective motion of many water molecules. A similar statement could be made about the molecules in the strings, plates, etc. The de Broglie waves, as far as we know, are *not* formed in this way. A *single* particle by itself has a de Broglie wave associated with it. These waves *of* matter are evidently part of the nature of matter itself and can pertain to the behavior of a single particle.

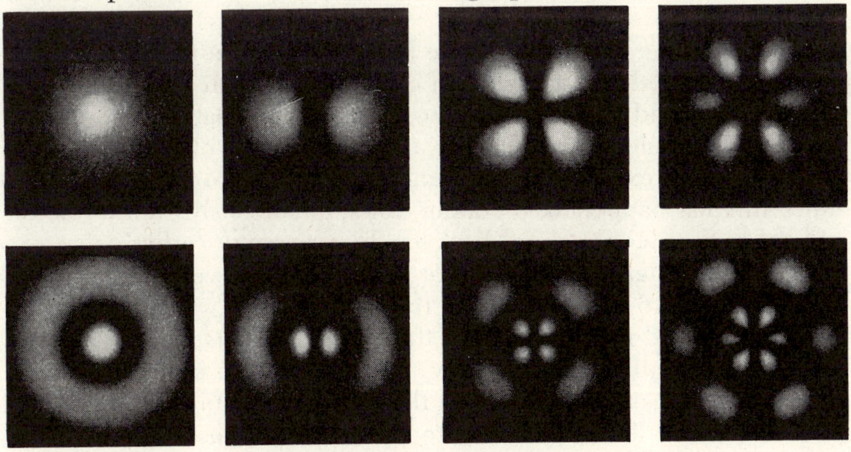

FIGURE 16-12 **Electron waves in the atom. (These are, of course, not actual photographs of the atom.) From Leighton, *Principles of Modern Physics*, McGraw Hill pp. 178–179.**

Because of the discoveries discussed above, matter and light presented the same problem. Both material particles (such as electrons) and electromagnetic field waves (such as light) showed wave *and* particle characteristics. There was thus a wave-particle duality problem with both matter and electromagnetic waves. There seemed to be no way to tell which was the more fundamental material reality, the particle or the wave (or field). Is it best to think of a field spread over space with waves traveling in it, or of localized particles moving through space? We have already seen (Chapter 15, p. 304) that both particle and wave can have mass and energy. In fact, because of the wave characteristics of the material particle, one can talk about a *matter field*, in which the matter waves exist. Or, if one is dealing with the field, the field must be lumped into particles, or *quantized* in order to explain some observations.

The Laser

An interesting application of quantum physics is the device called the laser. The laser produces a very straight, intense beam of light, which is very useful in a variety of applications (see, for example, Fig. 1-3, p. 5).

To explain the operation of one kind of laser, we recall that electrons are allowed only in states of definite energy in an atom. These states correspond to the "orbits" or standing-wave patterns described earlier. For an electron to jump to a state or "orbit" of higher energy, it must acquire exactly the right amount of energy to make the jump. Once this has happened, the atom can spontaneously emit a photon of light as the electron drops back to a lower-energy "orbit."

Let us consider an atom with at least three allowed electron "orbits" (see Fig. 16-13). Most of the time the electron we are interested in will be found in the lowest of these three "orbits," and the other two "orbits" will be empty. Now suppose that we shine a light on this atom that has photons of just the right energy to cause the electron in the first "orbit" to jump to the third. Suppose then that this electron, now in the third "orbit," immediately drops to the second "orbit." We cannot discuss the factors that influence the electron in making these changes, but we shall further assume that this drop from the third "orbit" to the second is far more likely than the drop

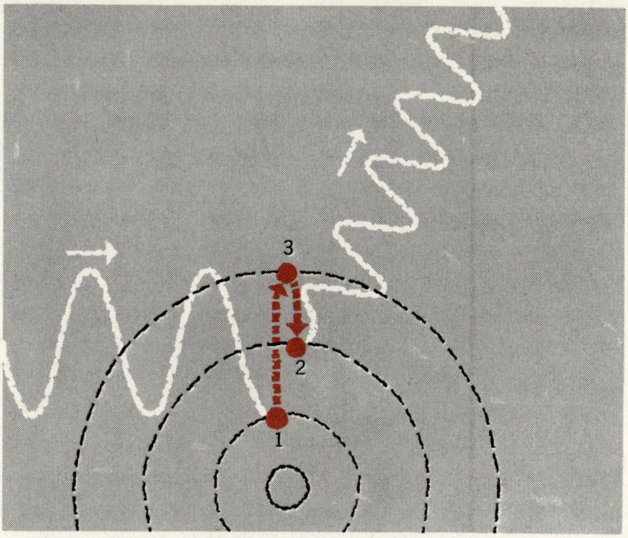

FIGURE 16-13 **An atom in a laser. The numbers refer to the three "orbits" discussed in the text.**

from the second "orbit" back down to the first. Thus, if the light keeps shining on a group of these atoms, electrons will keep rising to the third "orbit" and dropping back to the second. In the second "orbit," however, they will be more or less stuck. Electrons will drop into the second from the third much faster than they fall to the first. Therefore, most of the atoms have electrons in the second "orbit" under these conditions.

If some method could be found to get these electrons to drop back into "orbit" number one all at about the same time, there would be a beam of light from the emitted photons. We have already pointed out that an electron in an upper "orbit" can be stimulated so that it will fall to a lower "orbit" and emit a photon. This stimulation is produced by another photon of the same energy coming from outside the atom (see Fig. 16-10*b*). If we wait long enough, one of the electrons in the second "orbit" of one of the atoms will spontaneously fall back to the first "orbit." This is an unlikely event, but with enough atoms present it is bound to happen. This photon can then stimulate the same electron drop in another atom. Then we have two photons, each of which can cause an electron drop in other atoms, and so on until the atoms have all radiated (see Fig. 16-14). This explains the laser action except for one point.

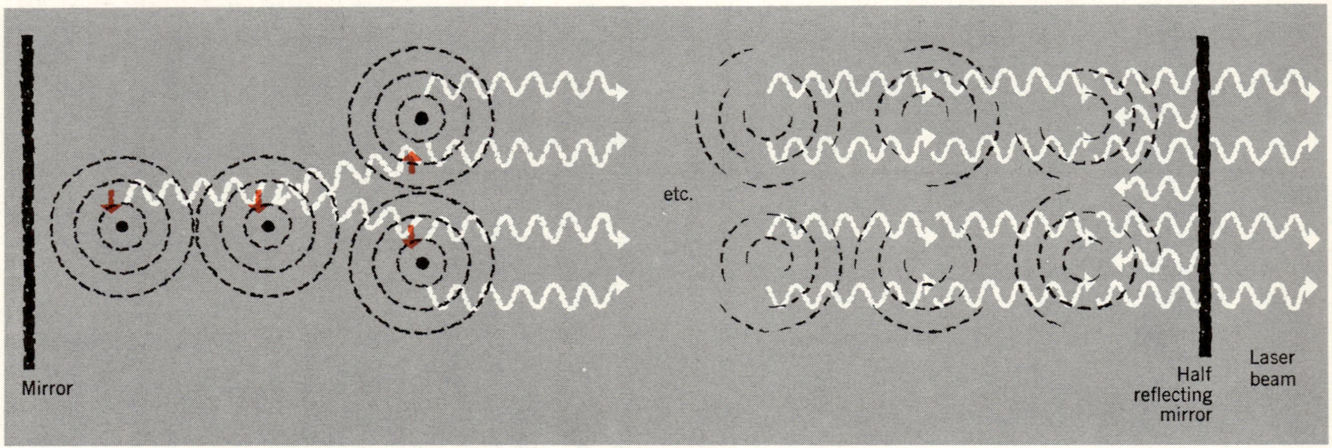

FIGURE 16-14 The action of a laser.

You will notice that there are mirrors installed at each end of the sample of laser material in Figure 16-14. If a photon is emited in the direction of one of the mirrors, it can bounce back and forth like the light in a light clock. If a photon is emitted in another direction, so that it does not strike one of the mirrors, it may move out of the apparatus before it stimulates many atoms to emit. The photon bouncing back and forth between the mirrors, however, stays in the apparatus for a while and continues to stimulate the emission of other photons. These emitted photons are in the same direction as the photon that stimulated their emission. Therefore, the beam of photons moving back and forth between the mirrors builds up to a very intense beam in both directions. If one of the mirrors is only partly reflecting, part of this intense beam comes through and can be used outside the apparatus. The other part of the beam is reflected back and continues to stimulate the emission of other atoms.

FUNDAMENTAL IDEAS

I. FUNDAMENTAL LAWS AND PRINCIPLES

- **A.** WAVE-PARTICLE DUALITY of both matter and electromagnetic radiation.
  1. Particle nature of electromagnetic radiation shown experimentally by:
     - a. Waves in a cavity.
     - b. Photoelectric effect.
     - c. Compton effect.
  2. Wave nature of material particles shown experimentally by:
     - a. Diffraction of electrons.
     - b. Bohr orbits as standing waves.
- **B.** EXCLUSION PRINCIPLE.

II. FUNDAMENTAL CONSTANT

- **A.** PLANCK'S CONSTANT: $h = 6.63 \times 10^{-34}$ joule second.

III. PARTICLES

- **A.** THE PHOTON, A fundamental particle
- **B.** THE ATOM
  1. Composition: positive nucleus and cloud of electrons.
  2. Structure: the Bohr atom: electron orbits.

3. Structure: the Schrodinger waves: three dimensional electron waves.

IV. APPLICATIONS

A. THE LASER

QUESTIONS

Particles of Light

1. Outline the photoelectric experiment described on p. 323. What was measured? What result was expected? What was observed? How did Einstein explain his result?
2. Why do we think of light as a wave?
3. Why do we think of light as a particle?
4. Can you account for the occurrence of the work function in the equation on p. 325.
5. Why must the rest mass of a photon be zero?

Waves of Matter

6. Why do we think of an electron as a particle?
7. Why do we think of an electron as a wave?
8. What is the difference between an electron and a photon?
9. Why do we think of the de Broglie waves as associated with a single particle instead of with a whole group of particles, as is the case for sound or water waves?

The Atom

10. Why was it thought that the stability of the atom contradicts the laws of electricity and magnetism?
11. How was the problem posed by the stability of the atom solved?
12. Why does an atom emit certain definite frequencies of light?
13. Does the fact that an atom can emit a photon of light violate the conservation-of-energy law? Explain.
14. What is the importance of Planck's constant?

The Laser

15. How can stimulated emission cause a laser to operate?
16. Why is the light that comes from a laser different in frequency from the light we shine on the material to make the laser operate?

17 The Uncertainty Principle and the Meaning of Quantum Physics

. . . There was a time when the newspapers said that only twelve men understood the theory of relativity. . . . There might have been a time when only one man did, because he was the only guy who caught on, before he wrote his paper. But after people read the paper a lot of people understood the theory of relativity in some way or other, certainly more than twelve. On the other hand, I think I can safely say that nobody understands quantum mechanics. So do not take the lecture too seriously, feeling that you really have to understand in terms of some model what I am going to describe, but just relax and enjoy it. I am going to tell you what nature behaves like. If you will simply admit that maybe she does behave like this, you will find her a delightful, entrancing thing. Do not keep saying to yourself, if you can possibly avoid it, "But how can it be like that?" because you will get 'down the drain', into a blind alley from which nobody has yet escaped. Nobody knows how it can be like that.

Richard P. Feynman[1]

[1]Richard Feynman, *The Character of Physical Law*, Cambridge: Massachusetts Institute of Technology Press, 1967, p. 129.

FIGURE 17-1

342 The Uncertainty Principle

In Newtonian physics the state of a system at one time determined its state at another time. If one knew the position and velocity of all the particles in a system and the forces between them, one could use the laws of motion to predict exactly the future behavior of the system, or even the past. Even if one did not know these positions and velocities, they did exist in principle and would themselves determine the future behavior of the system. There is, of course, some limitation on our knowledge of the position and velocity of a particle—the unavoidable experimental error. However, there was no reason to believe that this error could not be made arbitrarily small by improving our apparatus. The laws of Maxwell for the electromagnetic field were similar in the precision with which they described the field and predicted its behavior.

The principles of quantum physics directly contradict this earlier point of view. According to quantum physics, one *cannot* have precise knowledge of both the position and the momentum (and hence velocity) of a material particle. Therefore the predictions can be made only to a certain restricted degree of accuracy. This limitation arises not from any experimental error but from the laws of quantum physics themselves. We have not given these laws in their mathematical form, but we would like to discuss this consequence. It is called the *uncertainty principle* and was first stated by Werner Heisenberg.

According to the uncertainty principle, it is theoretically possible to know either the position or the momentum of a particle exactly, but we cannot know both exactly. The more we know of one, the less we can know of the other. No exceptions to this principle have been found. For large, ordinary objects, the amount of uncertainty is relatively small and unimportant. We can obtain enough information to predict the motions of the planets and the stars accurately for thousands of years. For electrons, however, there are conditions in which we cannot make predictions that are accurate for even a few seconds.

As an example of the uncertainty principle, we shall describe an attempt to prepare a particle in an accurately defined condition. Suppose that we want a narrow beam of electrons moving in a precisely defined direction. Figure 17-2 illustrates our attempt to obtain such a beam. We first obtain a beam of electrons all moving in the same direction and then try to narrow the beam by making it pass through a single slit. This method works up to a point, but we soon find that as we make the slit narrower, we actually make the beam

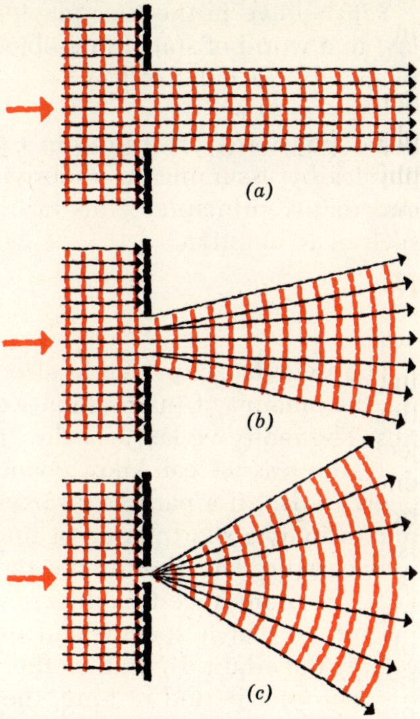

FIGURE 17-2 **Illustration of the uncertainty principle. As the position is determined more and more precisely, the momentum (in this case the direction of the momentum) becomes more and more uncertain.**

wider. The wave nature of electrons makes them diffract. We already know what to expect from a one-slit diffraction pattern. The narrower the slit, the wider the diffraction pattern (see Fig. 13-9, p. 251). As the slit is made narrower, more precisely defining the position of the electrons as they go through the slit, the direction of their momenta becomes more and more uncertain. The attempt to gain more information about the position of the particles by narrowing the beam thus results in a loss of information about their momenta.

As far as is now known, the uncertainty principle applies to all physical processes. Therefore, exact prediction is physically impossible, according to this principle. Some physicists have been uneasy about this uncertainty. Einstein in a letter to Max Born, expressed himself in these words.

You believe in the dice-playing God, and I in the perfect rule of law in a world of something objectively existing which I try to catch in a wildly speculative way.[2]

Some physicists are trying to explain the phenomena of quantum physics by assuming an underlying reality that does not involve an uncertainty principle. Thus far there is no experimental reason for such an assumption.

The Uncertainty Principle and Energy

The uncertainty principle also pertains to time and energy, limiting the amount of information we can have about these two quantities. The more we know of the energy of an object, or system of objects, the less we can know about the exact time at which it had that energy. Also, if a particle or a system remains in a particular condition for only a short period of time, then its energy in that condition is uncertain. The shorter the time interval in which the condition lasts, the more uncertain the energy. If the condition lasts for just an instant, so that it specifies an exact time, then the energy is completely uncertain. If, on the other hand, the condition lasts for an extremely long period of time, the energy may be known with great precision.

According to the uncertainty principle, therefore, a state of definite energy must last for a long time; the longer it lasts, the more certain we can be of the energy. We have reason to be interested in the states of a system that can exist unchanged for a long period of time. They should be involved in the long-lasting structures observed in the universe. We have already discussed the structure of the atom in terms of its energy states. The orbits in the Bohr theory of the atom, or the standing-wave patterns in the more accurate theory, correspond to states or conditions of definite energy for the atom. When we discuss the long-lasting structure of any system, be it atom, molecule, or nucleus, we are interested in its states of definite energy.

Most frequently, states of definite energy do not correspond to states of definite position and momentum. For example, the standing-wave states of definite energy in the atom do not correspond to

[2]Albert Einstein, letter to Max Born, November 7, 1944, published in Max Born. "*Natural Philosophy of Cause and Chance.*" New York: Dover, 1964, as quoted in V. Lawrence Parsegian, Alan S. Meltzer, Abraham S. Luchins, and K. Scott Kinerson, *Introduction to Natural Science, Part One: The Physical Sciences.* New York: Academic Press, 1968, p. 437.

states of definite position or momentum for the electron. The position of the electron is uncertain because the electron could be anywhere in the standing-wave pattern.

Informatiion and Planck's Constant

The amount of uncertainty in the position and momentum of an object is related to Planck's constant h.

$$(\text{Uncertainty in position}) \times (\text{Uncertainty in momentum}) \gtrsim h$$

The uncertainties of energy and time are related by a similar equation:

$$(\text{Uncertainty in time}) \times (\text{Uncertainty in energy}) \gtrsim h$$

(The symbol $\gtrsim$ means *greater than* or *roughly equal to.*) The product of the two uncertainties in each equation in roughly greater than or equal to h. This explains why the uncertainty does not bother us in large-scale operations. Planck's constant is a very small quantity. The uncertainty in both the position and the momentum of an object the size of a billiard ball is negligible in comparison with the size and momentum the ball would have under normal conditions. In dealing with very small particles such as electrons, Planck's constant is comparatively large, and the uncertainty may therefore be very important.

Because of its role in the uncertainty principle, Planck's constant determines the amount of information available in a physical situation. Therefore, it controls the maximum accuracy with which predictions of the future can be made on the basis of this information.

The Meaning of Quantum Physics

The Unobserved State[3]

Let us now try to deal with the wave-particle duality problem. We have two apparently necessary but apparently contradictory models for both matter and light—the particle model and the wave model. We need to examine the relationship between these models.

[3]Our discussion of the two-slit experiment here is similar to that of Richard Feynman in several places, for example, *The Character of Physical Law*, The M.I.T. Press, Cambridge, Massachusetts, and London, England, 1965, 1967, pp. 130-148.

Figure 17-3 diagrams the two-slit interference experiment described in Chapter 14. In this experiment we shall use electrons rather than light, although the main results would be the same for light. To the left of the slits we have a source of electrons. We shall assume that the electrons in the beam from this source all have the same velocity and are going in the same direction. At the right of the screen, some distance off, there is a photographic film. When an electron strikes this film, it exposes a tiny region marking the position at which it strikes.

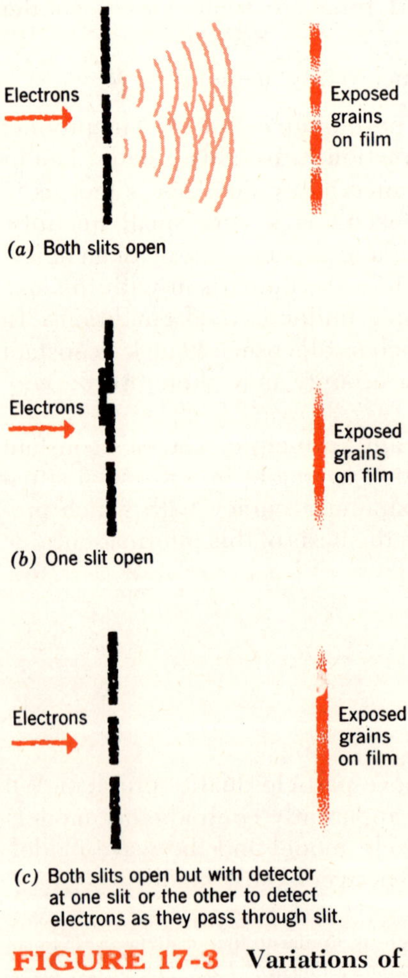

FIGURE 17-3 Variations of the two slit experiment with electrons.

The result of the experiment is the usual two-slit interference pattern on the film. This means that the electrons must somehow travel as waves through the slits. We also notice, on examining the film carefully, that the interference pattern on the film is caused by the exposure of a large number of individual grains or exposed spots on the film. Individual electrons strike the film at particular spots causing individual grains to be exposed. Thus the electrons must have struck the film as particles, not as waves.

This suggests a way out of the wave-particle dilemma that unfortunately turns out to be incorrect. Perhaps the wave behavior is not characteristic of single electrons but only of electrons in groups. Perhaps the wave behavior is a cooperative behavior, arising from the interactions among the electrons as they come through the slits. After all, the wave behavior is observed in the interference pattern, which cannot be formed unless many electrons strike the film. However, it turns out that this idea will *not* solve our problem, as we can see if we cut down the intensity of the electron beam striking the slits until there is a very low probability for more than one electron at a time to be traveling in the apparatus. Thus the great majority of the electrons go through the apparatus alone. We have to wait a long time for the pattern to be built up grain by grain, but when we get it, we see the same two-slit interference pattern we had before. Somehow, the behavior of each individual electron is affected by the wave property.

If each electron participates in the interference pattern, it must be affected by both slits. It is difficult to think of a single electron going through both slits. We can prove, however, that both slits are required to get the pattern. By closing one slit, we force the electron to go through one slit only. But as a result we get not the two-slit pattern but the one-slit pattern. By changing the experiment, we have obtained a different result (see Fig. 17-3*b*).

Does the electron go through one slit or both? We might devise the following thought experiment to try to answer this question. We set up a device near each slit that signals when an electron goes through that slit. We find from these signals that each electron does indeed pass through only one slit. When we look at the film pattern produced by these observed electrons, however, we find that *we no longer have the two-slit interference pattern.* We have two one-slit patterns instead, one from each slit (see Fig. 17-3*c*). By changing the experiment, we have again obtained a different result.

When we make a measurement to find out where the particle goes, we find that it goes through one slit or the other, never both. We fail, however, to obtain the two-slit pattern. Therefore we cannot answer the question in which we are interested: through which slit does each particle go when we get the two-slit pattern. We want to know where the particle is when we do not observe it, since when we do observe it we change the experiment and get a different result. Let us summarize this situation.

(a) When we do not experimentally determine through which slit the electron goes, we must think of it as going through both slits, since we get a two-slit pattern.

(b) When we *do* experimentally determine through which slit the electron goes, we find that it goes through only one slit and we get a one-slit pattern.

This same kind of problem arises in many other situations. Suppose that we wish to investigate a certain physical process. We make all the measurements we can on the system we are dealing with, without actually changing the process we wish to understand. Generally our measurements will not determine all the conditions of the system or the exact way in which the process takes place. In such circumstances we must think of the system as being (to some extent) in all the possible undetermined conditions at once. We must think of the process as occurring in all possible ways at the same time.

Thus, in our two-slit experiment we cannot say that the electron goes through one slit or the other unless it has been found to do so in some experiment or measurement. Furthermore, there is no way to determine through which slit it goes without disturbing the interference process we are studying. Therefore, we must think of the electron as being partially (or potentially) in all possible conditions at the same time, that is, as going through both slits in some sense.

When we *do* determine through which slit the electron travels, we actually change the process we are studying. This measurement, therefore, appears to give us information about a situation different from the one that existed before the measurement. The measurement interferes with the process.

The Copenhagen Interpretation of Quantum Physics

The conclusions we shall reach here represent the conventional interpretation of quantum physics today. It is the interpretation suggested by Max Born (1881–1970) and accepted by Bohr and others at Copenhagen. Some scientists, notably Einstein, Schrodinger, and de Broglie, were or are not happy with this way of looking at things. However, the so-called Copenhagen interpretation seems to agree with all the facts, and other interpretations are not yet required by experiment, as far as we know.

In the two-slit experiment, we found that we could not speak of a definite position for the electron unless we made a measurement that fixed it. If we do not make this measurement, then we must think of the electron as taking all possible paths or being in all possible places and conditions that are consistent with the measurements we *have* made. In the two-slit experiment there are two possible paths through the screen. One path leads through one slit, and another leads through the other slit. Since we have made it possible for the electron to go through either slit by leaving both slits open, we must consider that it does go through both. We must, however, consider it as going through as a wave. The waves from both slits interfere and produce an interference pattern. The pattern, however, cannot be detected by using only one electron. A single electron simply shows up as a single spot on the film. The pattern tells us where it is likely to appear, however, and this brings us to the conventional interpretation of quantum physics.

The waves are regarded as waves of probability. More accurately, the intensity of the wave at a point in space gives us the probability of finding a particle at that point. In the two-slit experiment, the waves must come through the slits. Then the waves interfere, setting up an interference pattern in space. The intensity of the wave at any point in space gives the probability of finding an electron there if a measurement is made. Finally, the waves come to the photographic film. It is very unlikely that the electron will go through the film without exposing it. The interposition of the film causes a measurement—it causes information to be created in an effect that we can observe. Thus the electron wave shows its particle behavior, giving us information in particle form. All we can find out from the wave is the probability of finding the particle at some point on the film. The probability is high where the interference pattern has a maximum. It is low where the interference pattern has a minimum. Apart from this probability the result is completely undetermined. The electron propagates as a wave and is observed as a particle. The wave gives the probability of finding the particle.

Suppose that we only set up a small piece of film that will cover only part of the interference pattern (see Fig. 17-4). Suppose that the electron does hit the film and is observed as a discrete particle. What happens to the rest of the continuous wave pattern that never touched the film? It vanishes instantly. If we know that the electron

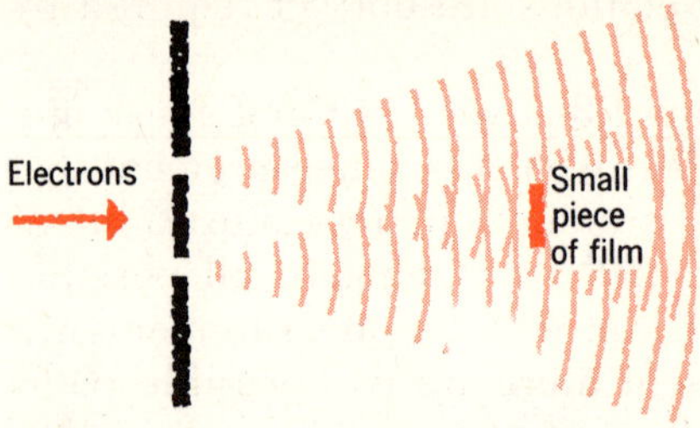

FIGURE 17-4 If the electron is detected by the film, the wave must "softly and suddenly vanish away".[4]

hit the film, then the probability of its being anywhere else is zero. The waves are waves of probability, so where the probability is zero, they must be zero also. When the particle is observed, the waves must vanish. We start over with a new wave, incorporating the new information that has been gained by observing the particle.

Applications of the Principles of Quantum Physics

The Zero-Point Energy of an Oscillator

The principles of quantum physics allow us to explain why the molecules and other small particles in material objects do not lose all their energy or cease all motion, even at the absolute zero of temperature. Let us consider an oscillator, for example the swing in Figure 17-5. Is it possible for this swing to have an energy state in which it is motionless at the lowest point of its path? If the swing remains at this point, then its position is exactly known. According to the uncertainty principle, therefore, its momentum must be completely uncertain. In such a circumstance, according to the principles of quantum physics, we must think of the swing as being at least partially in all these possible momentum states. Most of these states, however, are states of motion, and therefore the swing cannot be considered entirely motionless. Therefore, the swing will be swinging to some minute extent even in its lowest energy state.

[4]Lewis Carroll, *The Hunting of the Snark*, as quoted in Martin Gardner, *The Annotated Snark*, New York: Bramhall House, 1962, page 56.

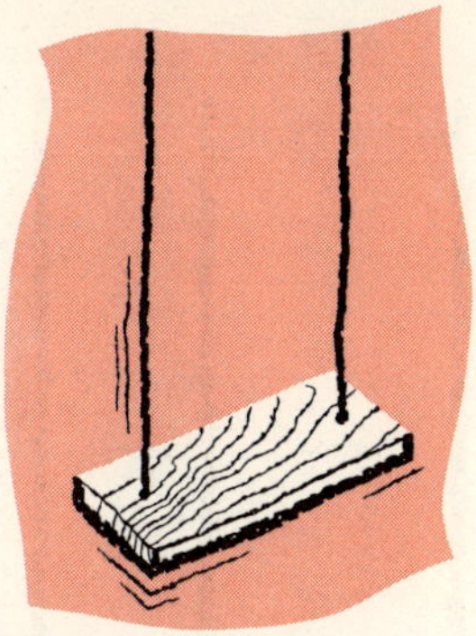

FIGURE 17-5 **Can the swing be completely at rest?**

Although this zero-point energy is insignificant for an object as large as a swing, it is important for the vibrating molecules, atoms, electrons, etc., in a material object. Even at absolute zero they would retain some oscillation because of this zero-point energy.

The Bound Electron

Using the uncertainty principle, we can explain why an electron resists being confined to an arbitrarily small volume. Suppose that we try to capture an electron at a certain position and hold it motionless. If we can do this, we will know the position of the electron exactly and also its momentum, since a truly motionless electron should have a momentum that is precisely zero. Figure 17-6 shows the device we will use in a thought experiment to try to perform this feat. One electron is inside the cylinder. The piston is slowly pushed in, decreasing the volume of space that the electron can occupy. The less volume the electron is allowed to occupy, the more information we have about its position. According to the uncertainty principle, however, the more information we have about the position, the less we are allowed to have about the momentum. As we continue to decrease the volume available to the electron, the momentum becomes more and more uncertain. The more uncertain the momentum of the electron becomes, the greater the likelihood that it is very high. As we have seen, we must consider the electron to be partially in all its possible states, and therefore we must consider the effect of these possible high-momentum states. That effect may be seen in Figure 17-6. The result is a bombardment of the piston by the electron, causing a resisting pressure on the piston. The farther the piston moves, the

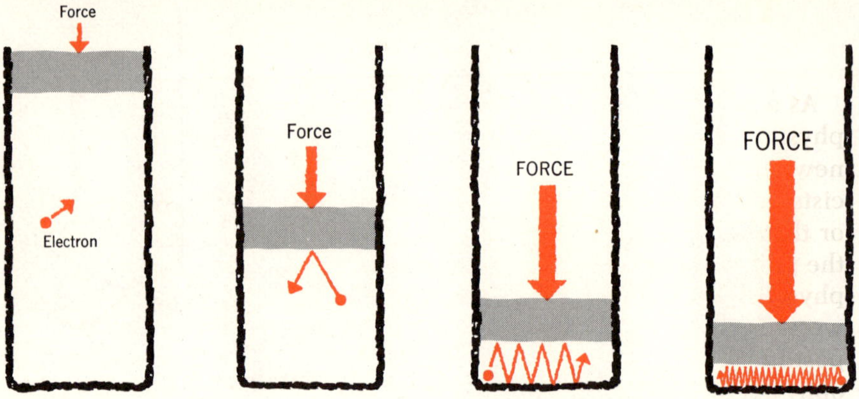

FIGURE 17-6 **How much space does an electron demand?**

more accurately we know the position of the electron, since it must be in a smaller and smaller volume. Therefore, the momentum becomes more and more uncertain, and the chance of high momentum for the particle becomes greater and greater. The greater the momentum, the greater the pressure resisting the motion of the piston, because the electron strikes it harder and more often. If the piston could finally trap the electron in one spot, the uncertainty of the momentum would have increased without any limit, and there would be an infinite force pushing back on the piston. No matter how hard the piston is pushed, therefore, the opposing force will eventually stop it.

Of course, we cannot actually observe the effect we have been discussing merely by using an ordinary piston and cylinder. The same principle, however, helps us to account for the size of an atom approximately. As we have seen, before the discovery of quantum physics, an electron moving around the nucleus would have been expected to radiate, lose energy, and finally fall into the nucleus itself. To have the electron located right at the nucleus with no kinetic energy and therefore no momentum, would violate the uncertainty principle, as our examples have indicated. There must be some state of lowest energy below which the electron cannot fall. Therefore, the lowest energy state must be a condition in which the electron is still moving around the nucleus with some momentum, and therefore some energy as well. This state, of course, corresponds to the lowest "orbit," or standing-wave pattern, in which the electron can exist in the atom.

The Philosophical Attitude of Modern Physics

As a result of the success of the theories of relativity and quantum physics, many of the basic ideas of science were questioned, and a new attitude and philosophy spread among physicists. Many physicists became very suspicious of entities that could not be observed, or that have no observable effects. This attitude is characteristic of the philosophy generally known as *positivism*. One early positivistic physicist-philosopher, who even disbelieved in atoms because they were unobservable, was Ernst Mach. A later physicist adopting a positivistic point of view was P. W. Bridgman, who called his philosophy *operationalism*. According to this philosophy, a physical quantity should be defined in terms of the actual physical operations necessary to measure it. Although there are many differences of opinion among those who adhere to this philosophy, it appeals to a great many physicists in one degree or another. It figured very significantly in the development of both relativity theory and quantum physics. Perhaps positivism should be considered to some extent an "antiphilosophy" rather than a philosophy. Many of those who were sympathetic to it held ordinary philosophy in some scorn. One of the prominent theorists of quantum physics is reported to have commented on a certain idea, "That is philosophy, and therefore nonsense."

Albert Einstein was very impressed with Mach's critical treatise on the history of physics. Despite the fact that Mach did not accept Einstein's theory, he probably did have an influence on Einstein's work. One of the results of the theory of relativity is to get rid of the ether, since it could not be observed and apparently had no effect that could be observed. Also, when Einstein examined space and time, he examined them not as absolute abstract entities, but simply as that which is measured by clocks and measuring rods—that is, he concentrated on the instruments and their operation.

In quantum physics we have abandoned the idea that a particle must be in a definite condition when we are not making an observation of that condition. The actual condition of such a particle may be not only unknown but in a sense nonexistent. This seems like an extreme application of positivistic principles, but it is widespread in physics today. This opinion results from the fact that we cannot assign a unique and definite condition to such a particle. The only significance of an unobserved particle is the effect it might have when it is finally observed. Thus it is the effect expected to be observed later, not some hypothetical present condition, that is important.

We have earlier presented the idea that the purpose of science is to make predictions. This view is also consistent with positivistic ideas and expresses a widespread attitude in physics today. Theories, assumed particles, fields, laws, etc., are merely abstract models that enable us to start with known present conditions and predict what will be observed later. Accordingly, the reality is in the observations; the theory merely connects present observations with future observations. Of course, because of the uncertainty principle these predictions cannot be made with absolute precision; we can merely give the probabilities for the events we expect to occur. (Sometimes the probabilities are so great as to amount to certainties, for all practical purposes.) The business of science is to state the probabilities as completely as possible for future events.

From this standpoint, relativity and quantum theory are extremely successful theories, the most successful known today for the very extensive range of phenomena that they cover. Of course, like any new theories, they must yield the same predictions as earlier theories for circumstances in which the earlier theories were correct. The theory of relativity gives the same results as Newton's laws when it deals with ordinary objects moving with speeds much less than that of light. Quantum theory gives the same results as Maxwell's theory of electricity and magnetism in many of the cases in which many electrons and many photons are involved. It also gives the same results as Newton's laws in many situations in which h can be regarded as very small and the wave nature of matter as unimportant.

Both the theory of relativity and the theories of quantum physics are capable of predicting the results of experiments to very great accuracy in many cases. No other theories, at once so comprehensive and so simple, can compete with them at present. Judging from the history of scientific theories, however, we can hardly expect them to be the last word.

FUNDAMENTAL IDEAS

I. FUNDAMENTAL LAWS AND PRINCIPLES

A. THE UNCERTAINTY PRINCIPLE, and its consequences.

B. THE COPENHAGEN INTERPRETATION OF QUANTUM PHYSICS, waves of probability.

The Uncertainty Principle and the Meaning of Quantum Physics

II. APPLICATIONS OF THE PRINCIPLES OF QUANTUM PHYSICS
   A. ZERO POINT ENERGY
   B. THE BOUND ELECTRON

III. METHODS AND PHILOSOPHY OF PHYSICS: Operationalism and other positivistic influences in (and of) physics.

QUESTIONS

The Uncertainty Principle

1. What does the uncertainty principle make us uncertain about?
2. What is the importance of Planck's constant?
3. Comment on the following statement. There can be no such thing as free will, because if I know the position and velocity of every particle and the forces between them, I can know everything that will happen.
4. If the uncertainty principle holds, why does a particle such as a baseball seem to have a definite and observable position and velocity, while an electron does not?
5. Some people claim that we did not have to wait for quantum theory to eliminate exact prediction in physics, since there was always some experimental error. What is the difference between this type of uncertainty and that expressed by the uncertainty principle?

The Copenhagen Interpretation of Quantum Physics and Applications

6. What is the Copenhagen interpretation of quantum physics? Why is it important?
7. Why can't we think of an electron as going through just one slit of the two-slit interference experiment? What happens if we detect the electron going through only one slit, or force it to do so?
8. What is an electron like when it is not being observed?
9. If matter propagates as a wave, why does an object such as a billiard ball appear to have a definite path and direction, and thus appear as a particle rather than a wave?
10. Can you think of a physical reason that an observed state of a system should be different from an unobserved state?
11. The eighteenth-century philosopher George Berkeley claimed that an object existed only by virtue of someone's perceiving it. How could

quantum physics, as illustrated in the two-slit experiment or otherwise, be used to advance, oppose, or modify his argument?

12. Why would it require a large force to confine an electron to a very small region of space?
13. According to Richard Feynman, "a philosopher once said 'It is necessary for the very existence of science that the same conditions always produce the same results'."[4] Was the philosopher right? Explain.
14. Describe the changes that have taken place in our understanding of the "field."
15. What are the similarities among waves on a string, waves a metal plate, and the electron waves in the atom? What are the differences?
16. In what way does quantum physics unify our understanding of the fundamentals of physics? In what way does it destroy a unity in our understanding?

[4]Richard Feynman, *The Character of Physical Law*, Cambridge: the M.I.T. Press, 1965,1967, p. 147.

18 The Nucleus and the Strong Interaction

I call heaven and earth to record this day against you, that I have set before you life and death, blessing and cursing: therefore choose life, that both thou and thy seed may live.

Deuteronomy 30:19[1]

Banesh Hoffman, in his popular exposition of quantum physics, ends his epilogue with a passage from the Book of Deuteronomy which we have quoted above. Why would a scientist quote from the Old Testament in a book on modern physics? The answer lies in the almost unbelievable amounts of energy available to mankind, for good or for evil, as a result of the nuclear forces. The photographs beginning this chapter illustrate two of the choices now open to mankind (see Figs. 18-1, 18-2). The choice that is not open is to return to the world that existed before nuclear energy was discovered. In this chapter we shall discuss the basic principles which explain how it is that this great energy is available.

[1]As quoted in the same context by Banesh Hoffman, *The Strange Story of the Quantum*, second edition, New York: Dover Publications, Inc., 1947, 1959, p. 232.

FIGURE 18-1 Hydrogen bomb explosion from fifty miles away. Courtesy of USAF.

FIGURE 18-2 The San Onofre Nuclear Generating Station supplies enough energy for a city of well over half a million people. Courtesy of Westinghouse.

Radioactive Nuclei

In 1896, the French physicist A. Henri Becquerel discovered the phenomenon of radioactivity. Radioactivity is the emission of high-energy particles (waves) from certain substances. The classic account is that Becquerel chanced to put some samples of the mineral pitchblende into the same drawer with some covered unexposed photographic plates. The radioactivity of the pitchblende exposed the plates. Becquerel investigated this phenomenon (as did such other scientists as Pierre and Marie Curie and Ernest Rutherford). The particles that had exposed the film proved to have come from specific kinds of atoms contained in the pitchblende. Investigation revealed that many other atoms could exist in forms that emit similar types of particles. Such atoms are called radioactive. Some examples are atoms of the metals radium, uranium, and thorium. Three types of particles were observed, called alpha, beta, and gamma particles. The beta particles proved to be high-energy electrons. The gamma particles proved to be photons, much like light, but of much shorter wavelength and much greater penetrating power. The alpha particles were shown to be helium atoms that had lost two electrons – in fact, just what we now call the nuclei of helium atoms.

Probability

The discovery of radioactive emission introduced a new type of process into physics—one that evidently involved probability instead of precisely predictable mechanical behavior. The exact moment at which a particle is emitted from a radioactive atom cannot be predicted; only the probability of emission can be specified. In the last chapter we saw how this probability concept was incorporated into quantum physics.

When a radioactive atom emits particles, a change takes place in its nucleus. The nuclear change is called a *disintegration* or *decay*. These terms are used regardless of whether the change in the nucleus is relatively minor or actually causes the nucleus to break up. In alpha and beta emission the nucleus is changed in such a way that the atom becomes a different kind of atom—an atom of a different chemical element. The great energy of the emitted particle originates in the nucleus, although leading scientists, including Lord Kelvin, once found it difficult to accept this fact.

There does not seem to be any way to predict when a particular nucleus will change by emitting a particle. One nucleus may change immediately while another, apparently just like it, may wait one-hundred years to change, or a million years. If we have a collection of identical nuclei, as for example in a sample of pure radium, we know how many of them *on the average* will disintegrate in a given time period. In the case of radium, we know that half of them will have changed within 1600 years, but we do not know which half, and at present there is no reason to believe that we will ever be able to tell which half. If we wait 1600 years half of the nuclei will have changed. If we wait another 1600 years half of those remaining will have changed, leaving one fourth of the original nuclei unchanged. If we wait still another 1600 years, half of the remainder will have changed (see Fig. 18-3). Thus, for a large sample of such atoms, we can determine that so-called *half-life*. The half-life is the period of time within which half of the nuclei will have disintegrated. Each type of nucleus that can disintegrate radioactively has a definite half-life. These half-lives can vary from billions of years to a few billionths of a second—in fact, there are no known limits as to how long or short a half-life can be. The disintegration rate, and thus the half-life, does not seem to be changed appreciably by the normal forces and chemical changes the atom is subjected to. The disintegration rate can be changed, however, by collisions with radioactive parti-

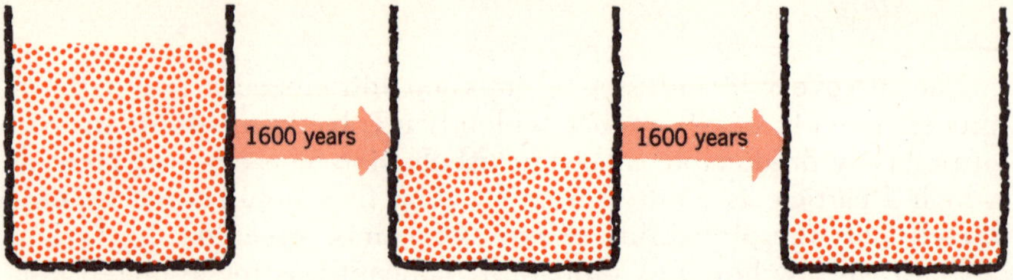

FIGURE 18-3 **The half-life of radium. We have drawn in only the radium atoms remaining. Those which have decayed have changed into other atoms, but we have not drawn these other atoms.**

cles, cosmic rays, and high-energy particles from cyclotrons and other particle accelerators.

It should be kept in mind that the half-life refers to average behavior. We can use it to predict the behavior of large groups of radioactive atoms. If we only have one atom, however, we can give odds as to how long it will remain unchanged, but we cannot tell exactly.

The Discovery of the Nucleus

Rutherford's assistants, at his suggestion, allowed a beam of high-energy alpha particles to pass through a very thin gold foil (see Fig. 18-4). The great majority of these particles passed through the foil without much deflection, but a few of them were deflected through large angles, some of them even coming straight back. This surprised Rutherford:

> It was quite the most incredible event that has ever happened to me in my life. It was almost as incredible as if you fired a 15-inch shell at a piece of tissue paper and it came back and hit you.[2]

This result meant that most of the cross section of the material target must be empty or covered with light particles, such as electrons. In order to deflect the alpha particles through angles, however, there must be heavier particles present. (For example in order to bounce a 15-inch shell off a piece of tissue paper, there must be something embedded in the tissue paper that is very massive.) These heavy par-

[2]E. N. da C. Andrade, *Rutherford and the Nature of the Atom*, Garden City, N.Y.: Doubleday Anchor Books, 1964, p. 111.

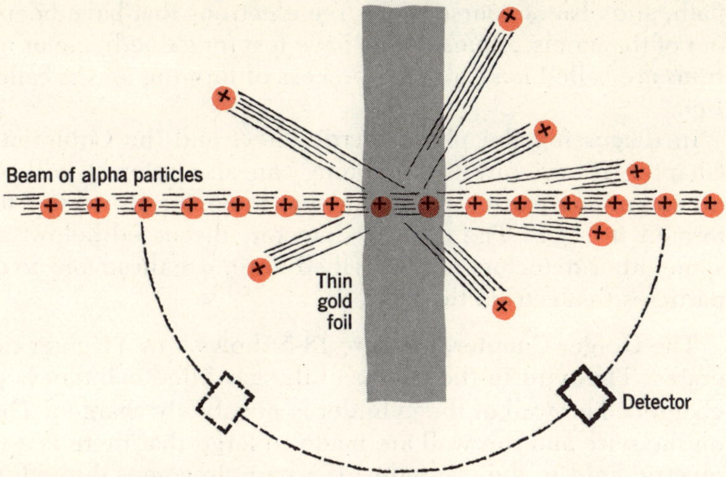

FIGURE 18-4 **Rutherford's experiment.**

ticles, according to Rutherford, must be small and highly charged. Thus the atomic nucleus was discovered. The atom consists of a surprisingly small, positively charged particle (about 10^{-15} meter in diameter) that contains nearly all the mass of the atom. Around it, out to a distance of about 10^{-10} meter, are the electrons. The alpha, beta, and gamma particles originate from the nuclei of certain kinds of atoms.

Most nuclei have less charge compared to their mass than they would have if they contained only protons. Therefore another kind of particle must also be contained in the nucleus. This fundamental particle, the *neutron,* was identified by J. Chadwick in 1932. The neutron, with no net charge, has a rest mass, m_n, equal to 1.67470×10^{-27} kilogram, which is slightly greater than the mass of the proton. Protons and neutrons are called *nucleons*, because together they make up the nuclei of atoms.

Particle Detectors

When alpha or beta particles pass through a gas, these high-energy particles interact with the molecules of the gas. Since these particles are electrically charged, they can interact with the electrons in the electron cloud outside the nucleus of the atom. They can strike these electrons and remove them from the atom. Thus, as the particle goes through the gas, it leaves gas atoms with missing electrons along its path, and also of course some free electrons that have been knocked out of the atoms. Atoms which have lost (or gained) one or more electrons are called *ions*, and the process of forming ions is called *ionization*.

In discussing the photoelectric effect and the Compton effect in Chapter 16 we saw that photons can also interact with electrons. Since gamma particles are photons, they can also cause ions to be formed in a gas. The particle detectors discussed below, as well as some other detectors not described here, use these ions to detect the particles that caused them.

The Geiger Counter. Figure 18-5 shows how a Geiger counter operates. The wire in the center of the gas-filled cylinder is positively charged. The wall of the cylinder is negatively charged. The charges on the wire and the wall are made so large that there is a very large electric field in the cyclinder. If a particle comes through the cylinder and creates ions and free electrons in the gas, the field causes these ions and free electrons to move. The electrons, in particular, are pulled so strongly toward the central positive wire that they strike gas atoms, knocking off other electrons. Thus an avalanche of electrons is formed, which, when it reaches the positive wire, creates a detectible current pulse through the Geiger counter. For each particle passing through the cylinder, a pulse is observed. Thus the Geiger counter is able to count the particles that come through it.

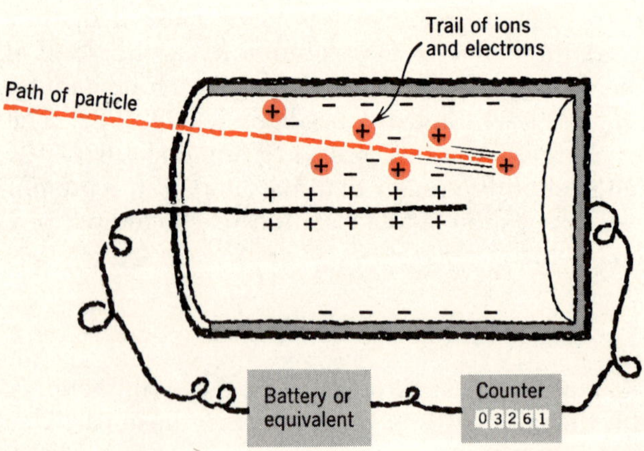

FIGURE 18-5 **Diagram showing the operation of a Geiger counter.**

The Cloud Chamber. The cloud chamber contains a gas (usually air) which itself contains water molecules. When such a gas is cooled, the water tends to condense, forming clouds of small water droplets. This process is what happens in the atmosphere when a cloud is formed. Warm air, with evaporated water in it, expands and cools as it rises in the atmosphere. As it cools below a certain temperature, the water molecules form drops, and a cloud is formed. In order for a cloud to form, however, the water molecules must come together into tiny drops. It is difficult for these drops to form unless there is some dust for them to condense on. The drops will also condense on ions. Thus, if a cloud is formed in clean air, the droplets will first form around the ions present in the air. If a moving particle has left ions along its path, this path will be visible because of the tiny droplets condensing on each of the ions. Each ion will become visible because of the drop that forms on it.

We cool the gas in the cloud chamber by suddenly expanding it (see Fig. 18-6). When it cools, droplets form on the ions left by any particle that has just passed through the chamber. These droplets lie along the path of the particle. Figure 18-6*b* is a photograph of particle paths in a cloud chamber.

A bubble chamber works in a way similar to the cloud chamber, except that instead of a gas condensing to form droplets, the bubble chamber contains a liquid (as, for example, liquid hydrogen) boiling to form bubbles of gas. These bubbles readily form on the ions in the liquid caused by passing particles, thus making particle tracks visible. Figure 19-1 on p. 382 is a photograph of particle tracks in a bubble chamber.

Nuclear Structure

We have pointed out that the nuclei of the various atoms are made up of two kinds of fundamental particles, neutrons and protons. The total number of protons in the nucleus determines the total charge on the nucleus. (We recall that the neutron has no net charge.) The electrons in the cloud outside the nucleus are held in the atom because they are attracted to the positively charged protons in the nucleus. In an electrically neutral atom, there are just as many electrons in the cloud outside the nucleus as there are protons in the nucleus.

The number of protons inside the nucleus is called the *atomic*

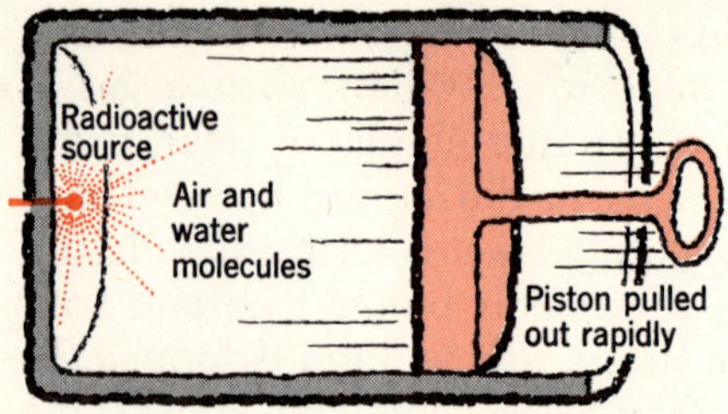

(*a*)

FIGURE 18-6a **Diagram illustrating the operation of a cloud chamber.**

number of that particular atom or nucleus. All atoms of the same chemical element have the same atomic number; in fact, it is the atomic number that determines the chemical element. Thus hydrogen, the simplest element, has an atomic number of one. The common form of hydrogen has one proton (and no neutron) in the nucleus, and one electron in the cloud. The most common form of helium, of atomic number two, has an alpha particle for its nucleus, consisting of two protons and two neutrons.

The *mass number* is the sum of the number of protons and the number of neutrons in the nucleus. Since the neutron and the proton have about the same mass, this number is very nearly proportional to the mass of the nucleus. The mass of the nucleus is very nearly the total mass of the atom, since the electrons are much less massive than the nucleons. The common form of hydrogen has a mass number equal to its atomic number, namely one, since it has only one nucleon in its nucleus (see Fig. 18-7*a*). The common form of helium has a mass number equal to four, because it has two protons and two neutrons in its nucleus (see Fig. 18-7*d*). For all known atoms except the common form of hydrogen, the atomic mass number is greater than the atomic number.

We have discussed the common form of hydrogen, which has a nucleus composed of one proton. There are two other forms of hydrogen. One of these, called *deuterium*, has a nucleus composed of one proton and one neutron (see Fig. 18-7*b*). Thus it has a mass number of two and is approximately twice as heavy as the common form of hydrogen. It is still the same chemical element hydrogen, however, because it still has the same number of protons in the nucleus. The third form of hydrogen, called *tritium*, has one proton and two neutrons in the nucleus (see Fig. 18-7*c*). All nuclei that have the same

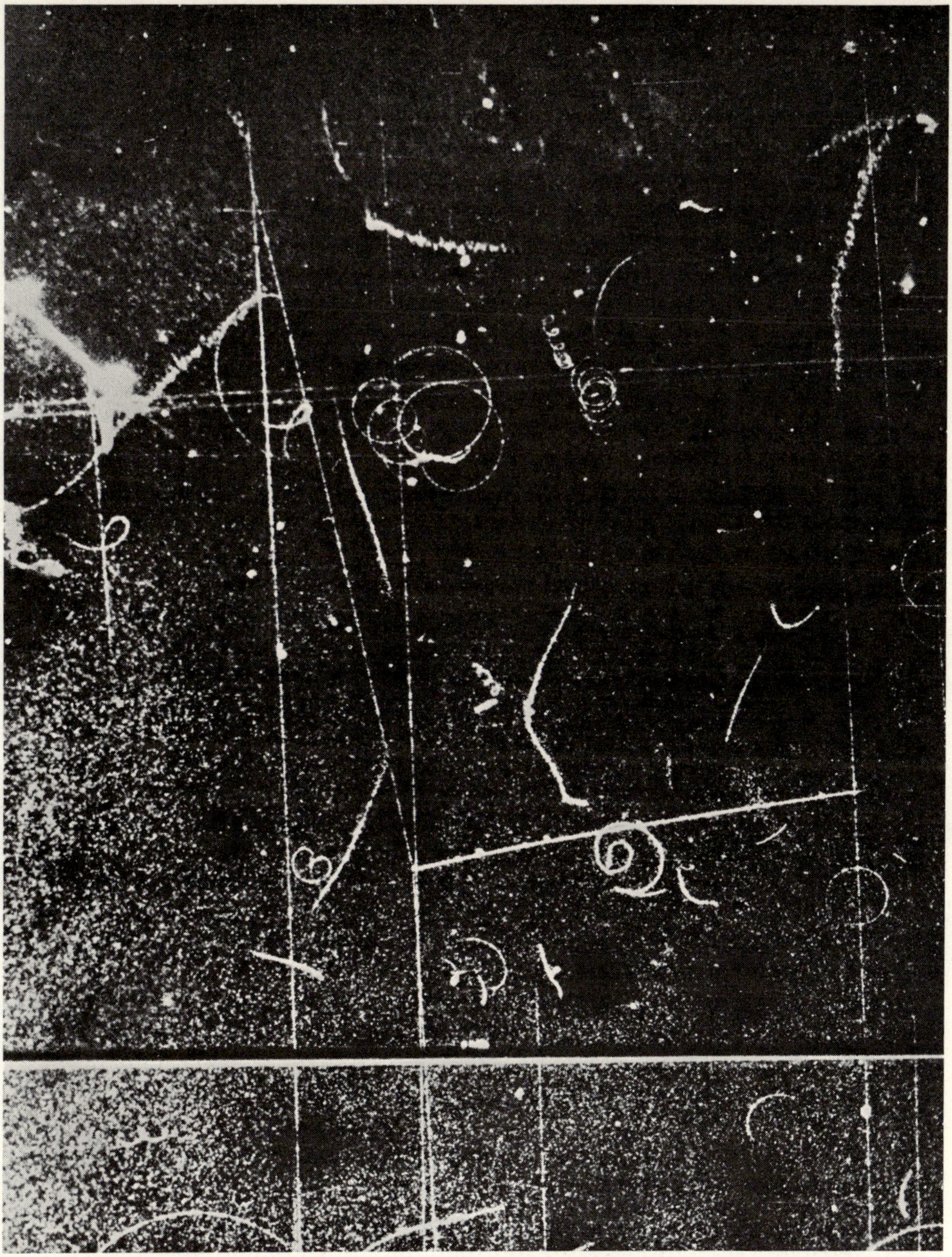

FIGURE 18-6b Particle tracks in a cloud chamber.

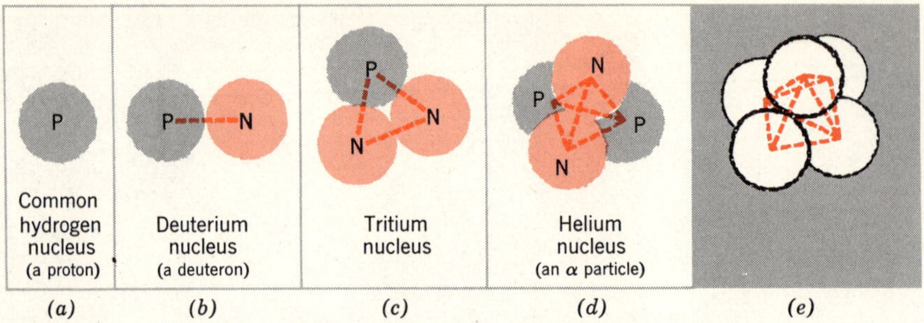

FIGURE 18-7 **Atomic nuclei are made up of one or more nucleons. (A nucleon is a proton or a neutron.)**

number of protons but different numbers of neutrons are called *isotopes* of each other. Hydrogen, as we have seen, has at least three isotopes. Most elements have several isotopes.

The Strong Interaction

One characteristic of the strong interaction is that it operates only between certain kinds of particles. We have considered two of the fundamental particles that exert or are subject to the strong interaction—the neutron and the proton. The forces of the strong interaction hold the nucleons together in the nucleus.

Another characteristic of the strong interaction is that it is very strong indeed, the strongest of all known interactions. It takes considerable energy to break up something that is held together by the strong interaction. This energy is so large that is has an appreciable amount of mass. For example, the mass of the alpha particle is about seven hundredths of a percent less than the sum of the masses of two protons and two neutrons. One might expect, therefore, to gain or lose a lot of energy in transformations involving the strong interaction. The atomic and hydrogen bombs are evidence that this does occur.

A third characteristic of the strong interaction is the very short range of the forces to which it gives rise. It is most effective at nuclear distances—less than about 10^{-15} meter, roughly the "radius" of a nucleon. Therefore it is strongest when acting between neighboring nucleons and is substantially less effective when the nucleons are separated by a distance greater than the radius of one nucleon. The strong interaction is about one hundred times stronger than the electromagnetic interaction when the nucleons are close together. Therefore we may often ignore electromagnetic forces between neighboring nucleons. The strong forces between protons and protons, protons and neutrons, and nuetrons and neutrons are all approximately equal in strength. The electromagnetic force is stronger than the strong force at long ranges, because the strength of the strong interaction falls off much more rapidly with increasing distance than that of the electromagnetic interaction.

At present the exact law governing the strong interaction is unknown.

Nuclear Binding Energies

The strong interaction and the electromagnetic interaction together account for the main features of nuclear binding energies. The binding energy (see Chap. 8, p. 159) is the energy that must be supplied in order to separate particles that are held together by a force. The nucleons that make up a nucleus have less mass when they are combined into a nucleus than when they are separated. Since energy must be supplied to the particles to separate them, mass must also be supplied because energy has mass. Therefore the nucleons will have more mass when separated than when combined in the nucleus.

We shall be concerned here with the total binding energy for the nucleus divided by the total number of particles in it—that is, the binding energy per particle. The actual binding energy of two nucleons is about 3.6×10^{-13} joule, corresponding to a binding mass of 4×10^{-30} kilogram. We shall call this mass B.

Figure 18-7*b* shows a proton and a neutron that have come together to form a *deuteron*, which is the nucleus of the second form of hydrogen, deuterium. The mass of this particle is less than the sum of the masses of the proton and neutron when separated because of the binding energy. The mass loss due to this binding energy is approximately B. The mass loss per particle, then, is $B/2$, since there are two particles making up the deuteron.

Figure 18-7*c* shows a proton and two neutrons that have come together to form the nucleus of tritium, the third isotope of hydrogen. There are now three particles, all attracted to each other so that they bind closely. Thus there are three "bonds," represented by the dotted lines in Figure 18-7*c*, since each particle is in close contact with each of the others. Therefore, we expect that there will be a mass loss of about $3B$. Since there are three particles, the mass loss per particle is $3B/3$, or B. The binding mass per particle is greater in this case than in the last.

Now consider the case of four particles, which will be arranged in a pyramid as shown in Figure 18-7*d*. (This kind of pyramid, with four corners, is called a tetrahedron.) Now there are six bonds and four particles; therefore we expect that the mass loss will be about $6b$ and the mass loss per particle will be $6B/4$, or $(3/2)B$. Thus the mass loss per particle has again increased. (Actually this treatment is much oversimplified for the alpha particle.)

One might think that this could go on forever, with the lowest energy state per particle being achieved when all the neutrons and protons in the world are amassed into one big nucleus. However, there are two other effects to consider. One of these limits the mass loss per particle, and the other actually decreases it.

To see the effect that limits the mass loss per particle consider the nucleus with five particles, as shown in Figure 18-7*e*. There is no way in three-dimensional space by which all of these particles can touch all of the others. Although each of four particles can all touch all of the others, each of five cannot. The fifth particle must be farther away from some of the particles than from others. This is very significant because the strong force is a short-range force. We will assume that it is effective only when the particles are "touching" although of course this is not exactly correct. In Figure 18-7*e*, there are five particles and nine bonds where the particles are touching one another. The mass loss per particle is now $(9/5)B$. This is still larger than in the last case, but the amount of increase is going down.

What will be the maximum mass loss per particle? Let us consider a particle in the middle of a large nucleus. Even in the closest possible packing, it can have no more than twelve nearest neighbors touching it. (Try this with a stack of oranges—see how many other oranges can touch a single orange.) This means 12 bonds, each of which is shared with another particle. Thus each particle has the equivalent of six bonds for itself. Therefore the mass loss per particle is $6B$ for a particle in the middle of the nucleus. A particle on the edge of the nucleus has fewer nearest neighbors, and therefore the mass loss per particle will be even smaller. The mass loss per particle cannot be greater than $6B$ according to this analysis.

Larger nuclei contain more protons. These protons repel one another because of their electric charge, and thus act against the attractive strong force. The electric force is much weaker than the strong force, but it acts at long range. Therefore each proton is bound by the strong force to its nearest neighbors, but it is repelled by every other proton in the nucleus. Despite the greater strength of the strong force, there is a maximum number of protons that can be housed together in a stable nucleus. These protons, because of their charge, act against the strong force; for large nuclei, the mass loss per particle actually decreases somewhat as the nucleus increases in mass number rather than becoming greater. Therefore, there is a size of nucleus for which the mass loss per particle is greatest. The simplified argument we have given here suggests a graph similar to that shown in Figure 18-8*a*. The actual graph is given in Figure 18-8*b*. The two graphs are somewhat different because of several effects not discussed here.

The Release of Nuclear Energy

An examination of Figure 18-8*b* helps us to understand how nuclear energy can be released. The greatest binding energies per particle occur for nuclei containing about 56 nucleons, such as the nucleus of iron. The fact that the graph has a maximum corresponding to the greatest binding energy per particle means that we could obtain energy in two ways by transforming nuclei.

1. If we could get the light nuclei at the left-hand side of the graph to join together, they would become the bigger nuclei that are farther to the right on the graph. Thus the binding energy per particle in these nucleons would increase. This means that the particles would be more strongly bound together than they were before, and therefore energy would have to be released. This process is known as fusion. Fusion occurs in the explosion of a hydrogen bomb, and in the sun.
2. If we could get the heavy nuclei at the far right of the graph to split approximately in half, the two pieces would be lighter nuclei, found at about the middle of the graph. Here again the binding energy per particle would be increased, although not as much as in the fusion process, as can be seen from the graph. Here also, then, energy would be released, but not as much as in the fusion process. This process, called nuclear fission, occurs in ordinary atomic-bomb explosions and in many nuclear reactions. In both fission and fusion new nuclei are formed, and some of the rest mass of the original nuclei is released as other forms of energy.

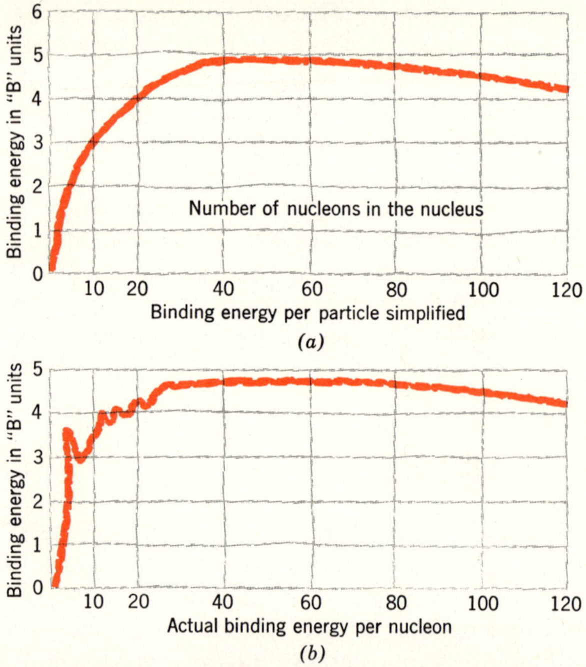

FIGURE 18-8 The binding energy per nucleon for various atomic nuclei. (*a*) Simplified graph corresponding to the discussion in the text. (*b*) More accurate graph including the results of some effects not discussed in the text.

Nuclear Fission: The Atomic Bomb

One kind of atomic bomb uses the type of nucleus that has atomic number 92 and mass number 235—the well-known uranium-235. This element appears well to the right on the graph in Figure 18-8*b* and therefore is a candidate for fission. If the uranium-235 nucleus can be made to break up it will indeed release energy. Each uranium-235 atom has a certain probability of breaking up by itself, but we want to stimulate many nuclei to break up at the same time. The fission process can be stimulated by irradiating uranium-235 with slow neutrons. These neutrons have no net charge, and therefore they are not repelled by the nucleus, as a proton would be. They easily fall into the nucleus and become bound to the other nucleons. The binding energy released as the neutron is captured appears as kinetic energy of the neutron and the other particles in the nucleus.

This increased energy causes the nucleus to break up, releasing not only energy, but also two or three more neutrons. These neutrons can now help to break up even more uranium-235 nuclei. Thus the process will be self-sustaining. Unfortunately, the neutrons that are released are not slow neutrons but high-energy (fast) neutrons. Before they can be captured easily by other uranium-235 nuclei they must be slowed down to lower kinetic energies. The fast neutrons will become slowed down by repeated collisions and scattering by the atoms of the uranium until they are slow enough to be captured by the nucleus. However, if the piece of uranium-235 is too small, most of the neutrons produced will leave the material before they are slowed down, and the reaction will not build up. If the material is greater than some critical size, however, the neutrons from one fission will be slowed down so that they start two or more other fissions. Then we have a chain reaction—on the average, each fission stimulates more than one more fission, and the reaction rapidly builds up to explosive rates (see Fig. 18-9). This causes an atomic bomb explosion. To explode the bomb it is necessary to take two pieces of uranium-235 below critical size and quickly put them together to make a piece above critical size.

Nuclear Fission: The Nuclear Reactor

If we do not want a fission bomb but wish to obtain nuclear energy from fission at a more controlled rate for less destructive purposes,

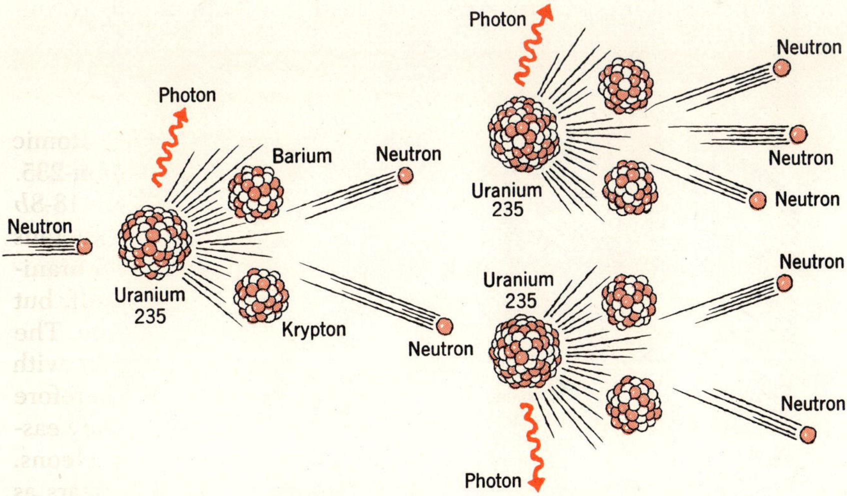

FIGURE 18-9 **A chain reaction.**

we use what is called a nuclear reactor. Figure 18-10 is a diagram of one kind of reactor. The "fuel" for this reactor is not pure uranium-235 but uranium that may contain about 20 percent uranium-235. Most of the rest of this uranium fuel is another isotope of uranium, with mass number 238—uranium-238. The uranium-238 will absorb most of the fast neutrons produced by the fission of the uranium-235 before these neutrons can be slowed down enough to be captured by other atoms of uranium-235. To reduce this effect, so that there will be enough slow neutrons to cause fission, the uranium rods are separated by a *moderator*, such as graphite. The moderator slows down the neutrons emerging from a uranium rod without absorbing too many of them. Then these neutrons will have the right speed to cause fission when they again enter a uranium rod.

To make sure that there are not too many slow neutrons, which might cause an uncontrolled chain reaction, control rods of some neutron-absorbing material, such as boron or cadmium, are inserted into the reactor. The deeper these control rods are inserted, the more neutrons they will absorb; thus they afford a method of controlling the process.

Fusion

The problem of attaining a practical fusion reaction, in which light nuclei come together to form more massive ones, is more difficult than the similar problem for fission. In a hydrogen-bomb explosion, deuterons combine in processes that lead, for example, to tritium

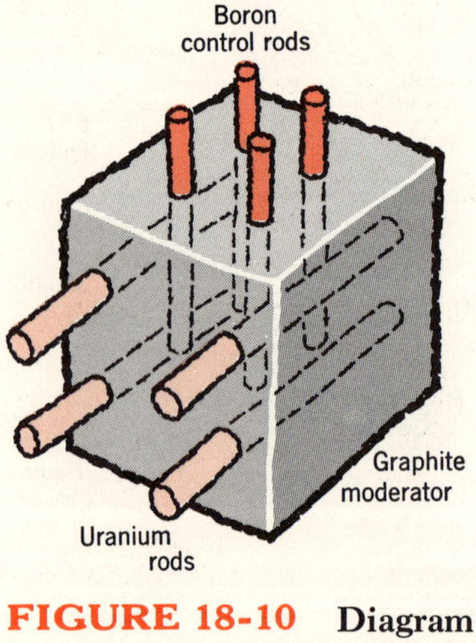

FIGURE 18-10 **Diagram of a nuclear reactor.**

nuclei and alpha particles. The deuterons, however, are positively charged, and therefore repel each other. They must have considerable kinetic energy so that they will collide and get close enough together for the strong force to overcome the electrical repulsion. They will have this high kinetic energy if they are at a high temperature.

In stars, such as the sun, this high temperature does occur. Presumably the sun's energy comes largely from a fusion process. These high temperatures have also been achieved in the hydrogen bomb, which requires an ordinary fission bomb to first obtain the necessary energies for fusion. Controlled release of power, however, presents a formidable engineering problem, since the high temperatures required to sustain the process will destroy any containing structure. At such high temperatures, however, the atoms are nearly all ionized and thus are electrically charged. Such a collection of mostly charged particles is called a *plasma*. In the cyclotron (see Chapter 12, p. 233), a magnetic field is used to keep the charged particles from escaping from the instrument too soon. More complex types of magnetic fields can be used to contain a plasma for very short times. There is some possibility that a plasma can be contained long enough to permit a controlled fusion reaction. Research is still being done on this problem (see Fig. 18-11).

Applications

Radioactived Dating

Lord Kelvin had estimated the age of the earth as no more than twenty or forty million years by calculating the time it must have taken for the earth to cool to its present temperature. This time span was an embarrassment to Darwin and other proponents of organic evolution, for it did not appear to be long enough for evolutionary processes. The apparent conflict of evolutionary theory with the laws of physics hampered the acceptance and development of the theory for over fifty years.[3]

Rutherford, in a paper presented to the Royal Institution, pointed out that radioactivity in the earth was a source of heat and that Kelvin's argument was therefore no longer valid:

[3]Loren Eiseley, *Darwin's Century: Evolution and the Men Who Discovered It*, Garden City, N.Y.: Doubleday Anchor Books, 1961, Chap. 9.

FIGURE 18-11 **The "2-X" nuclear fusion device at the Lawrence Radiation Laboratory at Livermore, California. Courtesy of USAEC, San Francisco Operations Office.**

I came into the room, which was half dark, and presently spotted Lord Kelvin in the audience and realized that I was in for trouble at the last part of my speech dealing with the age of the earth, where my views conflicted with him. To my relief, Kelvin fell asleep, but as I came to the important point, I saw the old bird sit up, open an eye and cock a baleful glance at me! Then a sudden inspiration came, and I said Lord Kelvin had limited the age of the earth, *provided no new source* [*of heat*] *was discovered.* That prophetic utterance refers

[4]From *Rutherford and the Nature of the Atom,* by E. N. da C. Andrade. Copyright ©, 1964 by Educational Services Inc. (Science Studies Series). Reprinted by permission of Doubleday & Company, Inc., p. 80.

to what we are now considering tonight, radium! Behold! the old boy beamed upon me:[4]

It was, however, a few years before Kelvin accepted radium as a new chemical element.

The discovery of radioactivity not only made Kelvin's argument irrelevant but also provided a means of making meaningful estimates of the age of the earth and certain structures and strata in it. These methods depend on the fact that the rate of radioactive disintegrations is not appreciably influenced by normal changes of temperature and pressure, or by normal chemical processes. Therefore, if we can measure or estimate how much of a sample of radioactive material has already disintegrated, we can find out how long the sample has been in its present condition. We must, of course, know the half-life of the radioactive material in order to do this.

Carbon-14 Dating. Carbon-14 is produced from nitrogen-14 in the atmosphere when the nitrogen-14 is stuck by neutrons, themselves produced by cosmic rays. The reaction that produces the carbon-14 is

$$N^{14} + n \rightarrow C^{14} + p^{+}$$

(N^{14} and C^{14} represent nitrogen-14 and carbon-14, respectively. Nitrogen-14 has a mass number of 14 and an atomic number of 7. Carbon-14 has a mass number of 14 and an atomic number of 6. n is a neutron and p^{+} a proton.) The carbon-14 thus formed becomes mixed with the ordinary carbon in the air and water and thus finds its way into living organisms. Carbon-14, however, is itself radioactive. Thus it is continually disintegrating, with a half-life of about 5730 years.[5]

The net result of this process is that there is a certain amount of carbon-14 in the air and water. If the rates of creation and disintegration are the same, then the amount of carbon-14 in the air and water will come to some constant value and remain there.

Living plants and animals take in carbon-14 from the air and water as part of their total carbon intake. When a plant or animal dies, however, it no longer takes in any more carbon. The carbon-14 it has already taken in, however, continues to disintegrate. As time goes on, therefore, there is less and less carbon-14 in the remains of the plant or animal. From the fraction of carbon-14 remaining, we can calculate how long the plant or animal has been dead. Wood, for example, can be dated in this way. Because of its relatively short half-life, carbon-14 is very useful in dating relatively recent artifacts, up to about 30,000 years old.

[5]D. T. Goldman and J. R. Roesser, *Chart of the Nuclides*, 9th Ed., Schenectady, N.Y.: General Electric Company, July 1966.

The use of the C^{14} method requires an estimate of the fraction of carbon-14 in the environment at the time the plant or animal died. The usual assumption is that this fraction remained the same until about 1850, when industrial processes began adding large quantities of carbon to the atmosphere. Recent evidence, however, indicates that the rate of formation of carbon-14 differs from the rate of disintegration, and that therefore the amount of carbon-14 is at present changing. Tree-ring data indicate that there have been changes in the fraction of carbon-14 in the air and water.

Alpha Emission

Since nuclei are positively charged, they repel other positively charged particles. When the positively charged alpha particles are directed toward the nucleus, as in Rutherford's experiment (p. 362), it is this electric repulsion that deflects the alpha particles. If an alpha particle has sufficient energy, it may be able to overcome this repulsion and reach the nucleus.

The electric repulsion acts like a barrier, keeping the alpha particles out of the nucleus. Figure 18-12 shows this barrier schematically. The height of the barrier at any point represents the energy that an alpha particle must have to reach that point. (The height of the barrier is the potential energy that an alpha particle would have at that point because of two fields of force acting on it, one caused by the strong force of attraction and the other by the repulsive electric force.) The top of the barrier is the point at which the attraction due to the strong force is equal in magnitude to the repulsion due to the electric force. At points closer to the nucleus than the top of the barrier, the strong force of attraction dominates the electric force. Farther away from the nucleus than the top of the barrier, the repulsive electric force dominates.

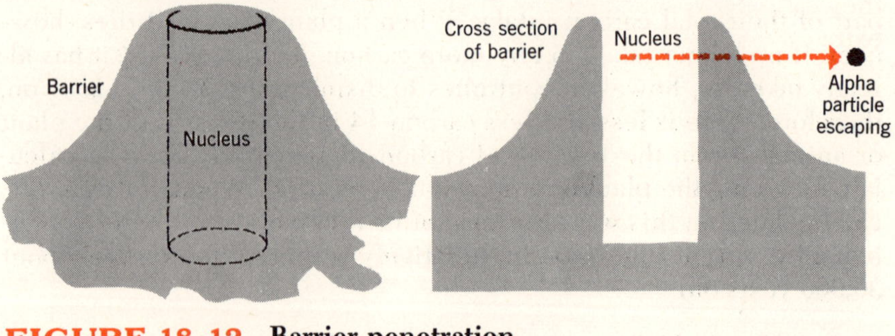

FIGURE 18-12 **Barrier penetration.**

In the radioactive process of alpha emission, alpha particles are coming out of the nucleus rather than being directed into it. One would expect that such alpha particles would have to have at least as much energy as that represented by the top of the barrier in order to get out. However, alpha particles that do not have this much energy are often emitted from nuclei, changing the nuclei in the process. Somehow the alpha particle gets through the barrier without having enough energy (apparently) to do so. This is one example of a process known as "barrier penetration," which is possible for a material particle according to quantum physics but not according to the earlier Newtonian physics.

According to Newtonian physics, the top of the barrier is inaccessible to the alpha particle because the particle does not have enough energy to reach it. It does have enough energy to reach the regions outside the varrier, however, if some way could be found for it to get across the barrier. According to the laws of quantum physics, however, a particle can "leak" through an energy barrier between one region and another. This can occur if the particle has enough energy so that both regions would be accessible to it except for the barrier. The higher and wider the barrier, the lower the probability for such leakage. If the barrier is infinitely high or infintely wide, leakage is impossible.

The uncertainty principle that relates time and energy is sometimes used to explain why it is possible for some of the alpha particles to penetrate into and through the barrier region. Over a long time the energy of the alpha particles corresponds to a certain precise energy state within the nucleus. For a very short time period, however, their energies can be very uncertain. This uncertainty, for very short times, allows them to have higher energies, and with these higher energies they can move into the barrier region. The distance through which they penetrate the barrier depends on how fast they can move with their additional energy during the short time the extra energy is allowed to them by the uncertainty principle.

FUNDAMENTAL IDEAS

I. FUNDAMENTAL LAWS AND PRINCIPLES

A. The strong interaction—supplies strong, short-range forces between nucleons (and some other particles). Exact force law unknown.

II. PARTICLES

A. FUNDAMENTAL PARTICLES

1. The neutron: much like a proton, but with no net charge.

B. THE NUCLEUS

1. Radioactivity.
   a. Alpha emission – alpha particle is a helium nucleus.
   b. Beta emission – beta particle is an electron.
   c. Gamma emission – gamma particle is a high-energy photon.
   d. Half-life and probability.
2. Nuclear structure.
   a. Atomic Number.
   b. Mass Number.
   c. Isotopes; the deuteron and deuterium.
   d. Nuclear binding energies.
3. Nuclear fission.
4. Nuclear fusion.

III. APPLICATIONS AND INSTRUMENTS

A. INSTRUMENTS

1. Geiger counter.
2. Cloud chamber.
3. Bubble chamber.

B. APPLICATIONS

1. Radioactive dating techniques.
   a. Carbon-14 dating.
2. Nuclear explosives using fission and fusion; nuclear reactor.

QUESTIONS

Radioactivity

1. What is radioactivity?
2. What is meant by the term "half-life"?
3. How does radioactive emission differ from the processes governed by Newton's laws of motion?
4. What was surprising about the experiment that sent alpha particles into a gold foil? What conclusions were drawn from it and why?

5. What is ionization?
6. Explain the operation of a cloud chamber.
7. Explain the operation of a Geiger counter.
8. Of what use are cloud chambers and Geiger counters?
9. How do you suppose one could tell that alpha particles are positively charged, beta particles negatively charged, and gamma particles uncharged?
10. A particle track in a cloud chamber appears to be very straight. How do you know that the path shown by the droplets is not straight "microscopically"?
11. Why does the gas in a cloud chamber cool down when it is forcibly expanded?

Nuclear Structure and the Strong Interaction

12. What are the particles that make up an atom? How are they arranged?
13. What are the main properties of protons and neutrons?
14. What are isotopes?
15. A proton is a hydrogen nucleus. So is a deuteron. Explain.
16. Define and distinguish between mass number and atomic number.
17. What is a nucleon?
18. The rest mass of the alpha particle is smaller than the sum of the rest masses of the two protons and two neutrons of which it is made. Explain how this is possible.

Fission and Fusion

19. Explain the main features of the curve in Figure 18-8*a*.
20. What does the binding energy per particle have to do with the energy release in the atomic and hydrogen bombs?
21. How can nuclear fusion and fission both be used to produce energy when they are opposite processes?

General

22. What did new discoveries in nuclear physics have to do with our knowledge of the age of the earth?
23. Why might it be easier to shoot neutrons into a nucleus than to shoot protons into a nucleus?
24. Sometines it is possible to break up deuterons by shooting them into a nucleus. The proton is stripped off the deuteron. Why?

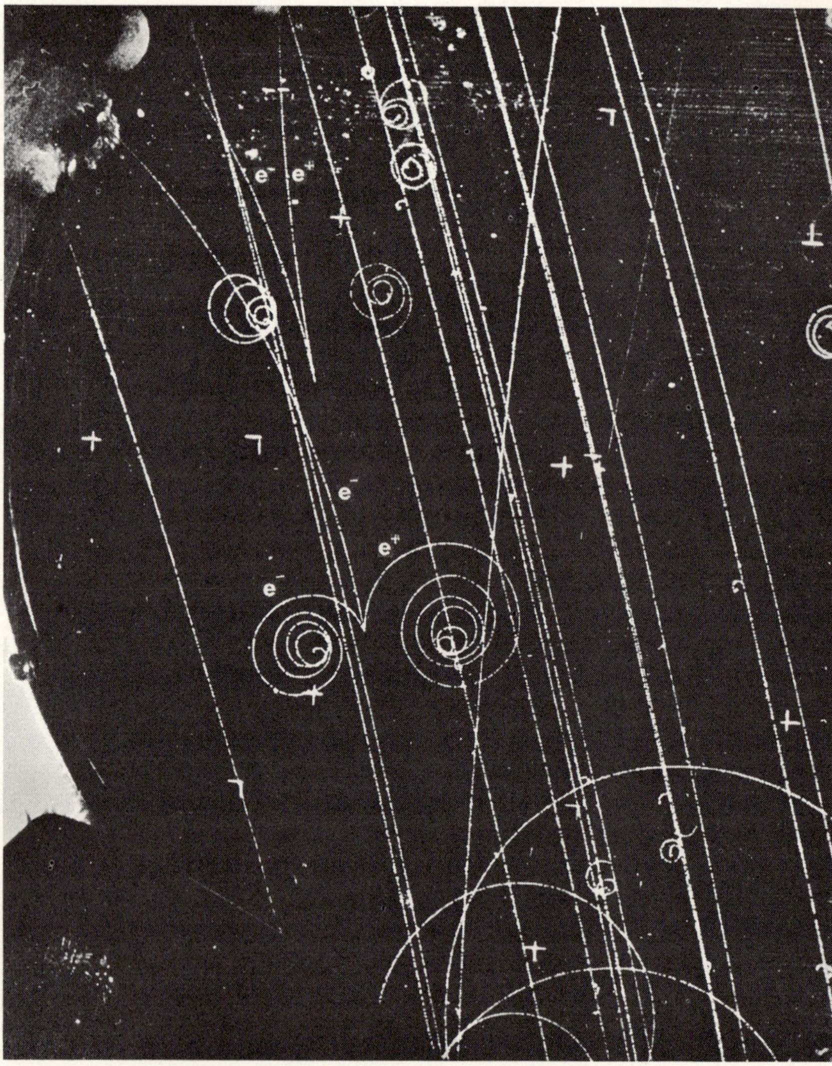
e^- e^+
e^-
e^+
e^-

19 The Elementary Particles and Their Interactions

*"You boil it in sawdust: you salt it in glue:
You condense it with locusts and tape:
Still keeping one principle object in view—
To preserve its symmetrical shape."*
Lewis Carroll[1]

What is everything made of? We raised this question in Chapter 9 when we discussed the atoms and molecules of which matter is made. Here we shall discuss particles which are even more fundamental, the so-called elementary particles. Everything is made of the elementary particles, which of course also have their field or wave aspect.

Some of these particles we have already introduced. In the last chapter we discussed the nucleons—the comparatively massive proton and neutron. We have also introduced the photon, the electron, and the muon. The muon is one of the mysteries of the universe—muons are apparently just like electrons except that they happen to have about 207 times as much rest mass.

We shall not discuss all of the elementary particles, nor shall

[1]Lewis Carroll, *The Hunting of the Snark*, as quoted in Martin Gardner, *The Annotated Snark*, New York: Bramhall House, 1962, p. 73. Thanks also to R. F. Streater and A. S. Wightman for quoting it also in a similar context in *PCT, Spin and Statistics, and All That*, New York: W. A. Benjamin, Inc., 1964, p. 96.

FIGURE 19-1 Tracks of elementary particles in a bubble chamber. Courtesy of Lawrence Radiation Laboratory.

we discuss all their properties. Rather we are interested in giving some idea of the very surprising behavior of these invisible entities whose interactions presumably underlie all phenomena. The study of the elementary particles reminds us that the dust has not yet settled on the second scientific revolution; many of our fundamental ideas may still be upset in the future.

Besides the elementary particles themselves, we are also interested in their interactions. There are only four fundamental kinds of interactions that are known. Three of them we have already studied. In order of increasing strength they are the gravitational interaction, the electromagnetic interaction, and the strong interaction. There is another interaction which is stronger than the gravitational interaction, but weaker than the other two. It is called the weak interaction, and we shall discuss it later in the chapter. It has the shortest range of any of the interactions. Our present understanding of these interactions is very imperfect. The force law for the strong interaction is unknown. The weak interaction also is not well understood. The gravitational interaction is so weak that we do not know its properties in elementary particle interactions where quantum physics is important. The theory of the electro-magnetic interaction is well developed and can be used to predict accurately the behavior of interacting particles. Even this theory, however, has some disturbing aspects.

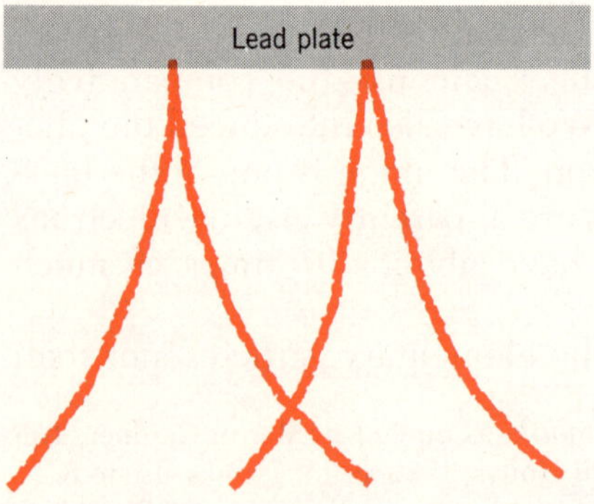

FIGURE 19-2 **Electrons and positrons curve in different directions in a magnetic field. A cloud chamber photograph of this same situation is shown in Figure 19-3.**

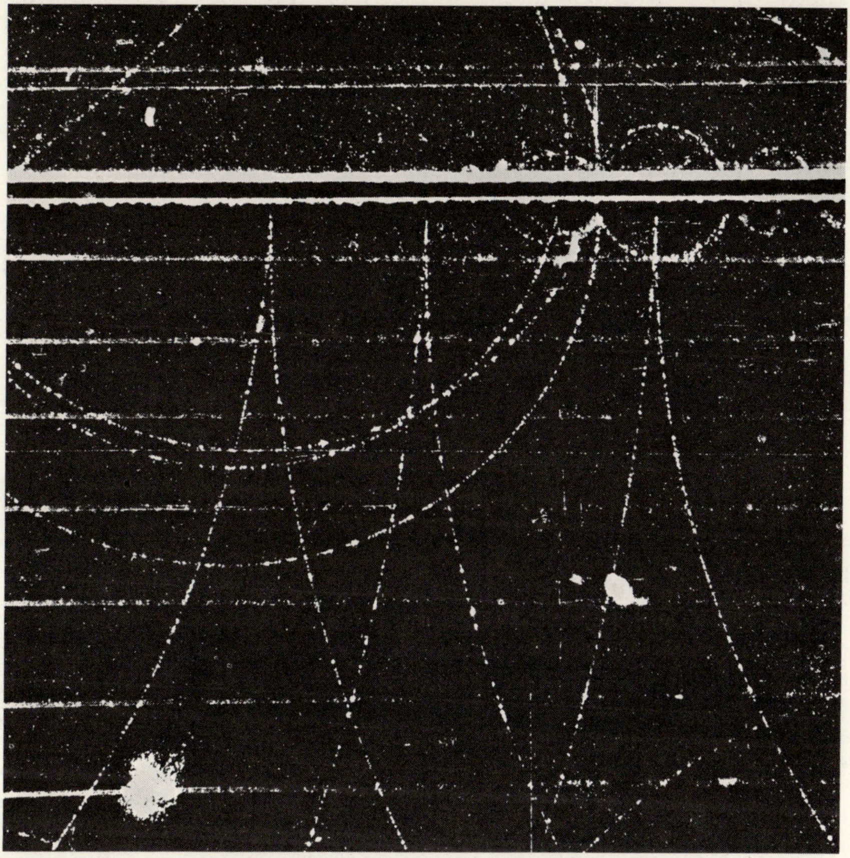

FIGURE 19-3 Cloud chamber tracks of electrons and positrons. (See Figure 19-2 also.) Photograph courtesy of Lawrence Radiation Laboratory, University of California.

Antimatter

The Discovery of Antimatter

In 1930, the American physicist C. D. Anderson announced the discovery of a new particle—the positive electron, also known as the *positron.* This particle has the same mass as the electron but it has a positive instead of a negative charge. The charge has precisely the same magnitude as that of the electron but it is opposite in sign. The positron appears to behave like an electron except for this difference in charge. Cloud-chamber tracts of an electron are curved in the presence of a magnetic field. The tracks of the positron are also curved, but in the opposite direction (see Figs. 19-2 and 19-3).

Today it is believed that for every particle there is an antiparticle, of the same mass as the particle itself but of opposite charge. Most of these antiparticles have been discovered. Thus there is the antiproton, with the same mass as the proton, but of negative charge. There is the antineutron, of the same mass as the neutron. Since the neutron has no net charge, the antineutron also has no net charge but it differs from the neutron in properties we have not yet discussed. The antiphoton is itself a photon; hence we say that the photon is its own antiparticle. There is also an antimuon, as is to be expected, since the muon behaves like an electron but has a larger mass. It is possible to have antiatoms made up of antiprotons, antineutrons, and positrons, rather than protons, neutrons, and electrons. We can even speculate about antiuniverses, made of antimatter. Perhaps some of the galaxies we see through the telescope are actually made up of antimatter rather than matter. Perhaps it is a coincidence that we ourselves are made of matter rather than antimatter.

Antiparticles figure very curiously in some modern physical theories. For example, in R. P. Feynman's very successful theory of the electromagnetic interactions, positrons appear as electrons moving backward in time.

The Annihilation of Matter

When a particle and its antiparticle come together they may annihilate each other; that is, they may both disappear, leaving two or more photons or perhaps other particles (see Fig. 19-4). The remaining photons ensure that energy and momentum will be conserved; they have the energy and momentum that were previously contained in the particles that were annihilated. In our part of the universe, when antimatter is produced, it soon finds some ordinary matter and is annihilated. Similarly, in certain processes of sufficient energy, particles and antiparticles are created in pairs. This process is called *pair production*. Of course, there must be at least enough energy available to make up the rest mass of the pair of particles before this can take place.

The annihilation of matter is an excellent example of the conversion of rest mass into other forms of mass-energy. When an electron

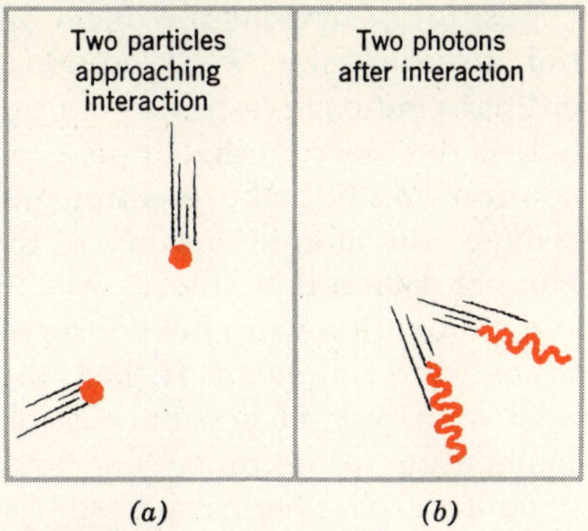

FIGURE 19-4 **The annihilation of matter.**

and a positron annihilate each other to form photons, their rest mass completely disappears, since photons have no rest mass. The rest mass of the electron and positron is completely transformed into the kinetic energy (mass) of the photons.

The Weak Interaction

When the neutron is separated from other particles, it is unstable. It emits two particles and changes into a proton, according to the following process:

$$n \rightarrow \mathrm{p}^{+} + \mathrm{e}^{-} + \bar{\nu}_e$$

This "equation" describes a particle reaction. The two sides of the reaction are separated by an arrow. To the left of the arrow are represented the particles that start the reaction, in this case the neutron, n, alone. To the right are represented the particles that are formed by or result from the reaction. In this case they are the proton, p^{+}, the electron, e^{-}, and a new particle, $\bar{\nu}_e$, the *antineutrino.* The antineutrino is the antiparticle of another new particle, the neutrino, which is represented by the symbol ν_e. (We often represent an antiparticle by putting a bar over the symbol representing the particle.)

The neutrinos and antineutrinos are particles with no charge and no rest mass. These particles apparently travel at the speed of light. They are very hard to detect, since they can pass through solid matter a tenth the size of the galaxy before being absorbed.

Besides describing the decay of an isolated neutron, the above reaction also describes what happens to a neutron in the nucleus in the ordinary radioactive process of beta emission. The emitted beta particle is the electron that appears on the right side of the reaction. The neutron emitting the electron changes into a proton. This process changes the atomic number of the nucleus, but not its atomic mass number. When this process was first studied, it appeared that energy and momentum were not conserved becaused neutrinos and antineutrinos were unknown. In fact, these particles were first assumed to exist as a way of accounting for the apparent loss of energy and momentum in the process of beta emission. Since then the detection of neutrinos has been reported in some very difficult experiments.

The reaction described above involves a kind of interaction which we have not yet discussed. The forces of this interaction are responsible for the emission of the electron and the antineutrino from the neutron, as it changes to a proton. This interaction is called the weak interaction, because its forces are weaker than those of either the strong interaction or the electromagnetic interaction. (They are, however, stronger than the forces of the gravitational interaction.)

In some nuclei another kind of beta emission can occur in which a proton changes to a neutron while emitting a positron (represented by the symbol e^+) and a neutrino:

$$p^+ \rightarrow n + e^+ + \nu_e$$

This reaction also involves the weak interaction. It cannot occur when the proton is alone, separated from other particles, because the rest mass of the proton is smaller than that of the neutron. In such a case the reaction would violate the law of conservation of mass-energy if it occurred spontaneously. In certain nuclei, however, the difference between the binding energy of the neutron and that of the proton may be great enough to permit release of the necessary energy when the proton changes to a neutron.

The symbols for the neutrino and the antineutrino in the reactions described above have "e" as a subscript. This subscript is used to differentiate the neutrino and antineutrino associated with the electron (the so-called electron neutrino and electron antineutrino) from another neutrino and antineutrino associated with the muon (called the muon neutrino and muon antineutrino). Thus there are two kinds of neutrinos and antineutrinos. The muon neutrino is sometimes called a neutretto.

Families of Particles

The elementary particles can be grouped into families as shown in Table 19-1. The baryon family contains the nucleons, and also other "heavy" particles, a few of which are listed in the table. The photon, and another particle, the graviton, are listed in a family by themselves. The photon, we recall, is a particle with zero rest mass that is associated with the waves in the electromagnetic field, and thus with the electromagnetic interaction. In fact, the electromagnetic interaction between charged particles may be thought of as an exchange of photons between the charged particles—the photons in a sense carry the interaction. There should presumably be a particle associated in the same way with the gravitational interaction. Although it has not yet been discovered, it has been named the graviton. (There has been serious experimental evidence recently presented for the existence of waves in the gravitational field.) The members of the meson family, particularly the pions, carry, or at least help carry, the strong interaction. The particle which carries the weak interaction is yet undiscovered.

All particles have mass—even if they have no rest mass they still have mass because of their energy. Therefore they should all participate in the gravitational interaction. All the particles (except presumably the graviton) participate in the weak interaction. All particles with charge participate in the electromagnetic interaction, as of course does also the photon, and indeed all the particles in the table except for the neutrinos (and antineutrinos) and the graviton. The members of the meson and baryon families participate in the strong interaction.

Members of the baryon, electron, and muon families can be assigned family numbers which are conserved in all particle reactions (see Table 19-1). We shall illustrate this for the baryons:

The proton by itself has enough rest mass to change into many electrons and other lighter particles. Why does it not do so? We do not know, but we can invent a principle to account for it. Both protons and neutrons (and the other baryons in the table) are assigned the baryon family number (or simply baryon number) of one. Their antiparticles are assigned the baryon family number of minus one. All lighter particles are assigned the baryon family number of zero. The baryon number of a collection of particles is just the sum of all the baryon numbers of the individual particles. The baryon number is conserved in all physical processes. This conservation law can be

Table 19-1 Some Properties of Some Fundamental Particles

Particle Family	Particle	Symbol	Mass[b]	Charge	Antiparticle[a]	Family Numbers: Electron No.	Muon No.	Baryon No.	Average Lifetime in Seconds
Photon	Photon	γ	0	0	γ Itself	0	0	0	Stable
Graviton	Graviton	g	0	0	g Itself	0	0	0	Stable
Electron	Electron	e^-	1	−1	e^+ Positron	1	0	0	Stable
	Electron neutrino	ν_e	0	0	$\bar{\nu}_e$ Electron antineutrino	1	0	0	Stable
Muon	Muon	μ^-	206.77	−1	μ^+ Positive muon	0	1	0	2.20×10^{-6}
	neutrino	ν_μ	0	0	$\bar{\nu}_\mu$ Muon antineutrino	0	1	0	Stable
Meson	Pions	π^+	273.2	+1	π^-	0	0	0	2.55×10^{-8}
		π^0	264.2	0	π^0 Itself	0	0	0	1.8×10^{-16}
	Kaons	K^+	966.3	+1	K^-	0	0	0	1.23×10^{-8}
		K^0	974.6	0	K^0 Anti K^0 (no charge)	0	0	0	Double
Baryon	Proton	p^+	1836.12	+1	p^- Antiproton	0	0	1	Stable
	Neutron	n	1838.65	0	$\bar{n}$ Antineutron (no charge)	0	0	1	1.013×10^3
	Lamda	Λ^0	2182.8	0	$\overline{\Lambda^0}$ Anti Λ^0	0	0	1	2.62×10^{-10}
	Sigma	Σ^+	2327.6	+1	$\overline{\Sigma^+}$ Anti Σ^+	0	0	1	7.9×10^{-11}
		Σ^0	2333.5	0	$\overline{\Sigma^0}$ Anti Σ^0 (no charge)	0	0	1	10^{-20}
		Σ^-	2342.6	−1	$\overline{\Sigma^-}$ Anti Σ^-	0	0	1	1.6×10^{-10}

[a]The antiparticle has the same rest mass, average lifetime as the corresponding particle, and the opposite change and family number as the corresponding particle.

[b]The rest mass is given in terms of the electron rest mass.

used to explain why the free proton is stable—there is no other heavy particle that has less rest mass. The proton cannot change into lighter particles only, since the baryon family number must be conserved. Similarly, a nucleon in an atomic nucleus cannot change completely into lighter particles but must change into at least one other heavy particle, preserving the heavy nucleus.

The electron and muon family numbers work in a similar way. The electron and the electron neutrino are assigned an electron number of one, and the antiparticles are assigned an electron number of minus one. The assignment of muon numbers to the muon family is exactly like the assignment of electron numbers to the electron family. Both the electron number and the muon number are independently conserved in all particle reactions.

There is no family number conservation law for photons, gravitons, or mesons. (These are the particles associated with the electromagnetic, gravitational, and strong interactions.) Another interesting fact about these particles is that they do not obey the exclusion principle. These particles are called *bosons*. The members of the electron, muon, and baryon families, on the other hand, do obey the exclusion principle—that is, no two protons, or no two muons, can be in precisely the same state at the same time, etc. These particles are called *fermions*.

Particle Identity

How many ways are there to arrange the two ping pong balls in Figure 19-5 so that there is one ball in each of the two different glasses? A little reflection convinces us that the answer is two. We can put the first ball in the large glass and the second one in the small glass, or we can reverse the procedure and put the second ball in the large glass and the first one in the small glass. Calculations of this sort are not mere mental exercises—they have a real application to the world. A gambler who is throwing two apparently identical dice must know, or at least act as if he knows, that there are two ways to get a total of three: he can get a one on the first of the two dice and a two on the other, or a two on the first and a one on the second. If he acts contrary to this fact he will eventually lose money.

Now let us suppose that instead of dealing with ping pong balls and dice, or other large scale objects, we are dealing with elementary particles. For example, how many ways are there to arrange the two

FIGURE 19-5 How many ways can two ping pong balls be arranged in two different glasses so there is one ball in each glass?

electrons so that there is one electron in each of the two different, precisely defined atomic "orbits" or states illustrated in Figure 19-6?

There is only one possibility—an electron in the first state, an electron in the second state, two electrons in all. We cannot ascribe individuality to these electrons. We cannot legitimately call one the "first" electron and the "other" the "second" electron. This same situation holds for the other kinds of elementary particles. The kind of thinking that is valid for balls and dice and other "ordinary" objects is *not* valid for elementary particles.

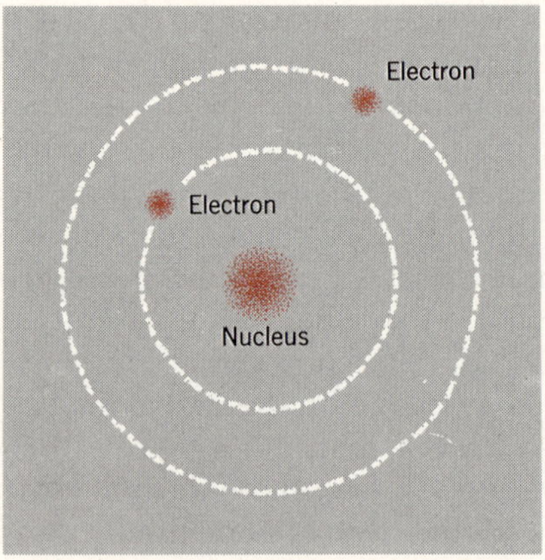

FIGURE 19-6 How many ways can two electrons be arranged in two different atomic "orbits"?

We may try to explain this curious difference between elementary particles and "ordinary" objects in the following way: electrons are truly identical in a way that ordinary objects are not. There is no way, for example, to put a tag or label on an electron to permanently identify it as it changes from state to state. Also, there is no way to follow the path of a particular electron. We cannot clearly distinguish its path from that of another electron—the uncertainty principle prevents us from precisely identifying its position without losing track of where it is going. If two electrons were to suddenly change places or states with each other, we would have no way of knowing that they had done so. This may help us understand why there is only one arrangement. What we have said for electrons is true for protons, photons, and other elementary particles. All protons are identical, all photons are identical, all muons are identical, etc.

This strange identity of elementary particles is not just a philosophical oddity, it has testable consequences in the real world. If we count the states of a many-particle system, we must take into consideration this identity. Our description of a physical system must not tell us which of a number of identical particles are in which positions or which states. Any two processes which differ only by the exchange of identical particles must be considered indistinguishable. If we violate these rules we find that we cannot correctly predict the results of experiments. Because of this identity we can state the following principle:

The laws of physics are not changed if identical particles are exchanged for each other.

The situation we have been discussing prompted Erwin Schrodinger to remark that "the elementary particle is not an individual."[2]

Symmetry and Elementary Particles

The principle stated in the previous paragraph is a new symmetry principle. It tells us that a certain operation may be performed, namely the exchange of identical particles, without changing the laws of physics.In the next paragraph we shall discuss another new

[2]Erwin C. Schrodinger, *Science, Theory, and Man*, New York: Dover, 1957, p. 194. Reprinted through permission of the publisher.

symmetry principle. In modern physics, symmetry principles are among the most important principles governing the behavior of elementary particles, and indeed, all physical phenomena.

Particle-Antiparticle Symmetry

The existence of antimatter appears to introduce another symmetry into the physical world. For every possible configuration of matter, we could get another by changing all the particles into antiparticles. This suggests that the laws of physics should be the same even if we were to change all the particles into antiparticles—that every possible process in the physical world would still be possible if the particles were antiparticles.

Violation of Reflection Symmetry and Particle-Antiparticle Symmetry

Processes that involve the weak interaction do not necessarily obey the principle of reflection symmetry discussed in Chapter 3 (p. 50). This means that processes are known to occur, such as beta emission, which when viewed in a mirror give processes that do *not* occur. The beta emission from cobalt-60 is a famous example. (The cobalt-60 nucleus has an atomic number of 27 and a mass number of 60.) Figure 19-7 shows how the spinning nucleus emits an electron in beta emission. The electron comes out in the opposite direction from the antineutrino, but the directions are usually in the same relation to the spin of the nucleus. If one looks at the mirror image, as indicated in the figure, the process does not look the same. The mirror image is not the same as the original process, just as the reflection of a right hand in a mirror looks like a left hand, not another right

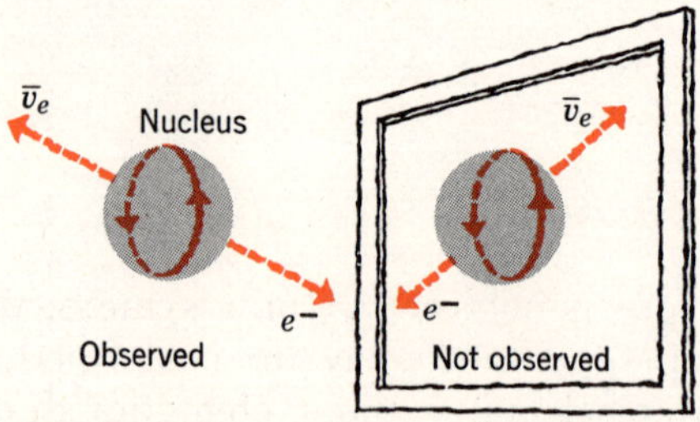

FIGURE 19-7 **Beta emission from cobalt-60 violates reflection symmetry.**

hand. This would be all right, except that the process that is observed in the mirror is not observed in nature—in reality the particles generally come out in the way indicated in the diagram and not as indicated in the mirror. For true reflection symmetry to hold, however, the reflected process should be just as likely as the unreflected process.

The weak interactions also violate the principle of particle-antiparticle symmetry. However, the weak interactions *almost* obey a combined symmetry principle. If we not only reflect the process illustrated in Figure 19-7 in a mirror, but also exchange all particles for their antiparticles, we obtain another symmetry operation which is a combination of reflection symmetry and particle-antiparticle exchange symmetry. There are some particle reactions involving the weak interaction which even break this combined symmetry principle, however.

If we not only perform the two operations described above, but also reverse time, we obtain a triple combination symmetry principle, which is referred to as PCT.[3] This combined symmetry principle is believed (or at least hoped) to hold universally. If it does hold, then, since the combined reflection and particle-antiparticle symmetry fails, the time reversal symmetry should also fail for the weak interactions.

Conservation Laws and Symmetry Principles

We have already mentioned that we do not know all the exact laws of interaction for fundamental particles. Even without these exact laws, however, we can understand a great deal about the behavior of fundamental particles by using the conservation laws and the symmetry principles. The picture that emerges is one of particles undergoing all possible kinds of reaction and transformations. Mass-energy, in the form of particles and kinetic energy, is constantly being transformed from one kind of particle to another, now bound up in the rest mass of particles, now appearing as photons or neutrinos with no rest mass whatever. This underlying chaotic behavior in all matter fully exploits the uncertainties of quantum physics. Through this chaos, limiting and regulating it, are the conservation laws and symmetry principles. According to these principles, some conceivable reactions and processes are forbidden. We have already seen

[3]The P stands for reflection (or rather for a quantity known as parity which is associated with it); the C stands for charge conjugation, which is the technical name for the particle-antiparticle exchange operation, and the T stands for time reversal.

that the disintegration of the proton is forbidden by the law of conservation of baryon number. The disintegration of the electron into less massive particles is forbidden by the law of conservation of charge. (All fundamental particles that have less mass than the electron are uncharged.)

Everything which is not forbidden is required. This may sound like an authoritarian nightmare, except that what is required is that there be some probability, however small, for a process unless it is forbidden. There is no need that it actually occur in a given particular case. Whenever we find some particle reaction that *never* occurs, however, we should look for some conservation law or symmetry principle which forbids it. It was in just such a way that a new conservation law was discovered. It was by studying some particle reactions that did *not* occur that Murray Gell-Mann and K. Nishijima discovered a new quantity, which they called *strangeness.* Strangeness is conserved in strong and electromagnetic interactions, but not in weak interactions. The reactions which did not occur could not have occurred without violating conservation of strangeness.

In this last paragraph we would like to mention a curious relationship between conservation laws and symmetry principles. We shall not try to explain why this relationship exists, but merely to state it.

What one would like to say is that every conservation law is equivalent to some symmetry principle and vice versa. What one can say is that many symmetry principles are equivalent to conservation laws. That is, if the symmetry principle holds, then the conservation law holds also. For example, according to our present understanding of physics, the law of conservation of momentum is a consequence of position symmetry. That is, the fact that the laws of physics do not depend on our location in space implies that momentum must be conserved. The law of conservation of energy follows from time symmetry. The law of conservation of angular momentum follows from directional symmetry. Thus we have a surprising equivalence between conservation laws and symmetry principles—some of the most important and fundamental laws and principles of modern physics.

FUNDAMENTAL IDEAS

I. FUNDAMENTAL LAWS AND PRINCIPLES

A. THE WEAK INTERACTION—Responsible for Beta Emission and Other Reactions.

B. CONSERVATION LAWS: Conservation of Electron, Muon, and Baryon Number.

C. SYMMETRY PRINCIPLES

1. Particle-antiparticle symmetry.
2. Double and Triple combined symmetries.
3. Breakdown of particle-antiparticle and mirror reflection symmetries and their combination for the weak interactions.
4. Particle exchange symmetries.
5. The equivalence of certain symmetry principles and conservation laws.

II. ELEMENTARY PARTICLES

QUESTIONS

Antimatter

1. What are the characteristics of an antiparticle?
2. Why would it be difficult for matter and antimatter to exist in close contact?
3. What difference is there between the cloud-chamber tracks of an electron and those of a positron? Explain this difference.
4. Can electrons be created or destroyed? Explain.
5. Can matter be created or destroyed? Explain.
6. In the annihilation of an electron with a positron, two charged particles disappear, and uncharged photons appear. Why does this not violate the law of conservation of charge?

Weak Interaction

7. Why was the neutrino invented?
8. How many kinds of neutrinos are there?
9. Why do we believe in neutrinos?
10. Does reflection symmetry really hold? Explain. What other symmetry principle does not hold?

Families of Particles

11. Describe the various family-number conservation laws.
12. Why is the proton stable?
13. From Table 19-1, find some characteristics that might be used to distinguish between the different families of particles.

Symmetry Principles and Conservation Laws

14. Is there any relation between symmetry principles and conservation laws?
15. What determines whether or not a particle will disintegrate into other particles?
16. What do we mean when we say that all electrons are the same? How does their identity differ from the similarity of two coins, for example?
17. What do you think Schrodinger meant by saying that a fundamental particle is not an individual?

General

18. Describe the fundamental forces or interactions of nature.
19. For each of the fundamental interactions, describe either a system that is held together by it or a reaction in which it is involved.

Appendix 1

Systems of Units

	Meter-Kilogram-Second[a]	Centimeter-Gram-Second	Foot-Slug-Second
Length	meter	centimeter = 10^{-2} meter	foot = .305 meter
Time	second	second	second
Mass	kilogram	gram = 10^{-3} kilogram	slug = 14.59 kilogram
Charge	coulomb	statcoulomb = 3.34×10^{-10} coulomb	coulomb
Force	newton = $\frac{\text{kilogram meter}}{\text{second}^2}$	dyne = 10^{-5} newton	pound = 4.45 newton
Energy	joule = $\frac{\text{kilogram meter}^2}{\text{second}^2}$	erg = 10^{-7} joule	foot-pound = 1.36 joule

[a]The meter-kilogram-second system is used throughout this book.

Appendix 2

Suggested Supplementary Reading

Andrade, E.N. da C., *Rutherford and the Nature of the Atom*, Anchor Books, Doubleday & Co., Garden City, New York, 1964.

Bronowski, J., *The Common Sense of Science*, Modern Library Paperbacks, Random House, New York, n.d.

Bronowski, J., *Science and Human Values*, Harper Torchbooks, Harper & Row, New York, Rev. Edition, 1965.

Burtt, E. A., *The Metaphysical Foundations of Modern Physical Science*, Anchor Books, Doubleday & Co., Garden City, New York, Rev. Edition, 1932. This book is more interesting than its title may suggest to those suspicious of metaphysics.

Cohen, I. Bernard, *The Birth of a New Physics*, Anchor Books, Doubleday & Co., Garden City, New York, 1960.

de Santillana, Giorgio, *The Crime of Galileo*, The University of Chicago Press, Chicago, 1955.

Drake, Stillman, *Discoveries and Opinions of Galileo*, Anchor Books,Doubleday & Co., Garden City, New York, 1957. This very entertaining book contains translations of Galileo's writings, including the letter to the Grand Duchess Christina and parts of *The Assayer*, in which Galileo replied to "Sarsi."

Eddington, Arthur S., *Space, Time, and Gravitation*, Harper Torchbooks, Harper & Row, New York, 1959. This book, originally published in 1920, begins with an interesting dialogue on the nature and measurement of space.

Einstein, Albert, and Leopold Infeld, *The Evolution of Physics*, Simon & Schuster, New York, 1938.

Fermi, Laura, and Gilberto Bernardini, *Galileo and the Scientific Revolution*, Fawcett Publications, Inc., Greenwich Conn., 1965. (Originally published by Basic Books in 1961.)

Feynman, Richard, *The Character of Physical Law*, M. I. T. Press, Cambridge, Mass., 1967. (Originally published in 1965.)

Ford, Kenneth W., *The World of Elementary Particles*, Blaisdell Publishing Company, New York, 1965. Very understandable account of elementary particles.

Galilei, Galileo, *Dialogue Concerning the Two Chief World Systems, Ptolemaic and Copernican,* trans. by Stillman Drake, University of California Press, Berkeley and Los Angeles, 1953.

Gamow, George, *One, Two, Three, . . . Infinity, Facts and Speculations of Science,* Mentor, The New American Library of World Literature, Inc.,New York, 1955. We would like to call special attention to Chapter VIII, The Law of Disorder.

Gardner, Martin, *The Ambidextrous Universe,* Mentor, The New American Library, Inc., New York, 1969.

Gardner, Martin, *Relativity for the Millions,* Pocket Books, New York, 1965. (Originally published by Macmillan in 1962.) A well-illustrated and interesting account of the special and general theories of relativity.

Gillispie, C. C., *The Edge of Objectivity,* Princeton University Press, Princeton, New Jersey, 1960. The development of science, with attention to its cultural and intellectual importance.

Hoffman, Banesh, *The Strange Story of the Quantum,* Second Ed., Dover Publications, Inc., New York, 1959.

Koestler, Arthur, *The Sleepwalkers,* Macmillan, New York, 1959. Sympathetic in his treatment of Kepler, less so in his treatment of Copernicus and Galileo.

Reichenbach, Hans, *From Copernicus to Einstein,* trans. by Ralph B. Winn, The Wisdom Library, The Philosophical Library, New York, 1942.

Schneer, Cecil J., *The Evolution of Physical Science,* Grove Press, New York, 1960.

Schrodinger, Erwin C., *Science, Theory, and Man,* Dover Publications, Inc., New York, 1957. (Originally published in 1935.) See especially Chapter IX on the nature of an elementary particle.

INDEX